Surfaces and Interfaces of Metal Oxide Thin Films, Multilayers, Nanoparticles and Nano-composites

Alejandro G. Roca • Paolo Mele
Hanae Kijima-Aoki • Elvira Fantechi
Jana K. Vejpravova • Martin Kalbac
Satoru Kaneko • Tamio Endo
Editors

Surfaces and Interfaces of Metal Oxide Thin Films, Multilayers, Nanoparticles and Nano-composites

In Memory of Prof. Dr. Hanns-Ulrich Habermeier

Springer

Editors
Alejandro G. Roca
MagneticNanostrutures Group
Institut Català de Nanociència i
Nanotecnologia
Barcelona, Spain

Hanae Kijima-Aoki
Creative Interdisciplinary Research Division
Tohoku University
Sendai, Japan

Jana K. Vejpravova
Department of Condensed Matter Physics
Charles University
Prague, Czech Republic

Satoru Kaneko
Kanagawa Institute of Industrial
Science & Technology
Ebina, Kanagawa, Japan

Paolo Mele
SIT Research Laboratories
Shibaura Institute of Technology
Saitama, Japan

Elvira Fantechi
Department of Chemistry and Industrial
Chemistry
University of Pisa
Pisa, Italy

Martin Kalbac
Department of Low-dimensional Systems
J. Heyrovsky Institute of Physical
Chemicals
Prague, Czech Republic

Tamio Endo
Japan Advanced Chemicals
Atsugi, Kanagawa, Japan

ISBN 978-3-030-74072-6 ISBN 978-3-030-74073-3 (eBook)
https://doi.org/10.1007/978-3-030-74073-3

This Springer imprint is published by the registered company Springer Nature Switzerland AG
The registered company address is: Gewerbestrasse 11, 6330 Cham, Switzerland

Editorial Note

The publication of this book is the result of 2 years' effort made by different researchers, having its origin at the IUMRS-ICEM Conference held in Daejeon (South Korea) in 2018. The book we published at the beginning of 2021 contains contributions from speakers and attendees at the Symposium S2D: *Surfaces and Interfaces of Thin Films, Multilayers and Nano-composites* and other colleagues from the field. We would like to express our gratitude to all the authors for their contributions and also the time spent reviewing the chapters and editing the book.

These two years have been a tough period, ridden with different obstacles, somber news from across the world, and a global pandemic. First, Prof. Daniel Niznansky passed away at the end of 2018. He was professor at the Charles University of Prague (Czech Republic) and has a high impact on the research careers of Jana Vejpravova and Alejandro G. Roca. Moreover, Prof. Hanns-Ulrich Habermeier passed away in 2019. He was an important physicist at world level. More than that, he was a person with a charming personality, coloring every conference dinner and presenting us with souvenirs. Their memories and work will remain alive in global science.

Lastly, we are still suffering from the COVID-19 pandemic, with millions of infections and more than one million deaths. COVID-19 has changed our lives since March of 2020. Indeed, it affected the writing of certain chapters of this book as authors suffered lockdowns at their workplaces. We hope with the finding of new vaccines we could end this terrible nightmare.

Scope of the Book and Short Description of the Chapters

Nanostructured metal oxides have gained a lot of attention during the last decades due to their appealing magnetic, electric, catalytic, and plasmonic properties. These properties mainly belong to the material itself; however, due to its nanostructuration, these properties can be affected and tuned on the basis of size, thickness, shape, crystallinity, defects, and interface of the material as well as interactions with other materials. Due to advance in the nanofabrication techniques/synthetic methods (please note that depending if we are talking about thin films or nanoparticles and if a "physical method" or "chemical method" the experts on this area give nomenclature to the procedure to obtain the material) and the intense study of these properties, nanomaterials have evolved and their performance has improved in widespread applications such as biomedicine, energy, environmental remediation, and data storage. Although there exist different types of nanomaterials depending on their dimensionality, nanoparticles and thin films are probably the most studied.

This book provides a general overview and current state of the art of different types of metal oxide nanomaterials, either in nanoparticles or thin film structure. It covers the development and optimization of different nanofabrication/synthesis techniques for nanostructures, which are currently under the attention of the research community, the study of the structure and interactions by different characterization techniques of heterostructured materials, and the final impact in different applications such as nanotherapy, data storage, super magnets, and high-frequency devices. This book is intended for experts, for consolidation of their knowledge, and also for students who wish to learn and understand the basics of nanostructured metal oxides in diverse forms.

The book is divided into four parts. The **first part** involves four chapters on metal oxide nanoparticles (Chaps. 1, 2, 3, and 4), ranging from the synthesis of magnetic iron oxide nanoparticles with anisometric morphology to other synthetic routes of more complex magnetic and plasmonic oxide nanoparticles. Structural and magnetic properties are also discussed using Mossbauer spectroscopy and other complementary techniques. The application of such metal oxide nanoparticles is discussed for widespread applications such as hyperthermia, data storage, microwave absorption, spintronics, and photoelectronics. **J. Vejpravova** presents an overview of the current understanding of the magnetism of spinel ferrites nanoparticles with different configurations using Mossbauer spectroscopy among other complementary techniques. Contribution of **A.I. Tovstolytkin** is related to the discussion of various approaches for self-controlled magnetic hyperthermia using complex magnetic oxide nanoparticles. **L. Gutiérrez** presents a general pathway on how to synthesize different anisotropic magnetic iron oxide nanoparticles and discusses the impact of anisotropy on different applications. **H. Matsui** describes the plasmonic properties of In_2O_3:Sn nanoparticles and their 3D assemblies at the near infrared range, and their use in thermal shielding applications

The rest of the parts are related to thin films, but a different material is the scope of each part. The **second part** is related to perovskites (Chaps. 5 and 6) for temperature-sensing devices and also for photocatalysis. **D.K. Goswami** introduces hexagonal barium titanate nanocrystals (h-BTNCs) as temperature-sensing materials for the fabrication of organic field-effect transistors (OFET) based temperature sensors. The chapter by of **J. M. Mora** is related to the study of different structural changes of semiconductor perovskites on the psychochemical, optical, and photoelectrocatalytic properties.

The **third part** is related to cuprates (Chaps. 7, 8, 9, 10, and 11) and their use as superconductors, ranging from thin films for data storage applications to heterostructures and interface engineering as well as and melt-textured materials. **A. Palau** shows the possibility of combining superconducting and ferromagnetic materials, the manipulation of magnetic textures, opening up new venues for energy-efficient information storage and manipulation. **N. Chikumoto** demonstrates that irradiation by neutrons and heavy ions is a robust method to improve the pinning properties in La214 and Y123. **A. K. Jha** presents the evolution of various interfaces in REBCO nanocomposite thin films through incorporation of secondary-phase nanoinclusions and their contribution to vortex pinning. **P.Mele** proposes the use of melt-textured NEG-123 +Gd-211 pellets, instead of sintered materials, as targets for the PLD laser ablation. **S. Miryala** demonstrates that the use of peritectic temperature liquid (Yb123+ $Ba_3Cu_5O_8$) and its mass are crucial for obtaining high-quality Y-123 through top-seeded infiltration growth process, which can be exploited for numerous industrial applications.

The **fourth part** is devoted to other metal oxide materials with thermoelectric properties and applicable to high-frequency devices (Chaps. 12 and 13). **H. Kijima-Aoki's** chapter is devoted to the evolution of the synthesis process, unique structure, and high-frequency soft-magnetic performance of nano-granular cobalt-(metal-oxide, metal-nitride) system thin films. **H. Assadi** presents some theoretical calculations done by density functional theory contribution about the thermoelectric properties of Na_xCoO_2 as a function of the type of cation substitution performed (Na or Co).

All the chapters have been reviewed by independent referees. For their invaluable refereeing assistance and expertise (some have served twice), we are grateful.

Memorial Note on Prof. Hanns-Ulrich Habermeier

The International Conference of IUMRS-ICEM-18 was held in Daejeon, Korea (International Union of Materials Research Societies – International Conference on Electronic Materials, August 19–24, 2018). Two symposia were organized by *Team Harmonized Oxides* (THO). THO is a unique scientist group which was established by Profs. Tamio Endo (Mie University) and Kazuhiro Endo (AIST) in Japan to develop materials sciences and technologies, and to promote international comprehensions and world peace. Under such policies, these symposia were conducted:

(1) ***Surfaces and Interfaces of Thin Films, Multilayers and Nano-composites***
(2) Carbon Related Materials in Honor of Nobel Laureate Prof. Suzuki

We invited two distinguished plenary speakers from Japan to the conference; Prof. Akira Suzuki (Nobel Laureate, Chemistry, 2010) delivered the Nobel lecture titled "Cross-Coupling Reactions of Organoboranes: An Easy Way for Carbon-Carbon Bonding." Akira Fujishima delivered a plenary lecture titled "TiO_2 Photocatalysis to Contribute Comfortable Atmosphere."

In **Symposium 1**, brilliant scientists who were very well known in their fields were invited. Principal invited speakers were **Hanns-Ulrich Habermeier**, Helmut Takahiro Uchida, Josep Nogués, Yukiko Yamada-Takamura, Hong Zhu, Axel Hoffmann, Boris Maiorov, and Oleksandr Tovstlytkin. All the participants presented excellent papers, and we had very active, exciting, and useful discussions in this symposium. A few scenes are shown in the photos below. Then we, organizers, agreed to publish a commemorative book of this symposium in order to share such valuable knowledge with the scientific community. We requested the participants and their affiliated scientists to contribute one chapter to this book in consonance with the symposium's scope. We are convinced that the papers presented by these authors are exquisite, thereby sharing profitable knowledge to readers. We would like to express our special thanks to all the chapter authors and reviewers of this book.

Incidentally, we published a book in relation to ***Symposium 2*** in 2020, which was organized in recognition of the Nobel lecture by Prof. Akira Suzuki.

Organizers of Symposium 1

Satoru Kaneko<Head> (Kanagawa Institute of Industrial Science and Technology)
Paolo Mele (Shibaura Institute of Technology)
Alejandro G. Roca (Catalan Institute of Nanoscience and Nanotechnology)
Takashi Nakajima (Tokyo University of Science)
Katherine Develos-Bagarinao (National Institute of Advanced Industrial Science and Technology)
Elvira Fantechi (Universita di Pisa)
Tamio Endo (Sagamihara Surface Laboratory)

Scope of Symposium 1

Recent material nanotechnologies have made the fabrication of nano-composites, ultra-thin and multilayered thin films, as well as nanoparticles possible, with a precise control of their properties. In the multilayers and nano-composites, the interfaces and grain-boundaries should give larger functional factors than inner parts of the layers and grains. It is difficult to understand grain-boundary roles due to their complexity. We first discuss the fabrication of multilayers and nano-composites, then understand their intrinsic nature arising from their peculiar structures. The ultimate goal is to elucidate commonalities and differences between the multilayer interfaces and nano-composite grain-boundaries to understand intricate roles of grain-boundaries. In the case of nanoparticles, the relation between the structural and magnetic properties on their performance in biomedicine is still a hot topic, such as magnetic hyperthermia, being one of the current alternatives to conventional therapies

Additionally, we would like to state that these two symposia organized by THO were held in conjunction with the meeting of **MRS Japan: Nano Oxides Materials Research Group**. Denomination of THO has changed to THM (Team Harmonized Materials, 2020) because the team will cover wider categories of materials henceforth.

Lastly, we would like to inform that Prof. Hanns-Ulrich Habermeier (Max-Planck, Stuttgart) has passed away on July 20, 2019, in Stuttgart, Germany. He acted as president of IUMRS (2015–2016) and president of E-MRS (2000–2003). He was a kind person, an international member of THO, and greatly helped us in our many international activities including conferences by IUMRS, MRS, E-MRS, MRS-J, and JSAP. He kindly delivered his fruitful keynote talk at Symposium 1, which was highly appreciated. The memorial photos are shown below. "Our sympathies are with his family!"

Editors

Alejandro G. Roca<Head> (Catalan Institute of Nanoscience and Nanotechnology, Barcelona, Spain)
Paolo Mele (Shibaura Institute of Technology, Saitama, Japan)
Tamio Endo (JapanAdvancedChemicals, Atsugi, Japan)
Elvira Fantechi (Universita di Pisa, Italy)
Hanae Kijima-Aoki (Tohoku University, Sendai, Japan)
Jana K. Vejpravova (Charles University, Prague, Czech)
Martin Kalbac (J. Heyrovsky Institute of Physical Chemistry, Prague, Czech)
Satoru Kaneko (Kanagawa Institute of Industrial Science and Technology: KISTEC, Ebina, Japan)

Tamio Endo (Japan Advanced Chemicals, Atsugi, Japan /
Professor Emeritus Mie University /
Honorary Professor South West Jiaotong University, Chengdu, China /
Representative MRS-J Research Group / Representative THO
Dec.1, 2020, Tsu, Japan

Contents

Part I
Metal Oxide Nanoparticles

Recent Progress in Mössbauer Studies of Iron-Based Spinel Oxide Nanoparticles

Jana K. Vejpravova

1 Introduction

Mössbauer spectroscopy (MS) is the γ-ray recoilless-free nuclear resonance spectroscopy named after Rudolph Mössbauer who received the Nobel Prize for Physics in 1961 for this discovery [1]. This method serves as a powerful tool for investigation of the oxidation state, local crystallographic environment and orientation of the magnetic moments in the absorbing sample through the interactions of the absorbing nuclei with its surrounding, called the hyperfine interactions [2]. The MS experiment has two options by means of a radioactive source: conventional radioactive isotope, which will be discussed further in detail, and synchrotron radiation [3, 4]. The energy-domain MS illuminates the nuclear hyperfine structure directly from the

Preface
When being asked to contribute to this book, my idea was to discuss the peculiarities in understanding magnetic behaviour of nanoscale oxide magnets, in particular the relevance of the classical and quantum pictures featuring their magnetic properties. However, in November 2018, my great friend and colleague Daniel Nižňanský passed away. He attracted me to the fascinating field of nanomagnetism, and in particular, he always contributed significantly with his enormous expertize in synthesis and Mössbauer spectroscopy of fine particles of magnetic oxides. In memory of his great contribution to the magnetic nanoparticle research, I will provide current state of the art in this field from the perspective of applying Mössbauer spectroscopy techniques to magnetic spinel oxide nanoparticles. At this point, I would like to express my great thanks to my outstanding students and colleagues: Barbara Pacáková, Simona Kubíčková and Alice Mantlíková who did an excellent job and substantial advancement of the field during their doctoral studies. Finally, my great thanks belong to Puerto Morales, Alejandro Gomez Roca, Gorka Salas, Carla Cannas, Marco Sana, Valentina Mameli, Dominika Zakutna and Anton Repko for unforgettable collaboration and precious inputs on various aspects related to magnetic nanoparticles.

J. K. Vejpravova (✉)
Department of Condensed Matter Physics, Charles University, Prague, Czech Republic
e-mail: jana@mag.mff.cuni.cz

A. G. Roca et al. (eds.), *Surfaces and Interfaces of Metal Oxide Thin Films, Multilayers, Nanoparticles and Nano-composites*,
https://doi.org/10.1007/978-3-030-74073-3_1

absorption/emission positions of the resonance line, while the time-domain MS reflects the hyperfine structure due to the coherent process of nuclear resonant forward scattering. The time-domain MS, also called the spectrum of quantum beats of nuclear forward scattering, is typically carried using synchrotron radiation. In 1986, Gerdau et al. introduced the nuclear resonant spectroscopy in the time domain [5], and since then, this experimental technique has been intensively developed [5, 6]. Because synchrotron radiation has the advantages of excellent collimation, high brilliance, pulsed structure and close-to-ideal polarization, the experiments can be carried in both the energy and the time domain [3, 5, 7–10]. A unique study of Ni nanoparticles using the time-domain MS may serve as an interesting example [11].

However, the conventional energy-domain MS is a very well-established tool in the magnetic NP research, and from last two decades of twentieth century, there have been many substantial contributions, e.g. [12–17].

This chapter aims to present a recent progress in the field within last ~ 10 years on a friendly level. In particular, the important results on spinel ferrite NPs with a complex internal structure, including core–shell NPs, are highlighted. After a brief introduction, an up-to-date insight into the core–shell phenomenon of fine magnetic NPs is given, followed by an overview of the MS of magnetic NPs. The multifaceted term—a "core–shell" NP—is discussed from the perspective of chemical composition, and structural and spin order, which can be addressed by the MS very efficiently. The uniqueness of the MS is also placed in context of other methods used for investigation of magnetic NPs. The next section summarizes a collection of recent works carried out on "magnetic", "structural" and "chemical" core–shell NPs. The importance of a collective response in NP ensembles and its impact on the MS will be also briefly addressed. Finally, future prospects and possible innovative applications of spinel oxide NPs are outlined.

2 Magnetic Nanoparticles—The Core–Shell Phenomenon

It is a well-known fact that magnetic domains are created in crystals with ferromagnetic or ferrimagnetic ordering in order to decrease the magnetostatic energy that is associated with the dipolar fields. The creation of the domains depends on the competition between the reduction of the magnetostatic energy and the energy required to form the domain walls separating the adjacent domains. The size of the domain wall is a balance between the exchange energy that tries to unwind the domain wall and the magnetocrystalline anisotropy with the opposite effect.

In magnetic NPs (literally speaking in close to spherical nanocrystals with a mean diameter below ~ 100 nm), the typical dimensions are comparable with the thickness of the domain wall, and thus below a critical diameter, d_{crit}, the NP becomes a single domain. The critical diameter was proposed by Ch. Kittel [18, 19] in the form:

$$d_{crit} = \frac{9\sqrt{A_{ex}K_u}}{\mu_0 M_S^{\ 2}} \tag{1}$$

where M_s is the saturation magnetization, A_{ex} and K_u are the exchange and uniaxial anisotropy constants, respectively. It is clear that the d_{crit} is material specific and ranges from few tens of nm in magnetic oxides to almost micrometre in hard rare earth-based magnets.

In magnetic NPs reaching the single-domain limit, magnetic properties are represented by a classical magnetic moment in the order of ~ 10^3–10^5 Bohr magnetons. Consequently, paramagnetic-like behaviour can be observed even below the T_c of the bulk material. The phenomenon is known as superparamagnetism (SPM) [20], while the magnetization reversal in the low temperature limit (in the so-called blocked state) has been introduced by Stoner and Wohlfarth [21]. Nevertheless, the macrospin approach neglects possible spin structure inside the NP. Decreasing the NP size, the numbers of atoms located at the surface (surface atoms) increases with respect to the all atoms, and at some defined size, surface atoms dominate the volume atoms. The atoms at the surface exhibit lower coordination number due to breaking of lattice symmetry at the surface proximity. Moreover, the exchange bonds are broken resulting in the spin disorder and frustration having consequences such as lowering of the saturation magnetization and/or lack of saturation in the high magnetic field [22, 23].

Already in 60-ties of the last century, the lower saturation magnetization of the NPs was identified, and several works tried to explain this feature considering the presence of a nonmagnetic layer (also termed magnetically "dead layer") [12] or presence of hydrogen in the lattice. In 1971, J. M. D. Coey [22, 23] suggested that the noncollinearity of the spins in NPs is responsible for the lower saturation magnetization and proposed the famous "core–shell" model of the NP.

In this case, the terms "core" and "shell" are related to a different spin arrangement in a model NP particle, which consists of a core with the bulk-like magnetic structure and a shell, where the spins are canted at random angles to the surface, giving rise to the so-called spin canting angle (please see Fig. 1, panel (a)). The spin canting angle depends on the number of the magnetic nearest neighbours connecting with the reduced symmetry and dangling bonds. This theoretical model was supported by the experimental evidence of the persistence of the 2nd and 5th absorption line in In-Field Mössbauer Spectroscopy (IFMS) pointing to the noncollinearity of some spins with the applied magnetic field. However, the same time he also pointed out that the possibility of some canting of the ions in the interior cannot be excluded.

In 1976, Morrish et al [24] tried to verify the Coey's model by inspection of a circular maghemite NPs with ^{57}Fe-enriched surface; however, the proof of enriching only the surface had not been given. Furthermore, it has to be noted that the enriching (thus further chemical treatment) usually induces additional disorder in the surface layer. They observed the broadening of the lines and the difference of the peak position of the 1st/6th line and 2nd/5th line, attributed to the lower effective field

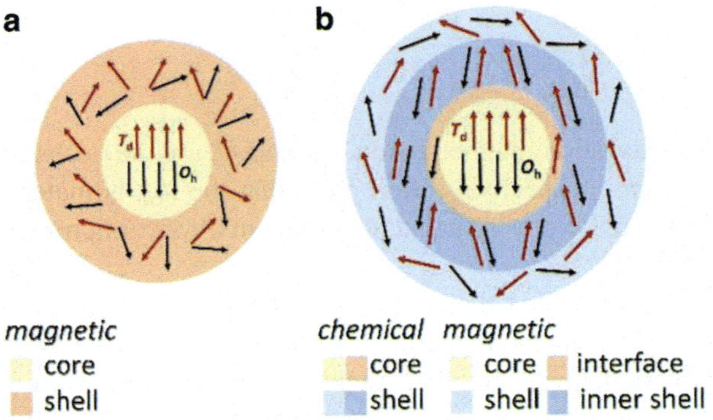

Fig. 1 Scheme of a core–shell model as suggested by Coey and termed "magnetic" in this chapter, adopted to a spinel ferrite NP (**a**) and a possible complex spin structure of a "chemical" core–shell NP of a spinel ferrite (**b**). Please note the interface between the inner and outer spinel phases. Illustration of a crystallographic (dis)order is omitted for clarity in both panels

that they connected with the surface spins. Up to date, there have been several studies with rather contrary results. Some studies suggested increase of the spin canting angle with decreasing particle size [17, 24]. Moreover, the same studies also investigated the influence of interparticle interactions (through the dilution of the samples in a polymer matrix) with no significant effect on the spin canting angle.

On the other hand, there are several works showing that the relative crystallinity of the NP is more important [24–27]. For example, Morrish et al [24] compared the maghemite NPs with different particle size and relative crystallinity—the well-crystalline 7 nm NPs and 95 nm NPs with high degree of disorder (the crystalline part had a mean diameter ~ 30 nm). Their results suggested that the NP internal structure is important and must be considered in any analysis of the surface spin canting as they observed quite a small difference in the spin canting angle in the two samples. Morales et al [25] studied 100 nm maghemite NPs with different vacancy ordering and found out that the spin canting angle varied with the vacancy ordering, suggesting the spin canting is a volume effect. So far, the volume or surface nature of the spin canting has been still debated.

There has been also several theoretical studies trying to solve the presence of the surface effects [13, 28, 29]. For instance, Pankhurst et al [29] proposed that the complete spin alignment cannot be achieved in the iron(III) oxide NPs due to the large magnetic anisotropy constant. After few years, this theoretical suggestion was questioned by Hendriksen et al in 1994 [13] who measured the IFMS of frozen ferrofluids. The study revealed that the spin canting angle is independent of the initial orientation of the magnetic moments in the NPs with respect to the external field; thus, the origin of the spin canting cannot be explained by enhanced magnetic anisotropy.

In past decade, the role of structure and morphology of NPs became a strong argument in explaining the complexity of spin arrangement in the magnetic NPs. It has been reported that the magnetic properties of NPs mostly result from their internal structure [14, 27, 30, 31], particle size distribution [32, 33] and/or interparticle interactions [15, 34].

Yet, correlation of the magnetic properties with the crystal structure or particle size is not as straightforward as it is generally believed. First, it is usually claimed that very small NPs exhibit high ratio of the surface to volume spins that lowers the resulting saturation magnetization, M_s [22, 23, 26]. However, several studies have shown smaller M_s for large NPs [35, 36]. Furthermore, it is a common practice to evaluate magnetic properties of NPs upon their size, which is a vague term when taking into account the fact that every experimental method such as X-ray Diffraction (XRD), Transmission Electron Microscopy (TEM) or magnetization measurements provides a "size" with a different physical meaning. For instance, the TEM determines only a specific projection of the physical size of a NP, d_{TEM}, without any deeper insight into its relative crystallinity. Thus, correlations of the d_{TEM} with the magnetic response of NPs is highly misleading, but it has been widely performed [35, 37].

Without doubts, the internal structural and spin arrangement play important role and their proper investigation is a crucial point in correlation of structural and magnetic parameters of the NPs with their magnetic properties related to their performance in applications.

With increasing availability of the advanced instrumental techniques and advancements in the data processing and analysis, the current trend is to drain maximum information from a specific technique taking into account its operation range by means of dimension and time (or frequency) as it is summarized in Fig. 2.

A great benefit of the MS is its elemental selectivity and a very local character of probing the sample. Nevertheless, when studying very small objects such as NPs, one has to consider that all sample parameters with a given distribution will introduce complexity to the resulting spectra. Vice versa, a complicated character of the sample can be tracked with the help of MS via the complex information encoded in the MS.

MS also has a very sharp location on the energy (frequency). Therefore, it is often impossible to use it as a single probe to structure of magnetism of the material. Besides the less common techniques operating on the same scale, such as nuclear magnetic resonance NMR [36] and neutron diffraction [38], the widely used methods for investigation of magnetic NPs are: transmission electron microscopy (TEM) including high-resolution TEM (HRTEM) and electron energy loss spectroscopy (EELS), X-ray diffraction (XRD), dynamic light scattering (DLS) and various magnetometries among which the volume methods (vibrating sample magnetometry—VSM, superconducting quantum interference device—SQUID) prevail. Some groups successfully applied X-ray Photoemission Spectroscopy (XAS) and Extended Absorption Fine Structure (EXAFS) to get more insight into the local coordination environment of the cations in the spinel lattice [39]. In addition, advanced studies based on small-angle X-ray and neutron scattering and X-ray Magnetic Circular Dichroism (XMCD) have been performed [40–44]. Magnetic microscopies and other microscale methods such as magnetic force microscopy (MFM) and micro-Hall or micro-SQUID magnetometries are rarely applied as they

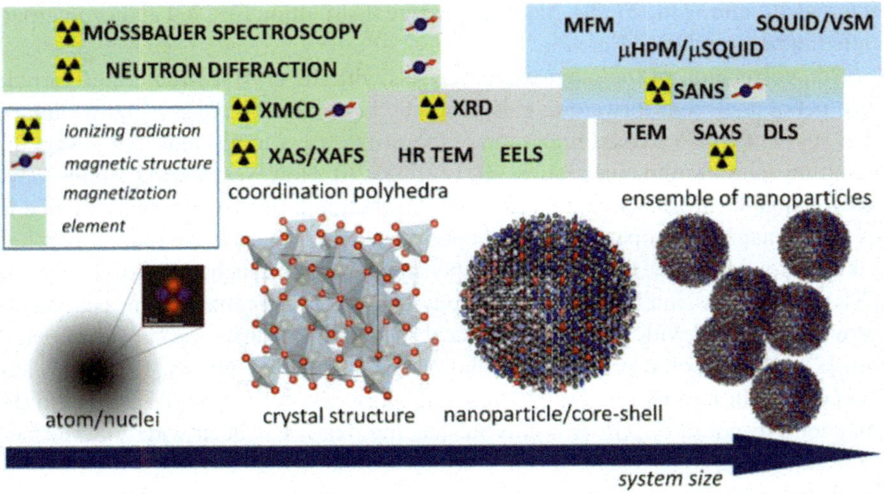

Fig. 2 Tentative diagram of the most common experimental methods used for investigation of magnetic nanoparticles in comparison to their typical operation range (given by the system size ranging from a single atom over a coordination polyhedron, unit cell, nanoparticle and ensemble of nanoparticles). The green blocks correspond to the methods, which are capable of elemental (chemical) selectivity, while the blue blocks mark the magnetometry-related methods, which enable detection of a certain sum of elementary magnetic moments. The grey blocks contain techniques with no specific elemental and magnetic selectivity; please note that for the spinel ferrites the Z values are very similar; thus, the electron density driven contrast is very low. Methods requiring ionizing radiation are marked with the "radiation" symbol. Methods which enable to resolve magnetic structure either on the level of a magnetic unit cell or on the level of a single particle are labelled by a "spin" symbol

require positioning of the NPs on a flat surface or a microprobe, respectively, and a signal from a single NP can be recorded [45, 46]. However, the techniques have very limited usage for understanding the complex structural and spin architectures within a single magnetic NP.

At this point, one has to revisit the meaning of the core–shell term. In the original concept by Coey, the core–shell is related to a different spin structure of the two parts of a single NP. In last few years, the so-called bimagnetic core–shell spinel oxide NPs have been explored intensively, namely due to their great performance in magnetic fluid hyperthermia [47]. In such NPs, one has to keep in mind that are in principle three different possibilities of understanding the core–shell phenomenon. First, the chemical composition varies across the NP volume; the core is composed of a different spinel ferrite than the shell, and a kind of "chemical core–shell" structure can be introduced. The bimagnetic core–shell NPs are typically prepared by a seed-mediated growth, which means that the second spinel ferrite phase (shell) is grown on a different spinel ferrite NP (core). However, the core has already a certain amount of structural disorder at the surface [30]. Consequently, a kind of internal interface within a single NP is formed and another structurally disordered layer forms in the proximity of the NP surface. Consequently, quite complex "structural core–shell" arrangement can be observed. Keeping in mind the already complex "structural core–shell" situation, the final arrangement of magnetic moments in a "magnetic core–shell" perspective may become very complex due to the structural disorder at the inner interface and at the outer shell (please see the Fig. 1, panel (b)). Coming back to the common characterization techniques, the information they provide is usually related to one of the possible core–shell concepts. In this context, the MS (except photoelectron spectroscopies, which have limited usage due to very low penetration depth) is the only one providing information about the Fe valence. Moreover, a great challenge nowadays is to understand the MS experiments on the bimagnetic core–shell NPs to give more insight into all types of core–shell arrangement. In this vein, advanced MS studies profiting from temperature- and magnetic field-dependent experiments are needed. The most important aspects will be summarized in the next section.

3 MS of Magnetic NPs—Fundamentals

The small NPs exhibit relaxation time in order of 10^{-9} s that is close to the time window of the MS (10^{-8} s). It allows us to study the relaxation of the NPs by means of MS. Furthermore, the big advantage of the MS is that it is not restricted to the well-crystalline samples, and thus, all kinds of spinel ferrite NPs can be investigated using MS. An overview of the essential information embedded in the parameters of the spectra (position, intensity and profile) is given in Fig. 3, panel (a).

Below the blocking temperature T_B, NPs are in blocked state, thus the relaxation across the energy barrier is negligible, and the magnetically split spectra (evolved in sextet) are observed in the MS spectra. However, the macrospin fluctuates around the direction of the magnetization easy axis resulting in the so-called collective magnetic excitations that reduces the value of B_{hf} to resulting B_{obs} according to the equation:

$$B_{\mathrm{obs}} = B_0\left(1 - \frac{k_B T}{2K_{\mathrm{eff}} V}\right), \tag{2}$$

where B_0 is the B_{hf} without the effect of the collective magnetic excitations. The maximum reduction is found to be around 5—15%; the higher reduction leads to collapse of the magnetically split spectra to a broad singlet or doublet [48]. The temperature dependence of the B_{hf} in the blocked state is usually used for the determination of the K_{eff} of the NPs [2, 48, 49].

With increasing temperature, the thermal fluctuations dominate, leading to the collapse of the magnetically split spectra. The T_B determined from the MS is defined as the temperature at which half of the spectra area is magnetically split (sextet) and half of the spectra is in a doublet and/or singlet form.

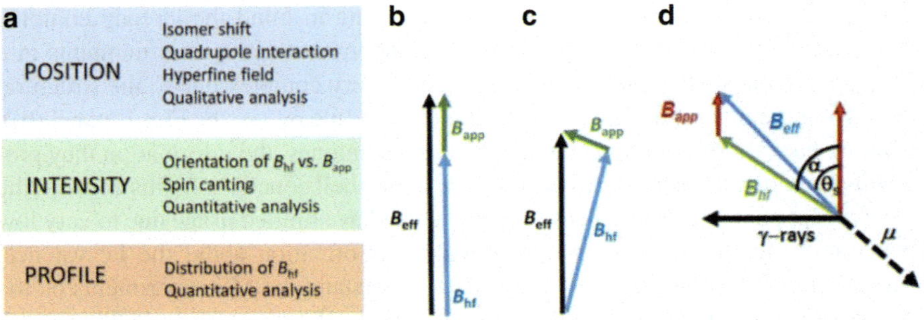

Fig. 3 Scheme of the parameters that can be derived from the refinement of MS (**a**). The simple sketch of behaviour of the B_{hf} in B_{app}: the parallel orientation of the B_{hf} with respect to B_{app} (**b**); the nonparallel direction of the B_{hf} with B_{app} that results in the smaller B_{eff} is depicted on the panel (**c**). The scheme of the spin canting angles derived from the perpendicular set-up of the IFMS (**d**). The θ_s is the angle between the B_{eff} and B_{app}; the α is the angle between the B_{hf} and the B_{eff}. The magnetic moment, μ, points antiparallel with the respect to the B_{hf} in the ferrite oxides

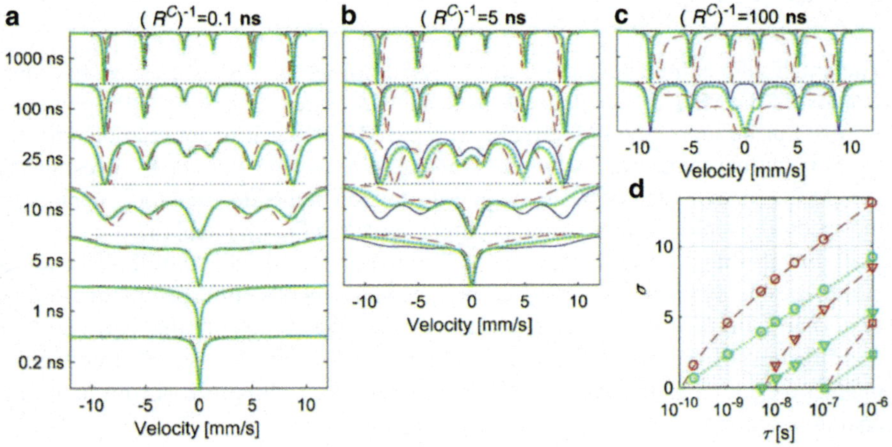

Fig. 4 Theoretical Mössbauer relaxation spectra for a hyperfine field switching between ±55 T and zero quadrupole interaction at SPM relaxation time calculated using the Blume–Tjon model [53] (blue, full lines), the three-level (S = 1) model (green, dotted lines) and the multilevel (S = 60) model (red, dashed lines). For the multilevel model, the relaxation time, τ, was calculated using the Brown model with the indicated values of R^{-1} ($\tau = R^{-1}f(\sigma)$;$\sigma = KV/k_BT$, where K is the anisotropy constant, V is the NP volume), and for the three-level model, it was calculated using the simplified Néel model. Panel (d) shows the results for the three-level model (green) and the multilevel model (red) for R^{-1} = 0.1 ns (circles), 5 ns (triangles) and 100 ns (squares). (Adopted from ref. [50])

A recent work by Fock et al [50] treated the problem of MS from the viewpoint of fluctuations of the direction of the magnetic hyperfine field by theoretical means. For noninteracting particles, the SPM relaxation results in the spectra consisting of a sum of a sextet and a doublet with a temperature-dependent area ratio. This is in accordance with the exponential dependence of the SPM relaxation time on the particle size and temperature. An alternative interpretation of these features given by some authors is a first-order magnetic transition from a magnetically ordered state to a paramagnetic state [51]. However, the mistaken base of this interpretation is even corroborated by the fact that the doublet component has been found to transform to a magnetically split component when relatively small fields are applied, which excludes the formation of a paramagnetic state. In other cases, the spectra of magnetic NPs consist of sextets with asymmetrically broadened lines without the presence of doublets. It has been suggested that such spectra can be explained by a multilevel model [52], according to which the relaxation takes place between a large number of states. The authors suggested that the spectra with asymmetrically broadened lines (at least in some cases) should be better explained by the influence of magnetic interparticle interactions on the magnetic fluctuations, which brings the importance of the so-called real effects (size distribution, interparticle interactions) into play again.

Figure 4 gives an overview of calculated spectra using Blume–Tjon [53] and Jones–Srivastava model [52] in the three-level (S = 1) and multilevel (S = 60) conditions. While the character of the spectra is somewhat comparable, the increasing S (number of levels) in the multilevel model leads to decrease of the B_{hf}.

In the real systems, the particle size distribution generally results in the distribution of the relaxation time. Due to the exponential dependence of the relaxation time, the broad distribution of the B_{hf} is observed in these systems [49, 50].

In order to obtain some information about the possible core–shell spin structure, represented often by the so-called spin canting angle (angle between the B_{app} and B_{hf}/B_{eff} of a NP) that is usually attributed to the presence of the surface spins, the in-field MS (IFMS) is usually performed. However, the computation of the spin canting angle is not so obvious in spinel ferrite and bimagnetic spinel ferrite NPs, and the precise determination of its value is in most cases impossible.

Generally, there are two approaches, how the spin canting angle can be determined. First, the spin canting angle can be derived from the relative areas of the 2nd and 3rd absorption lines. For an arbitrary angle between the γ-rays and magnetic field, the resulted equation is:

$$3:\frac{4\sin^2\theta_S}{1+\cos^2\theta_S}:1, \tag{3}$$

where θ_s is the angle between the B_{hf} and the direction of γ- rays. This formula is only valid for the transitions from $I = 3/2$ to $I = ½$ levels, that is the case of the nucleus of ^{57}Fe. In the case of powder samples, the probability (thus the relative intensities) is proportional only to the square of the Clebsch–Gordan coefficients resulting in the 3:2:1:1:2:3 ratio [2].

It has to be noted that this approach can be inaccurate due to several reasons. First, the area of the absorption lines is affected by relative thickness of the sample, and the thick sample results in the saturation of the lines, thus affecting ratio of the line areas [54]. Second, the θ_s is the angle between the B_{eff} and B_{app} that does not coincide with the orientation of the magnetic moment, μ, as is displayed in Fig. 3, panel (d).

In ferrite oxides, the μ of the nucleus is oriented antiparallel to the $B_{\rm hf}$ due to the negative Fermi contact term, others contributions can be neglected, and therefore, the size should be proportional to the value of the B_{hf}. If the external field is applied, the magnetic moment tends to align in parallel direction with the $B_{\rm app}$. The B_{hf} will rotate as to be antiparallel with the μ. Therefore, even in $B_{\rm app}$, the μ is antiparallel with the B_{hf}, not the $B_{\rm eff}$. Determination of the angle from the area of the lines is related to the angle between the quantization axis and the direction of the γ-rays that is usually parallel with the $B_{\rm app}$, as it is in spinel ferrites.

There are two possible experimental set-ups used for the IFMS. First, the $B_{\rm app}$ is applied parallel with the direction of the γ-rays (Fig. 3, panel (b)); the resulting equation is similar to the (2). Second, the $B_{\rm app}$ is applied perpendicularly to the direction of the γ-rays (Fig. 3, panel(c)), and the Eq. (2) converts to the:

$$4\left(\frac{2}{\sqrt{1+\sin^2\theta_S}}-1\right):1. \tag{4}$$

The second approach accounting for the nonzero angle between the B_{hf} and $B_{\rm app}$ that is physically relevant requires the measurements of both the zero-field spectrum together with the in-field spectrum, yielding the B_{hf} and $B_{\rm eff}$, respectively. The spin canting angle is then given by the following expression:

$$\cos\alpha = \frac{B_{\rm eff}^{2} - B_{\rm app}^{2} - B_{hf}^{2}}{2B_{\rm app}B_{hf}}. \tag{5}$$

When studying spinel oxide NPs, one usually treats a whole ensemble of NPs. It has been demonstrated many times that especially the dense ensembles of magnetic NPs experience strong dipolar fields and the magnetization reversal in such systems is determined by the strength of the mutual interparticle interaction [32, 34, 55].

Based on the stochastic approaches, a simplified three-level stochastic model taking into account the magnetic anisotropy, precession and diffusion of uniform magnetization of single-domain particles was proposed by Chuev [56] in order to describe the MS of magnetic NPs in a weak magnetic field; results of his calculations are shown in Fig. 5. The MS obtained for the different geometries show clear differences, which are originated by the magnetic anisotropy, precession and diffusion of uniform magnetization of the NPs. The obvious polarization effects present in the theoretical spectra point to the fact that a relevant description of the experimental data containing randomly or partly oriented NPs (by means of their easy axis to the applied magnetic field) requires description based on the spectral parameter distributions.

Another example can be found in the recent work of Majetich et al. [57]. They have shown that multiparticle correlations, both parallel and perpendicular to the applied field, exist in ordered assemblies of NPs. By atomistic simulation, they have

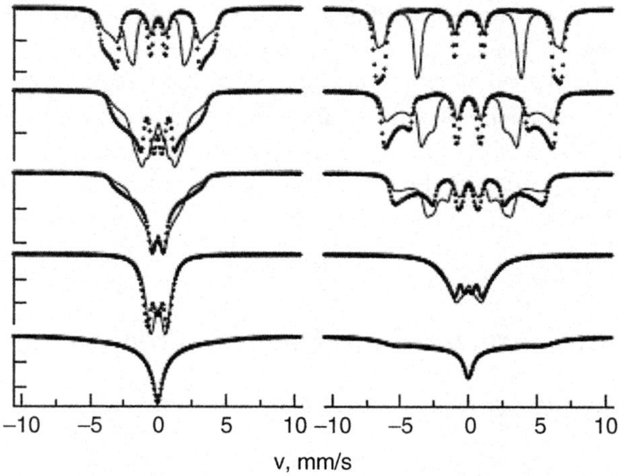

Fig. 5 IFMS of an ensemble of NPs for a transverse and longitudinal magnetic field geometry calculated using three-level relaxation model. The normalized field strength h = 0.01, 0.1, 0.5, 1 and 2 (from bottom to top). (Adopted from ref. [56])

revealed that the magnetic frustration in the shell, which may originate from Dzyaloshinskii–Moriya interactions (DMI) of Mn B site ions, leads to a modest amount of surface canting, which can act as a source of anisotropy.

Strong exchange coupling between the core and shell causes the core spins to cant, as well. In dense assemblies, magnetostatic interactions among the particles favour canting of the particle moments in the same direction. This coherent canting results in a so-called canted superferromagnet or canted supermagnet that collectively shows canted ferromagnetic behaviour.

In summary, the MS in principle provides a lot of information about the spin orientation in spinel ferrite NPs, and it is capable of reflecting real effects such as particle size distribution, internal and surface spin canting, and interparticle interactions. However, a careful analysis and a critical interpretation of the results are needed to obtain a realistic picture. Examples of exploring the chemical, structural and magnetic core–shell spinel ferrite NPs by MS are given in the next section.

4 Selection of Recent Studies

In this section, a selection of interesting works published within last ~ 10 years and going beyond the state of the art described previously is presented. Let me stress out that one can easily find few hundreds of research papers published within the last 10 years, which report on the application of MS on spinel ferrite NPs. However, most of the studies use ambient MS or temperature-dependent MS, providing only room temperature and 4.2 K or 77 K spectra. Majority of the works do not go beyond the knowledge established by the Mørup, Fiorani or Tronc and their successors, which means that the complex nature of the spectra due to the chemical, structural and magnetic disorder as well as nontrivial distribution of hyperfine parameters is not considered. The IFMS studies are also quite rare. Literally speaking, the MS is mostly used as a reliable tool for determination of the iron valence and a complementary technique of phase analysis, e.g. to the XRD. Therefore, the aim of this section is to highlight the works with substantial progress in applying the MS to the spinel ferrite NPs. First, the iron oxide NPs with a complex internal structure are discussed. Then, other spinel ferrite NPs including chemically doped systems are presented. Finally, several studies of nominally chemical core–shell spine ferrite NPs with a very complex scenario of spin order are included.

4.1 Structural and Magnetic Properties of Spinel Ferrite NPs

Iron Oxide NPs

The most studied spinel ferrite in the form of NPs is magnetite, and the MS is usually used to confirm the presence of Fe^{2+}. However, it has been demonstrated many times that the magnetite NPs undergo a continuous topotactic oxidation to maghemite, and consequently, the structural arrangement of a real NP gains complexity.

The groups of Vejpravova and Morales focused on the question of spin disorder and spin canting in a series of highly uniform iron oxide NPs [27, 30, 58].

Selected samples with different relative crystallinity were investigated, also by the IFMS with B_{app} increasing from 0 to 6 T with the step of 1 T. The evolution of hyperfine parameters was determined with emphasis on the calculation of the spin canting angle. Finally, the resulting parameters were correlated with internal structure of the NPs, which was probed by HR TEM and XRD, and corroborated by the magnetic measurements. It has been pointed out that the value of the spin canting angle is strongly dependent on the approach used and exact value of the spin canting angle is disputable. The value of θs determined from the area under a peak is strongly dependent on the refinement procedure used. The value of α is derived from the values of B_{hf}, B_{eff} and B_{app}, where the B_{hf} is determined with considerable experimental error due to the fully overlapped sextets; thus, the resulting α is only estimative.

The well-crystalline samples with different particle diameter served as great candidates to solely correlate the resulting spin canting angle with the increasing particle size. The effect of the spin disorder is expected to be negligible in these samples. On the other hand, imperfections in internal structure of the NPs with different origin were observed in the two samples with identical TEM size. One sample can be viewed as a magnetic core–shell NP (Coey-like), where the disordered spins are located at the surface. The other sample possesses stacking faults that split the highly crystalline NP into at least two crystalline domains; therefore, the disordered spins are located in the whole volume of the NP, namely at the internal interfaces. For the first time, IFMS study of these two samples allows to disentangle the origin of the possible spin canting, but only in context of other characterization techniques.

For example, the normalized IFMS at 6 T did not reveal any significant differences in the profile for both samples, the asymmetric broadening at the lower values of the B_{hf} is pronounced and comparable in both spectra, and the θs and α values are identical within the experimental error. It has been demonstrated that a realistic orientation of spins with respect to the B_{app} can be obtained by measuring the evolution of the values of B_{eff} with increasing B_{app} that gives the complex behaviour of the spins in individual sublattices [27, 58]. Please note that the quite popular refinement of the MS by the Coey-like core–shell model has to be performed carefully as the values of the B_{eff} for the disorder (shell) spins are dependent on the orientation of the spins of the individual sublattices with respect to the B_{app} [27].

Another complex study by means of magnetic, XRD and MS studies on uniform magnetite/maghemite core–shell NPs was reported by Iyengar et al [59]. A very interesting aspect of this work is the use of vibrational spectroscopies (Raman and Fourier Transform Infrared Spectroscopy, FTIR), which clearly confirmed presence of magnetite and maghemite phases with a distribution of coordination polyhedral deformation. The fit of the MS to a core–shell model (magnetite/maghemite) enabled determination of the shell thickness.

Multicore γ-Fe_2O_3 nanoparticles were studied by Kamali et al [31]. They focused on the formation of a core–shell structure, consisting of multiple maghemite NPs as the core and silica as the shell. Low-temperature MS reveals the presence of pure maghemite NPs with all vacancies at the B sites. Surprisingly, the multicore γ-Fe_2O_3 NPs show similar magnetic behaviour comparing to the isolated particles of the same size. However, this conclusion in fact illustrates the importance of taking into account interparticle interactions present in all NP samples studies as powders.

A very detailed MS study of core/shell Fe_3O_4/γ-Fe_2O_3 NPs has been published by Kamzin et al [60]. The phase composition, the structure of cores and shells, and the dependences of the shell thickness on the fabrication technique were determined for magnetite core with a fixed size (8 nm) and γ-Fe_2O_3 shells of a varying thickness (from one to five nm). It was found that the surface layer and the "bulk" of these shells had different magnetic structures. This difference was attributed either to the frustration of spin magnetic moments or to the formation of a canted spin structure in the surface layer. The intermediate layer between the core and the shell is likely to be in a spin-glass state. Both the Mössbauer spectra and the reconstructed B_{eff} distributions have characteristic features suggesting either a narrow size distribution of the studied NPs or the existence of relaxation processes in these NPs.

A work by Kalska-Szostko et al [55] pointed out to the importance of intra- and interparticle interactions as evidenced by a study carried out on ferrite magnetic NPs with core–shell structures obtained in two-step preparation process. The interesting aspect of this study is the comparison between two "inverse" core–shell structures: the magnetite shell on maghemite core and *vice versa*.

Beside the spinel iron oxide core–shell NPs, some studies report on structures composed of different iron oxides. In this case, the system is in principal homogeneous by means of elemental composition, but it can be viewed as a "structural" core–shell as the two iron oxide phases have a different lattice symmetry. Please note that the magnetite/maghemite core–shell NPs can be in principle viewed in the same way; however, both phases form a spinel structure, in spite of a different occupation of the lattice sites.

An example of such a formally "structural" core–shell system has been studied by Kamzin et al [61]. FeO/Fe_3O_4 nanoparticles were synthesized by thermal decomposition method and the MS of the phase composition of the synthesized nanoparticles clearly revealed the simultaneous presence of three phases: magnetite Fe_3O_4, maghemite γ-Fe_2O_3 and wustite FeO. This work also confirmed a scenario proposed for "simple" iron oxide NPs [30] that the saturation magnetization is not the only factor governing the SAR, and the efficiency of heating of magnetic FeO/Fe_3O_4 nanoparticles may be increased by enhancing the effective anisotropy.

Another example is the work by Lak et al [62]. Size-dependent structural and magnetic properties of FeO–Fe_3O_4 NPs prepared *via* decomposition of iron oleate were investigated. The authors applied many different experimental techniques (HR TEM, XRD, MS and magnetization measurements) and received a complete picture of the internal structural and spin order. Based on that the authors provided a model describing the phase transformation from a pure Fe_3O_4 phase to a mixture of Fe_3O_4, FeO and interfacial FeO–Fe_3O_4 phases with increasing NP size. The reduced magnetic moment in FeO–Fe_3O_4 NPs was attributed to the presence of differently oriented Fe_3O_4 crystalline domains in the outer layers and paramagnetic FeO phase. Interestingly, the exchange bias energy was found to dominate in the magnetization reversal mechanism, and the SPM blocking temperature in FeO–Fe_3O_4 NPs depends strongly on the relative volume fractions of FeO and the interfacial phase. This work is another great example of a system with a very complex structural and magnetic core–shell structure, however, composed of two elements only.

Other Spinel Ferrite NPs

The situation becomes even more complex when the spinel structure contains another element; the most common are cobalt and manganese ferrites. Nevertheless, the first complex MS studies have been reported since 90-ties of the last century. The most discussed point is the degree of the spinel inversion, which was addressed by many groups in cobalt, zinc and copper ferrite NPs [38, 63–70]. A general problem (already discussed in Sect. 2) is how to rigorously disentangle the effect of degree of inversion and the spin canting phenomenon. Though, the information about the internal structure of the NPs obtained from the XRD or TEM enables one to interpret the character of the MS and IFMS under the realistic approximations, as shown, e.g., in [63] (please see Fig. 6).

Even more complex results have been reported for spinel ferrites upon doping [71–74]. Cu-doped cobalt ferrite was explored by Batoo et al [71]. MS at room temperature shows two ferrimagnetically relaxed Zeeman sextets. The dependence of MS parameters such as isomer shift, quadrupole splitting, line width and hyperfine magnetic field on Cu^{2+} ion concentration is in line with the continuously changing magnetic parameters, such as effective anisotropy constant.

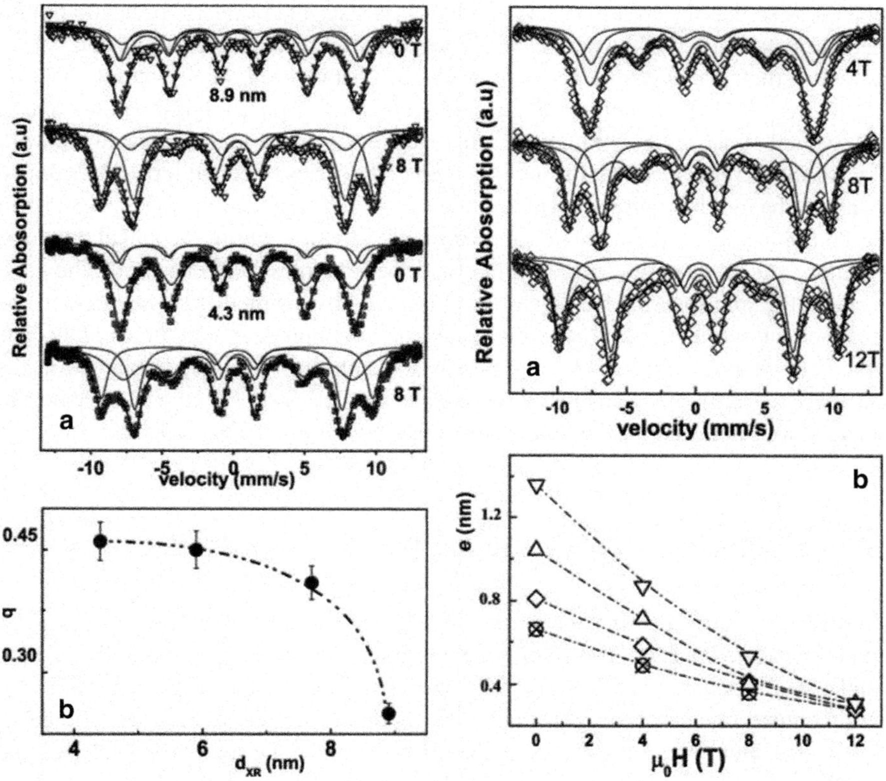

Fig. 6 (**a**) MS at 4.2 K of samples with a different particle diameter in zero and 8 T applied fields. All spectra are fitted with three magnetic sextets. (**b**) Mossbauer spectra at 4.2 K of the sample with diameter of 4.3 nm at the indicated fields. (**c**) Fraction of Fe ions belonging to the outer shell of canted spins as a function of the particle diameter. (**d**) Thickness of the disordered spins surface layer as a function of applied field for the samples with a different diameter. (Adopted from ref. [63])

Complex magnetic properties resulting from core–shell interactions in nanosized $Ni_{0.25}Co_{0.25}Zn_{0.5}Fe_2O_4$ synthesized by chemical coprecipitation method were treated by Lakshmi et al [72]. The authors observed exchange bias phenomenon, and they suggested that the effect is arising from the core–shell interaction. The observed variation in coercivity and exchange bias field suggested that only the core is affected by the cooling field. The authors used complementary IFMS regimes: zero-field-cooled and field-cooled and parallel and perpendicular orientation of the γ rays with respect to the B_{app}, respectively. Careful analysis of the spectra revealed that 70% of the spins are in the shell. The system can be modelled as an ordered core with conventional collinear arrangement of spins at the A and B sites and a canted, highly frustrated surface. The effects observed are comparable to those obtained in nanogranular systems comprising ferromagnetic particles embedded in an antiferromagnetic matrix. The large volume fraction of surface spins completely isolates the cores so that the entire ensemble behaves as a system of nearly perfectly noninteracting particles. However, the NPs were prepared by coprecipitation, and consequently, they have significant size distribution and level of structural disorder therefore all absolute values should not be considered on a too quantitative basis. The spinel lattice can accommodate other elements, such as lanthanides. In the work by Burianova et al [73, 74], the authors prepared a series of cobalt ferrite NPs doped with lanthanum by microemulsion method. Detailed IFMS studies were performed to determine spin canting angles and cation distribution within the spinel network. Canting angles up to 40 degrees in the La-doped samples were observed. The presence of the spin surface effects was also supported by magnetic measurement as the magnetization did not saturate even in considerably high magnetic fields (7 T). The observed features originated from the surface spin disorder were explained in the simplified frame of the magnetic core–shell model; however, the importance of local strains and structural disorder due to the La doping is also proposed.

4.2 Chemical Core–Shell Spinel Ferrite Nanoparticles

It was demonstrated in the previous section that the core–shell phenomenon is a much intricate concept outreaching the common view of a NP composed of a core and a shell, each of a different material or with a different spin alignment. In a general context, one can distinguish between chemical, structural and magnetic core–shell structures. With increasing number of elements in the spinel oxide phases and artificially formed "chemical" core–shell NPs with multiple internal regions with a different composition and structural/spin order, rigorous understanding limits to "mission impossible". As already mentioned, the MS has quite limited information ability when used as a single probe of chemistry, structure and magnetism. Therefore, realistic approximations profiting from a robust methodological approach must be considered.

A showcase study of bimagnetic core–shell NPs composed of cobalt ferrite core and iron oxide or manganese ferrite shell was reported by the group of Cannas and co-workers [75]. They obtained a clear evidence of the chemical, structural and magnetic core–shell formation indirectly by comparing the MS of the core–shell samples and an *ad hoc* mechanical mixture and directly by mapping the NP's chemical composition by EELS. Chemical-sensitive electron tomography revealed detailed three-dimensional images of the NPs with a subnanometre spatial resolution.

Another great example featuring a maximum synergy of the experimental and theoretical methodologies has been recently released by the York group [57]. In my opinion, the work represents the benchmark by means of exactness for all future studies on spinel ferrite NPs.

The study focuses on intraparticle and interparticle effects in Fe_3O_4/Mn-ferrite core/shell structures [57]. The authors demonstrated that strong DMI can lead to magnetic frustration within the shell and cause canting of the net NP macrospin. The chemical composition and structural composition explored with the help of

MS, XAS and XMCD were used to determine the configuration on the Mn sites. In addition, polarized SANS experiments were performed and revealed parallel and perpendicular magnetic correlations, suggesting multiparticle coherent spin canting in an applied field. Atomistic simulations have revealed the underlying mechanism of the observed spin canting (for a graphical representation of the model, please see Fig. 7). The results show that strong DMI can lead to magnetic frustration within the shell and cause canting of the net particle moment. Nevertheless, strong exchange coupling between the core and shell also causes the core spins to cant. In ensembles of NPs, the magnetostatic interactions cause the NP magnetic moments to cant in the same direction giving rise to a superferromagnet or canted supermagnet state. These results have illuminated how core–shell NP systems can be engineered for spin canting across the whole volume of the particle, rather than solely at the surface. In context of practical applications, the local strains and DMI could affect the response of NP spins to the high-frequency magnetic field in the magnetic fluid hyperthermia, as suggested previously [30].

5 Conclusions and Future Prospects

Within the last decade, the research on spinel oxide NPs has been mostly motivated by fine tuning of the NPs for biomedical applications. It has been demonstrated that the MS is a great tool for addressing various aspects of spinel oxide NPs. Nevertheless, the power of MS is even multiplied when the results are put in context of other probes operating on different structural and magnetic scales and corroborated by theoretical calculations. In particular, the complex nature of the structural and magnetic ordering within a single NP as well as the mesoscopic effects (size distribution, interparticle interactions) can be revealed by the advanced experiments

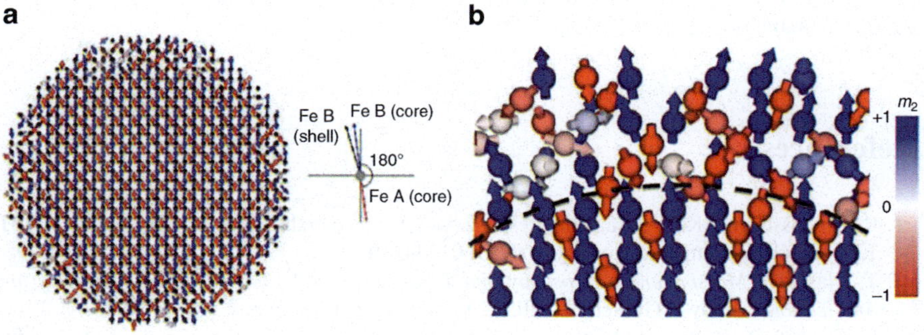

Fig. 7 (**a**) Visualization of the simulated spin configuration of a Fe_3O_4 core and $Mn(Fe_{1-x}Mn_x)_2O_4$ shell including DMI interactions on Mn B sites. The simulation temperature is set at 0 K in a 0.1 T externally applied field along the [001] crystal direction. (**b**) Enlargement of a region of panel (**a**), with dashed line to show the boundary between core and shell. Colour bar indicates direction of spin magnetization (blue, +1 or red, −1) on the spin sites. (Adopted from ref. [57])

and data analysis. Thanks to much deeper understanding of the complexity on the level of a single NP, some novel terms such as "intraparticle" interactions, "internal structure" and "nanointerface" have been introduced. In this chapter, the confusions in understanding the "core–shell" phenomenon were mitigated by definition of a chemical, structural and magnetic core–shell NPs, and the role of internal interfaces has been emphasized.

Certainly, the recent trends in understanding the fundamentals of NP magnetism include multifaceted experimental investigations in context of theoretical simulations capable of the complex system description. Such rigorous approach opens door to better understanding the mechanism underlying large variations in performance seen in magnetic NPs used for magnetic hyperthermia. For example, instead of tuning the common particle parameters (phase composition, size), the role of surface and internal spin frustration due to either DMI or local strains could affect the response of entire NP to alternating magnetic fields, which would impact the heat generation.

Being affected with the current situation of COVID-19 pandemic disease, the role of the tiny magnets in the detection and even therapeutic protocols should not be overlooked as they are widely present in numerous commercial kits for RNA and DNA separation. For example, a NP-based simple viral RNA detection for a RT-PCR molecule diagnosis has been reported very recently [76]. The development of high-impact methodologies based on the spinel oxide (ferrite) NPs can both speed up the detection procedures and make them very efficient by means of miniaturization. Therefore, the new wave of magnetic NP research may rise due to the great promises of their utilization in the revolutionary diagnostic technologies. Nevertheless, understanding the fundamentals of their magnetic performance is a must for a targeted development in this direction.

Acknowledgments This work was supported by the European Research Council (ERC-Stg-716265) and Ministry of Education, Youth and Sports of the Czech Republic under Operational Programme Research, Development and Education, project Carbon allotropes with rationalized nanointerfaces and nanolinks for environmental and biomedical applications (CARAT), number CZ.02.1.01/0.0/0.0/16_026/0008382.

References

1. R.L. Mössbauer, Kernresonanzfluoreszenz von Gammastrahlung in Ir191. Z. Phys. **151**, 124–143 (1958). https://doi.org/10.1007/BF01344210
2. T. Glaser, in *Mössbauer Spectroscopy and Transition Metal Chemistry. Fundamentals and Applications*, ed. by P. Gütlich, E. Bill, A. X. Trautwein, (Springer, Berlin, Heidelberg, 2011)
3. E. Gerdau, R. Rüffer, H. Winkler, et al., Nuclear Bragg diffraction of synchrotron radiation in yttrium iron garnet. Phys. Rev. Lett. **54**, 835–838 (1985). https://doi.org/10.1103/PhysRevLett.54.835
4. G.V. Smirnov, U. van Bürck, A.I. Chumakov, et al., Synchrotron Mössbauer source. Phys. Rev. B **55**, 5811–5815 (1997). https://doi.org/10.1103/PhysRevB.55.5811

5. M. Seto, R. Masuda, S. Higashitaniguchi, et al., Synchrotron-radiation-based Mössbauer spectroscopy. Phys. Rev. Lett. **102**, 217602 (2009). https://doi.org/10.1103/PhysRevLett.102.217602
6. T. Li, X. Zhang, The prime beat components extraction method for the time spectra analysis of nuclear resonant forward scattering. Materials, **12**(10), 1657 (2019). https://doi.org/10.3390/ma12101657
7. J.B. Hastings, D.P. Siddons, U. van Bürck, et al., Mössbauer spectroscopy using synchrotron radiation. Phys. Rev. Lett. **66**, 770–773 (1991). https://doi.org/10.1103/PhysRevLett.66.770
8. E. Gerdau, R. Rüffer, R. Hollatz, J.P. Hannon, Quantum beats from nuclei excited by synchrotron radiation. Phys. Rev. Lett. **57**, 1141–1144 (1986). https://doi.org/10.1103/PhysRevLett.57.1141
9. R. Masuda, K. Kusada, T. Yoshida, et al., Synchrotron-radiation-based Mössbauer absorption spectroscopy with high resonant energy nuclides. Hyperfine Interact. **240**, 1–6 (2019). https://doi.org/10.1007/s10751-019-1672-x
10. M. Seto, R. Masuda, S. Higashitaniguchi, et al., Mössbauer spectroscopy in the energy domain using synchrotron radiation. J. Phys. Conf. Ser. **217**, 012002 (2010). https://doi.org/10.1088/1742-6596/217/1/012002
11. R. Masuda, Y. Kobayashi, S. Kitao, et al., 61 Ni synchrotron radiation-based Mössbauer spectroscopy of nickel-based nanoparticles with hexagonal structure. Sci. Rep. **6**, 6–10 (2016). https://doi.org/10.1038/srep20861
12. E. Tronc, P. Prene, J.P. Jolivet, et al., Magnetic behaviour of γ-Fe2O3 nanoparticles by mössbauer spectroscopy and magnetic measurements. Hyperfine Interact. **95**, 129–148 (1995). https://doi.org/10.1007/BF02146310
13. P.V. Hendriksen, S. Linderoth, C.A. Oxborrow, S. Morup, Ultrafine maghemite particles. II. the spin-canting effect revisited. J. Phys. Condens. Matter **6**, 3091–3100 (1994). https://doi.org/10.1088/0953-8984/6/16/014
14. S. Mørup, E. Brok, C. Frandsen, Spin structures in magnetic nanoparticles. J. Nanomater. **2013** (2013). https://doi.org/10.1155/2013/720629
15. S. Mørup, M.F. Hansen, C. Frandsen, Magnetic interactions between nanoparticles. Beilstein J. Nanotechnol. **1**, 182–190 (2010). https://doi.org/10.3762/bjnano.1.22
16. E. Tronc, P. Prené, J.P. Jolivet, et al., Spin canting in γ-Fe2O3 nanoparticles. Hyperfine Interact. **112**, 97–100 (1998)
17. P. Prené, E. Tronc, J.P. Jolivet, et al., Mössbauer investigation of non-aggregated γ-Fe2O3 particles. Hyperfine Interact. **93**, 1409–1414 (1994). https://doi.org/10.1007/BF02072885
18. J. Stöhr, H.C. Siegmann, *From Fundamentals to Nanoscale Dynamics* (Springer-Verlag, Berlin, Heidelberg, 2006)
19. B.D. Cullity, C.D. Graham, *Introduction to Magnetic Materials* (Wiley, Hoboken, 2008)
20. W.F. Brown, Thermal fluctuations of a single-domain particle. Phys. Rev. **130**, 1677–1686 (1963). https://doi.org/10.1103/PhysRev.130.1677
21. E.C. Stoner, E.P. Wohlfarth, A mechanism of magnetic hysteresis in heterogeneous alloys. Philos. Trans. R. Soc. Lond. Ser. A Math. Phys. Sci. **240**, 599–642 (1948). https://doi.org/10.1098/rsta.1948.0007
22. J.M.D. Coey, Noncollinear spin arrangement in ultrafine ferrimagnetic crystallites. Phys. Rev. Lett. **27**, 1140–1142 (1971). https://doi.org/10.1103/PhysRevLett.27.1140
23. J.M.D. Coey, D. Khalafalla, Superparamagnetic γ-Fe2O3. Phys. Status Solidi **11**, 229–241 (1972). https://doi.org/10.1002/pssa.2210110125
24. A. Morrish, K. Haneda, P. Schurer, Surface magnetic structure of small γ-Fe2O3 particles. J. Phys. Colloques. **37**, C6-301–C6-305 (1976). https://doi.org/10.1051/jphyscol:1976663
25. M.P. Morales, S. Veintemillas-Verdaguer, M.I. Montero, et al., Surface and internal spin canting in γ-Fe2O3 nanoparticles. Chem. Mater. **11**, 3058–3064 (1999). https://doi.org/10.1021/cm991018f
26. A.G. Roca, D. Niznansky, J. Poltierova-Vejpravova, et al., Magnetite nanoparticles with no surface spin canting. J. Appl. Phys. **105**, 114309 (2009). https://doi.org/10.1063/1.3133228

27. S. Kubickova, D. Niznansky, M.P. Morales Herrero, et al., Structural disorder versus spin canting in monodisperse maghemite nanocrystals. Appl. Phys. Lett. **104**, 223105 (2014). https://doi.org/10.1063/1.4881331
28. H. Kachkachi, A. Ezzir, M. Noguès, E. Tronc, Surface effects in nanoparticles: application to maghemite -Fe O. Eur. Phys. J. B. **14**, 681–689 (2000). https://doi.org/10.1007/s100510051079
29. Q.A. Pankhurst, R.J. Pollard, Origin of the spin-canting anomaly in small ferrimagnetic particles. Phys. Rev. Lett. **67**, 248–250 (1991). https://doi.org/10.1103/PhysRevLett.67.248
30. B. Pacakova, S. Kubickova, G. Salas, et al., The internal structure of magnetic nanoparticles determines the magnetic response. Nanoscale **9**, 5129 (2017). https://doi.org/10.1039/c6nr07262c
31. S. Kamali, E. Bringas, H.-Y. Hah, et al., Magnetism and Mossbauer study of formation of multi-core gamma-Fe2O3 nanoparticles. J. Magn. Magn. Mater. **451**, 131–136 (2018). https://doi.org/10.1016/j.jmmm.2017.10.102
32. B. Pacakova, A. Mantlikova, D. Niznansky, et al., Understanding particle size and distance driven competition of interparticle interactions and effective single-particle anisotropy. J. Phys. Condens. Matter **28**, 206004 (2016). https://doi.org/10.1088/0953-8984/28/20/206004
33. H.S. Dehsari, V. Ksenofontov, A. Moeller, et al., Determining magnetite/maghemite composition and core-shell nanostructure from magnetization curve for iron oxide nanoparticles. J. Phys. Chem. C **122**, 28292–28301 (2018). https://doi.org/10.1021/acs.jpcc.8b06927
34. D. Fiorani, D. Peddis, Understanding dynamics of interacting magnetic nanoparticles: From the weak interaction regime to the collective superspin glass state. J. Phys. Conf. Ser. **521**, 012006 (2014). https://doi.org/10.1088/1742-6596/521/1/012006
35. G. Salas, C. Casado, F.J. Teran, et al., Controlled synthesis of uniform magnetite nanocrystals with high-quality properties for biomedical applications. J. Mater. Chem. **22**, 21065–21075 (2012). https://doi.org/10.1039/c2jm34402e
36. S. Belaïd, S. Laurent, M. Vermeech, et al., A new approach to follow the formation of iron oxide nanoparticles synthesized by thermal decomposition. Nanotechnology **24**, 055705 (2013). https://doi.org/10.1088/0957-4484/24/5/055705
37. M. Estrader, A. Lopez-Ortega, I.V. Golosovsky, et al., Origin of the large dispersion of magnetic properties in nanostructured oxides: FexO/Fe3O4 nanoparticles as a case study. Nanoscale **7**, 3002–3015 (2015). https://doi.org/10.1039/c4nr06351a
38. V. Blanco-Gutierrez, E. Climent-Pascual, M.J. Torralvo-Fernandez, et al., Neutron diffraction study and superparamagnetic behavior of ZnFe2O4 nanoparticles obtained with different conditions. J. Solid State Chem. **184**, 1608–1613 (2011). https://doi.org/10.1016/j.jssc.2011.04.034
39. A. Kuzmin, J. Chaboy, EXAFS and XANES analysis of oxides at the nanoscale. IUCrJ **1**, 571–589 (2014). https://doi.org/10.1107/S2052252514021101
40. P. Strunz, D. Mukherji, G. Pigozzi, et al., Characterization of core-shell nanoparticles by small angle neutron scattering. Appl. Phys. A Mater. Sci. Process. **88**, 277–284 (2007). https://doi.org/10.1007/s00339-007-4008-7
41. S. Mühlbauer, D. Honecker, É.A. Périgo, et al., Magnetic small-angle neutron scattering. Rev. Mod. Phys. **91**, 1–75 (2019). https://doi.org/10.1103/revmodphys.91.015004
42. M. Bersweiler, P. Bender, L.G. Vivas, et al., Size-dependent spatial magnetization profile of manganese-zinc ferrite M n0.2 Z n0.2 F e2.6 O4 nanoparticles. Phys. Rev. B **100**, 1–10 (2019). https://doi.org/10.1103/PhysRevB.100.144434
43. M. Bonini, A. Wiedenmann, P. Baglioni, Small angle polarized neutrons (SANSPOL) investigation of surfactant free magnetic fluid of uncoated and silica-coated cobalt-ferrite nanoparticles. J. Phys. Chem. B **108**, 14901–14906 (2004). https://doi.org/10.1021/jp049286a
44. S. Brice-Profeta, M.-A. Arrio, E. Tronc, et al., Magnetic order in γ-Fe2O3 nanoparticles: a XMCD study. J. Magn. Magn. Mater. **288**, 354–365 (2005). https://doi.org/10.1016/j.jmmm.2004.09.120
45. R. Russo, E. Esposito, C. Granata, et al., Magnetic nanoparticle characterization using nanoSQUID based on niobium Dayem bridges. Phys. Procedia **36**, 293–299 (2012). https://doi.org/10.1016/j.phpro.2012.06.162

46. L. Angeloni, D. Passeri, S. Corsetti, et al., Single nanoparticles magnetization curves by controlled tip magnetization magnetic force microscopy. Nanoscale **9**, 18000–18011 (2017). https://doi.org/10.1039/C7NR05742C
47. M. Kim, C.S. Kim, H.J. Kim, et al., Effect hyperthermia in CoFe2O4@MnFe2O4 nanoparticles studied by using field-induced Mossbauer spectroscopy. J. Korean Phys. Soc. **63**, 2175–2178 (2013). https://doi.org/10.3938/jkps.63.2175
48. S. Mørup, Mössbauer effect in small particles. Hyperfine Interact. **60**, 959–973 (1990). https://doi.org/10.1007/BF02399910
49. S. Mørup, H. Topsøe, B.S. Clausen, Magnetic properties of microcrystals studied by mössbauer spectroscopy. Phys. Scr. **25**, 713–719 (1982). https://doi.org/10.1088/0031-8949/25/6A/015
50. J. Fock, M.F. Hansen, C. Frandsen, S. Mørup, On the interpretation of Mössbauer spectra of magnetic nanoparticles. J. Magn. Magn. Mater. **445**, 11–21 (2018). https://doi.org/10.1016/j.jmmm.2017.08.070
51. I.P. Suzdalev, Magnetic phase transitions in nanoclusters and nanostructures. Russ. J. Inorg. Chem. **54**, 2068 (2009). https://doi.org/10.1134/S0036023609130038
52. D.H. Jones, K.K.P. Srivastava, Many-state relaxation model for the Mössbauer spectra of superparamagnets. Phys. Rev. B **34**, 7542–7548 (1986). https://doi.org/10.1103/PhysRevB.34.7542
53. M. Blume, J.A. Tjon, Mössbauer spectra in a fluctuating environment. Phys. Rev. **165**, 446–456 (1968). https://doi.org/10.1103/PhysRev.165.446
54. U. Gonser, F. Aubertin, S. Stenger, et al., Polarization and thickness effects in Mössbauer spectroscopy. Hyperfine Interact. **67**, 701–709 (1991). https://doi.org/10.1007/BF02398222
55. B. Kalska-Szostko, M. Cydzik, D. Satula, M. Giersig, Mossbauer studies of core-shell nanoparticles. Acta Phys. Pol. A **119**, 15–17 (2011). https://doi.org/10.12693/APhysPolA.119.15
56. A. Chuev, Mössbauer spectra and magnetization curves of nanoparticles in a weak magnetic field. J. Phys. Conf. Ser. **217**, 8–12 (2010). https://doi.org/10.1088/1742-6596/217/1/012011
57. S.D. Oberdick, A. Abdelgawad, C. Moya, et al., Spin canting across core/shell Fe3O4/MnxFe3−xO4 nanoparticles. Sci. Rep. **8**, 3425 (2018). https://doi.org/10.1038/s41598-018-21626-0
58. B. Pacakova, S. Kubickova, A. Mantlikova, et al., Spinel ferrite nanoparticles: Correlation of structure and magnetism, in *Magnetic Spinels- Synthesis, Properties and Applications*, (InTech, London, 2017)
59. S.J. Iyengar, M. Joy, C.K. Ghosh, et al., Magnetic, X-ray and Mossbauer studies on magnetite/maghemite core-shell nanostructures fabricated through an aqueous route. RSC Adv. **4**, 64919–64929 (2014). https://doi.org/10.1039/c4ra11283k
60. A.S. Kamzin, I.M. Obaidat, A.A. Valliulin, et al., Mossbauer studies of the structure of core/shell Fe3O4/-Fe2O3 nanoparticles. Tech. Phys. Lett. **45**, 426–429 (2019). https://doi.org/10.1134/S1063785019050079
61. A.S. Kamzin, A.A. Valiullin, H. Khurshid, et al., Mossbauer studies of core-shell FeO/Fe3O4 nanoparticles. Phys. Solid State **60**, 382–389 (2018). https://doi.org/10.1134/S1063783418020129
62. A. Lak, M. Kraken, F. Ludwig, et al., Size dependent structural and magnetic properties of FeO-Fe 3O4 nanoparticles. Nanoscale **5**, 12286–12295 (2013). https://doi.org/10.1039/c3nr04562e
63. E.C. Sousa, H.R. Rechenberg, J. Depeyrot, et al., In-field Mossbauer study of disordered surface spins in core/shell ferrite nanoparticles. J. Appl. Phys. **106**, 93901 (2009). https://doi.org/10.1063/1.3245326
64. S. Kubickova, J. Vejpravova, P. Holec, D. Niznansky, Correlation of crystal structure and magnetic properties of $Co_{(1-x)}Ni_xFe_2O_4/SiO_2$ nanocomposites. J. Magn. Magn. Mater. **334**, 102 (2013). https://doi.org/10.1016/j.jmmm.2013.01.005
65. A. Repko, D. Nižňanský, J. Poltierová-Vejpravová, A study of oleic acid-based hydrothermal preparation of CoFe2O4nanoparticles. J. Nanopart. Res. **13**, 5021–5031 (2011). https://doi.org/10.1007/s11051-011-0483-z
66. J. Vejpravová, V. Sechovsky, J. Plocek, et al., Magnetism of sol-gel fabricated Co Fe_2O_4Si O_2 nanocomposites. J. Appl. Phys. **97**, 124304 (2005). https://doi.org/10.1063/1.1929849

67. G. Concas, G. Spano, C. Cannas, et al., Inversion degree and saturation magnetization of different nanocrystalline cobalt ferrites. J. Magn. Magn. Mater. **321**, 1893–1897 (2009). https://doi.org/10.1016/j.jmmm.2008.12.001
68. M. Siddique, N.M. Butt, Effect of particle size on degree of inversion in ferrites investigated by Mössbauer spectroscopy. Phys. B Condens. Matter **405**, 4211–4215 (2010). https://doi.org/10.1016/j.physb.2010.07.012
69. J. Kurian, S.P. John, M.M. Jacob, et al., Mössbauer studies of nanocrystalline ZnFe2O4 particles prepared by spray pyrolysis method. IOP Conf. Ser. Mater. Sci. Eng. **73**, 012032 (2015). https://doi.org/10.1088/1757-899X/73/1/012032
70. N. Moumen, P. Bonville, M.P. Pileni, Control of the size of cobalt ferrite magnetic fluids: Mössbauer spectroscopy. J. Phys. Chem. **100**, 14410–14416 (1996). https://doi.org/10.1021/jp953324w
71. K.M. Batoo, D. Salah, G. Kumar, et al., Hyperfine interaction and tuning of magnetic anisotropy of Cu doped CoFe2O4 ferrite nanoparticles. J. Magn. Magn. Mater. **411**, 91–97 (2016). https://doi.org/10.1016/j.jmmm.2016.03.058
72. N. Lakshmi, H. Bhargava, O.P. Suwalka, et al., Magnetic properties resulting from core-shell interactions in nanosized Ni0.25 Co0.25 Zn0.5 Fe2 O4. Phys. Rev. B Condens. Matter Mater. Phys. **80**, 1–6 (2009). https://doi.org/10.1103/PhysRevB.80.174425
73. S. Burianova, J. Poltierova Vejpravova, P. Holec, et al., Surface spin effects in La-doped $CoFe_2O_4$ nanoparticles prepared by microemulsion route. J. Appl. Phys. **110**, 073902 (2011). https://doi.org/10.1063/1.3642992
74. S. Burianova, J.P. Vejpravova, P. Holec, et al., Observation of surface effects in La-doped CoFe2O 4/SiO2 nanocomposites. IOP Conf. Ser. Mater. Sci. Eng. **18**, 022015 (2011)
75. M.S. Angotzi, A. Musinu, V. Mameli, et al., Spinel ferrite core-shell nanostructures by a versatile solvothermal seed-mediated growth approach and study of their nanointerfaces. ACS Nano **11**, 7889–7900 (2017). https://doi.org/10.1021/acsnano.7b02349
76. Z. Zhao, H. Cui, W. Song, et al., A simple magnetic nanoparticles-based viral RNA extraction method for efficient detection of SARS-CoV-2. *bioRxiv*, 518055:2020.02.22.961268 (2020). https://doi.org/10.1101/2020.02.22.961268

Nanoscale Heat Mediators for Magnetic Hyperthermia: Materials, Problems, and Prospects

Alexandr Tovstolytkin, Anatolii Belous, Yaryna Lytvynenko, Yuliia Shlapa, Serhii Solopan, and Larissa Bubnovskaya

1 Overview

1.1 Motivation and Significance of the Problem Under Study

Among the pressing issues of our time, a special place belongs to the problem of the fight against cancer diseases, in particular with deeply located tumors. Diseases of this type cause fear and association with the death fate, although in reality it is only a diagnosis, and most cancers in the early stages of detection are treatable. The incredibly disturbing is the fast increase in cancer incidence around the world. According to the data of the World Health Organization (WHO) [1] in the absence of new methods of treatment, the number of cancer cases can grow from 19 million people in 2020 to 29.5 million until 2040. At the same time, the cancer death rate is growing too. As claimed by the same prediction of WHO [1], the estimated number of deaths due to oncological diseases during this period will increase by 71.5%: from 10.1 million people in 2020 to 16.4 million in 2040. Until 2040 a significant growth of the mortality rate is expected in African countries (>100%) and Asia (~77%), and slightly less in North America (~59%) and Europe (~32%). Therefore,

A. Tovstolytkin (✉) · Y. Lytvynenko
Institute of Magnetism of the NAS of Ukraine and MES of Ukraine, Kyiv, Ukraine
e-mail: atov@imag.kiev.ua

A. Belous · Y. Shlapa · S. Solopan
V.I. Vernadsky Institute of General and Inorganic Chemistry of the NAS of Ukraine, Kyiv, Ukraine
e-mail: belous@ionc.kiev.ua

L. Bubnovskaya
R.E. Kavetsky Institute of Experimental Pathology, Oncology and Radiobiology of the NAS of Ukraine, Kyiv, Ukraine

A. G. Roca et al. (eds.), *Surfaces and Interfaces of Metal Oxide Thin Films, Multilayers, Nanoparticles and Nano-composites*,
https://doi.org/10.1007/978-3-030-74073-3_2

the instant task is to improve already existing methods of diagnosis and treatment of the oncological diseases, particularly those with deeply located tumors, as well as developing the new ones.

1.2 Progress in Cancer Treatment and Current Challenges

Nowadays, there are powerful methods of fighting oncological tumors, such as surgery, chemotherapy, and radiation therapy (radiotherapy). The two of the last methods are characterized by high efficiency, but at the same time, they are "aggressive" with many side effects that negatively affect the normal functioning of the body and the quality of life in general. In the case of chemotherapy, the injected drugs are spread throughout the body and it is necessary to apply "huge" doses to achieve the desired concentration of the drug in the tumor area. This has a devastating effect not only on malignant formations but also on healthy organs, which significantly suppresses their vital activity. For this reason, quite often a patient must stop the drug treatment only to save the body from full destruction. At the same time, the radiation doses received during radiotherapy affect both the tumor itself and healthy tissues in its vicinity. This brings up the important questions: "Is it possible to focus the therapeutic effect only on a tumor? And how can harmful side effects be reduced?". The technological progress provides an answer to these questions—magnetic hyperthermia (MHT) is one of the promising and safe methods with the opportunity to localize the effect only on the malignant formation and is gradually being introduced into world medical practice.

The principle of MHT is based on the well-known fact that cancer cells are more sensitive to the heat than healthy body tissues. In particular, it is known that heating to 42–45 °C increases the sensitivity of the affected cells to chemotherapy or radiotherapy. In this case, the doses of drugs or irradiation can be significantly lowered (twice or even three times compared to the independent use of these methods); consequently, MHT allows reduction of the side effects. Also, cancer cells can be destroyed by heating to 46–50 °C (in medicine it is known as thermoablation).

MHT consists of three important stages: the first—injection of magnetic nanoparticles (magnetic fluids) and their localization in the tumor area; the second—nanoparticle heating in an external alternating magnetic field (AMF); and the third—the use of chemotherapy or radiotherapy when reaching and maintaining a temperature of 42–45 °C, or further heating of the tissues to the temperature of the destruction of cancer cells at 46–50 °C (thermoablation). Thus, magnetic nanoparticles play the role of heat mediators, which locally heat their environment (tumor) to the required temperature providing the desired therapeutic effect.

There are several advantages of magnetic hyperthermia over other methods of cancer treatment. Firstly, the location of the nanoparticles can be effectively controlled by the magnetic field that allows them to concentrate and held in the tumor area. Secondly, the use of magnetic hyperthermia makes it possible to treat deeply located tumors with little or no surgery. Third, the benefits include almost complete absence of the side effects of treatment.

2 Self-Controlled Magnetic Hyperthermia: Ideas and Approaches

Magnetic hyperthermia still has many unresolved issues on its way to wide usage. Important among them is the problem of reliable temperature control in the tumor area and beyond (to avoid overheating of healthy tissues), as well as achieving a uniform temperature distribution in the tumor volume.

The optimal temperature interval for the destruction of malignant tumors is 42–45 °C (46–50 °C for the case of thermoablation) [2, 3]. The problem of uneven heating of the tumor in the absence of the risk of overheating and damage to healthy tissues of the body can be solved when used as an heat mediator of MHT substances for which the phase transition from magnetic to nonmagnetic state (Curie temperature T_C) slightly exceeds the above ranges. In this case, when it reaches a temperature close to T_C, heat dissipation in the tumor containing magnetic nanoparticles is automatically stopped. This eliminates the need for precise control of the tumor temperature (since the placement of temperature sensors in the tumor is often simply impossible), as well as the power of electromagnetic radiation. Such approach also allows avoiding the risk of overheating and the associated with it subsequent damage to healthy tissues, as well as skin burns over the tumor. If necessary, this will make it possible to increase the "heat dose" by increasing the time of hyperthermal exposure. As a result, the tumors are evenly warmed up, including deeply located ones.

Currently, the most common magnetic materials for the heating of tumors and targeted drug delivery are magnetite (Fe_3O_4) and maghemite (γ-Fe_2O_3), which crystallize in a spinel structure and whose biocompatibility is well established. However, they have relatively high Curie temperatures (near or above 500 °C), which prevents them from being used in self-controlled MHT [4–7].

To uniformly heat the tumor in the absence of risk of overheating and damage to healthy body tissues, there are very promising nanoparticles developed based on strontium-substituted lanthanum manganite (La, Sr)MnO_3, which are characterized by the structure of deformed perovskite and ferromagnetism with Curie temperature T_C in the range 30–77 °C, where the latter can be tuned by changing the chemical composition and synthesis conditions. Also, magnetic materials with relatively low Curie temperature can be obtained from solid solutions with spinel structure, for example, nickel–zinc ferrite $Ni_{1-x}Zn_xMn_2O_4$.

Thus, there are several groups of nanomaterials, which are currently attracting the attention of scientists:

1. Materials *requiring external temperature control* at the time of MHT:

1.1. Ferrimagnetic spinels AFe_2O_4 (A = Fe, Ni, Mn, Co, Zn, etc.), which may have a wide range of magnetic properties due to variations in chemical composition and method of preparation [6, 8–10]. The advantages of spinel ferrites are the relatively high saturation magnetization and high chemical stability, as well as the opportunity of making an almost unlimited number of solid solutions.
1.2. Composite nanostructures, in particular those that combine soft and hard magnetic materials in core/shell nanoparticles. By combining the high saturation magnetization of soft material with the high coercivity of a hard material, it is possible to achieve the parameters overcoming those of the individual counterparts [11–14]. In this context, attention should be paid to the spinel ferrite materials due to the same crystallographic structure and almost negligible lattice mismatch, and a wide range of magnetic properties.

2. Materials with *automatic preset temperature control*, which include the following:

2.1. Solid solutions based on the ferrites with spinel structure and relatively low Curie temperature, particularly nickel–zinc ferrites $Ni_{1-x}Zn_xMn_2O_4$ [15–17].
2.2. The lanthanum–strontium manganites (La, Sr)MnO_3 with structures of perovskite. These materials have a moderate coefficient of heating in an alternating magnetic field and exhibit an easily controlled Curie temperature near the range optimal for magnetic hyperthermia [18–21].

3 Techniques for the Synthesis of Magnetic Nanoparticles

3.1 Basic Characteristics of the Heat Mediators for the Hyperthermia

Magnetic nanoparticles have to satisfy a number of requirements to be used as the heat mediators of MHT. First of all, they must meet biomedical requirements (non-toxicity, biocompatibility with living organisms, etc.). In particular, they should have certain parameters regarding cytotoxicity, genotoxicity, antioxidant, and antiviral activity [10, 22]. The size of the nanoparticles also plays a significant role: It should be in the range from 10 to 100 nm—the smaller particles are quickly removed by renal clearance, while the bigger ones are taken up by liver and spleen [23–26].

Along with the need to meet the biomedical requirements, the MHT heat mediators should have high heating efficiency under the influence of AMF with certain parameters. A quantitative measure of heating efficiency is specific loss power (SLP; also referred to as the specific absorption rate, SAR). SLP represents the energy loss of the AMF per unit time, normalized per unit mass of magnetic nanoparticles (MNPs) [18, 27].

It should be noted that some limitations also apply to the magnetic field parameters used to heat the nanoparticles. On the one hand, the heating efficiency of the nanoparticles increases with the increase in amplitude and frequency of the AMF. But on the other hand, medical restrictions are imposed on the values of these parameters for the safe effect of the field on normal tissues of the body. Thus, the frequency f should not exceed several MHz, and the product of the magnetic field amplitude by the frequency, $H_{max} f$, should be less than 5×10^9 $A \cdot m^{-1} \cdot s^{-1}$ [10, 18].

As a result, the magnetic nanoparticles used as heat mediators in the hyperthermia must meet the stringent criteria: (1) They have to be made (or coated) of non-toxic biocompatible material, (2) not stick together to ensure homogeneous distribution in the tumor and prevent thrombus formation, (3) be less than 100 nm in size for effective cell penetration, and, at the same time, (4) be characterized by efficient heating in an alternating magnetic field.

3.2 *Methods of Synthesis: Advantages and Drawbacks*

To prepare the nanoparticles (NPs), it is necessary to choose the methods of synthesis, which would satisfy the previously presented requirements. However, different materials acquire different properties at the same conditions of synthesis; therefore, *there are no sole (universal) methods of synthesis*, which would be able to allow preparing the nanostructures with desired (predefined) properties.

One of the features, which plays an important role during choosing the method and conditions of synthesis, is the energy of the formation of crystalline structure. According to the literature data [28], the energy of formation for the spinel structure equals approximately 30–60 kJ/mol that is much less than for the perovskite structure, for example. Due to this fact, nanosized weakly agglomerated crystalline NPs of the materials with the spinel structure can be obtained during the synthesis process *at relatively low temperatures*. It allows the *crystalline* NPs of materials with the spinel structure during the synthesis *without additional heat treatment*.

At the same time, such a method of synthesis does not allow obtaining the crystalline NPs of perovskites since the energy of the formation of the crystalline structure for them is much higher [29]. Therefore, an *amorphous* precipitate is formed during the synthesis of NPs with the perovskite structure in solutions independently on the method; to obtain *the crystalline* particles, it is necessary to carry out *additional heat treatment*, which results in the growth of NPs and their significant agglomeration.

Structure and properties of (La, Sr)MnO_3 manganite NPs, which crystallize in the perovskite structure, are sensitive to the conditions and methods of synthesis, and it is necessary to perform the synthesis in solutions to obtain them in nanoscale form. It is known that the temperature of formation of single-phase crystalline product is higher than 1000 °C during the synthesis of manganite NPs in aqueous solutions that leads to the growth and agglomeration of the particles [30]. Therefore, the methods of synthesis with using *the nonaqueous solutions and organic compounds* can be promising because they can allow blocking the interactions between individual particles in contrast to the NPs synthesized in an aqueous medium.

Consequently, taking into account the features of the crystalline structure, *several methods* were chosen for the synthesis of magnetic NPs with the spinel and perovskite structures: synthesis via cryochemical method, precipitation in microemulsions and nonaqueous solutions, hydroxide precipitation, and solgel method.

3.3 Synthesis of Weakly Agglomerated Nanoparticles of Ferrispinels AFe_2O_4 (A = Fe, Ni, Co, Mn, Zn)

To synthesize the NPs with the spinel structure by the precipitation methods, aqueous solutions of metal salts $FeCl_3$, $FeSO_4$, $Ni(NO_3)_2$, and $Zn(NO_3)_2$ were used as the starting reagents and solutions of ammonia or sodium hydroxide as the precipitators. Oleic acid was used for the stabilization of synthesized NPs.

Synthesis of Fe_3O_4 NPs via cryochemical method was performed in the closed system in the argon atmosphere according to the technique described in [31]. The solution, which contained the stoichiometric amount of $FeSO_4$ and $FeCl_3$ salts, was frozen using the liquid nitrogen up to the formation of ice, which was placed into the aqueous solution of ammonia under the intensive stirring to the formation of the black mixture. The obtained mixture was stirred for 20 min at 20–30 °C. Formed NPs were separated and washed with the bidistilled water. Prepared product was dispersed in the ethanol; oleic acid was added to the solution, and it was stirred for 2 h at 50 °C. NPs coated with the oleic acid were decantated, washed with ethanol, and dispersed in the necessary volume of water.

As it was established by TEM microscopy, synthesized NPs were weakly agglomerated with the average size of 10–15 nm and narrow size distribution.

Synthesis of Fe_3O_4 NPs in the reverse microemulsions was performed according to the scheme presented in [31, 32]. To prepare the reverse microemulsions, it used the necessary surfactant (Triton X–100, Brij–35 or CTAB), n-butanol as an additional surfactant, cyclohexane as a solvent, and corresponding aqueous phase. The aqueous phase of the first microemulsion consisted of the necessary amount of iron salts and the aqueous phase of the second one–the solution of precipitator (ammonia). To perform the synthesis, microemulsions of salts and precipitators were mixed for 1 h with further intensive stirring and heating at 70 °C for 3 h. Precipitated NPs were separated by centrifugation and washed with ethanol and bidistilled water.

TEM microscopy data showed that such a method of synthesis allowed obtaining nanosized weakly agglomerated particles, which sizes could be varied in the range from 4–8 to 10–20 nm by using different surfactants for preparation of the microemulsions.

Synthesis of NPs with the spinel structure in the solution of diethylene glycol (DEG) was performed as described in Refs. [33–36]. For synthesis, stoichiometric amounts of metal salts ($FeSO_4$ or $A(NO_3)_2$) and $Fe(NO_3)_3$ were dissolved in DEG and stirred for 1 h; simultaneously, the solution of NaOH in DEG was prepared. The alkaline solution was added by drops to the mixture of salts with constant stirring for 4 h. The obtained solution was heated at 200–220 °C for 60 min; after this, the solution of oleic acid in DEG was added to the reaction mixture and stirred for 1 h. Formation of the NPs of AFe_2O_4 compounds was observed during the cooling the mixture to the room temperature, and they were washed from the impurities by dispersing in ethanol with further centrifugation. NPs were saved in the solution or dried in the air at 30–50 °C for further investigations.

As it was established by TEM, this method of synthesis provided the obtaining nanosized weakly agglomerated particles with size of 4–9 nm.

3.4 Synthesis of Core/Shell Structures $Fe_3O_4/CoFe_2O_4$

Synthesis of core/shell structures $Fe_3O_4/CoFe_2O_4$ was performed according to the technique presented in [14, 37]. Fe_3O_4 synthesized by the precipitation in diethylene glycol without the addition of the oleic acid was used as a core. The synthesis of the shell was performed in the argon atmosphere. Necessary amounts of cobalt and iron nitrates were dissolved in DEG and stirred for 1 h. The solution of precipitator NaOH in DEG was prepared separately, and it was added to the solution of salts with stirring for 2 h. Previously synthesized Fe_3O_4 NPs were put into the solution and stirred for 2 h. The reaction mixture was heated up at 200–220 °C for 90 min reaching the desired temperature for 1.5 h. When the heat treatment finished, oleic acid was added with further stirring of the reaction mass up to cooling. Synthesized NPs were washed with the water–alcohol solution; the precipitate was separated by centrifugation. During the synthesis of core/shell structures, it was considered that Fe_3O_4 NPs (core) had a spherical shape with an average size of 6.3 nm [33], on which different amounts of $CoFe_2O_4$ was precipitated. The volume of the coating $CoFe_2O_4$ (shell) was calculated according to [14] using the equation $V = 4/3*\pi*((R_2)^3-(R_1)^3)$ (R_1 and R_2—radii of starting and coated spherical particle, respectively) and taking into account that $m = \rho \cdot V$ (m—mass, ρ—density of the material of core or shell).

3.5 Synthesis of Solid Solutions of $Ni_{1-x}Zn_xMn_2O_4$ with the Spinel Structure

Synthesis of NPs of solid solutions of $Ni_{1-x}Zn_xMn_2O_4$ ($x = 0$; 0.25; 0.50; 0.75; 0.80; 0.85; 0.90; 0.95; 1.00) by precipitation in the aqueous solutions was performed according to the technique described in [15]. As it was shown, the amorphous precursors formed during the synthesis process at 80 °C; therefore, these precursors were subjected to the heat treatment at 800 °C for 2 h. The main advantages of this method are its simplicity, low cost, and the possibility to synthesize a large amount of the final product. However, high-temperature treatment of the powder leads to the growth and agglomeration of NPs that is the drawback for their practical application, particularly in medicine.

3.6 Synthesis and Properties of Nanosized Particles Based on (La, Sr)MnO_3

Synthesis of (La, Sr)MnO_3 (LSMO) NPs by precipitation in the nonaqueous solutions. A solid solution of (La, Sr)MnO_3 manganite is quite complex system since it consists of the ions of three different metals—La, Sr, and Mn. As it is known, all of them precipitate in the aqueous solutions at different pH values that make the synthesis more complicated [30]. Therefore, the synthesis of LSMO NPs by precipitation in the nonaqueous solutions can be more satisfactory in terms of the choice of the conditions for synthesis. Usually, polyhydric alcohols, namely glycols (for example diethylene glycol DEG), are chosen as a medium for synthesis. It is due to several important factors: (1) application of glycols allows performing the coprecipitation of the metal cations, which have a large difference in pH values for the precipitation in the aqueous solution; (2) as the polyhydric alcohol, DEG can react with the metal cations; therefore, it can affect the synthesis process and stabilization of NPs; (3) DEG has the relatively high viscosity that allows avoiding the interaction between NPs during the precipitation process.

Synthesis of manganite NPs by precipitation in DEG solution was performed according to the technique described in detail in [38]. The solution, which contained the stoichiometric amounts of metal nitrates in DEG, was heated up to 200 °C in the argon atmosphere; then, the solution of NaOH in DEG was added by drops to the reaction mixture with constant stirring. The obtained reaction system was stirred and heated up on the oil bath at 200–220 °C for 1 h. After the heating was finished, the solution of oleic acid in DEG was added to the system, and it was left for cooling to the room temperature with stirring. Particles were separated by centrifugation and dispersed in ethanol. An obtained precursor was dried in the air at 30–50 °C, and to obtain crystalline NPs, this precursor was subjected to the heat treatment at 800 °C.

Synthesis of $La_{1-x}Sr_xMnO_3$ NPs by precipitation in reverse microemulsions. Synthesis of manganite NPs by precipitation in the reverse microemulsions was performed as described in [39]. It is known that the structure and nature of the surfactant can significantly affect the size and properties of synthesized NPs [40]. One more important thing for obtaining LSMO NPs is that it is necessary to choose surfactants, which do not contain any metal ions. It is connected with the fact that such ions can additionally incorporate into the structure during the heat treatment of the nonmagnetic precursor and it can result in a significant change of the properties (particularly, magnetic ones) of the final crystalline product. Therefore, microemulsions were prepared based on three different surfactants: Triton X-100, CTAB, and Brij-35.

To synthesis LSMO NPs via this method, it was prepared two microemulsions (M1 and M2), which consisted of corresponding surfactant (CTAB, Triton X-100 or Brij-35), n-butanol as cosurfactant, cyclohexane, and aqueous solution. The aqueous phase of M1 contained metal salts, and the aqueous phase of M2 contained the solution of the precipitator (NH_3). M2 was added to M1 dropwise for 1 h with further heating the mixture at 70 °C for 1 h. Precipitated NPs were separated by centrifugation, and then they were washed with ethanol and bidistilled water for several times by desperation and centrifugation. To obtain crystalline NPs, corresponding amorphous precursors were subjected to the high-temperature treatment at 800 °C for 2 h.

When LSMO NPs were synthesized via a solgel method, water-soluble salts $La(NO_3)_3$, $Mn(NO_3)_2$, and $Sr(NO_3)_2$ were used as the starting reagents [41–47]. Necessary molar amounts of salts were dissolved in bidistilled water, and gel-formed agents, citric acid and ethylene glycol, were added to this solution. The obtained mixture was heated up at 80 °C with stirring, polyesterification reaction took place, and polymer gel was formed. Pyrolysis of the gel was at 200 °C, and precursor powder was obtained as a result of pyrolysis [48]. This precursor was subjected to further mechanochemical treatment using different approaches: (1) heat treatment at 800 °C for 2 h; (2) grinding with alcohol and heat treatment at 800 °C for 2 h; (3) grinding with alcohol and further three-stage heat treatment at 400 °C, 600 °C, and 800 °C for 2 h; (4) grinding in the mill and heat treatment at 800 °C for 2 h; (5) effect of ultrasonication for 3 h with further heat treatment at 800 °C for 2 h; (6) pressing into the tablet with further heat treatment at 800 °C for 2 h; (7) heat treatment at 600 °C for 10 h; (8) grinding with the ammonium nitrate with further heat treatment at 800 °C for 2 h.

It is important to note that the application of different methods of mechanochemical treatment of the precursor after solgel synthesis allows obtaining NPs with different size and size distribution. However, considering some methodological features of synthesis and treatment of obtained particles [49], this method allows decreasing the size of NPs to 35 nm and obtaining relatively weakly agglomerated manganite NPs.

As it was shown by X-ray diffraction data, nonmagnetic amorphous precursors formed after synthesis via all three methods. It was established that when manganite NPs were synthesized by methods using organic compounds and nonaqueous medium, single-phase crystalline structure began to form in one stage at 600° and it completely formed at 800 °C.

Results of microstructural studies of manganite NPs showed that their size and morphology significantly depended on the method and conditions of synthesis. It was established that manganite NPs synthesized via three different methods had the average sizes in the range of 20–40 nm with narrow size distribution [39].

It follows from the above-considered data that there is *no unique method* to synthesize MNPs with reliably tunable and reproducible parameters. On the contrary, each specific material with certain properties needs a thoroughly adapted fabrication technique. The main characteristics of the techniques we employed in this work for the fabrication of MNPs with the spinel and perovskite structural types are presented in Table 1.

4 Materials Requiring External Temperature Control when Used in MHT

4.1 Ferrimagnetic Spinels AFe_2O_4 (A = Fe, Ni, Mn, Co, Zn, etc.)

Synthesis of nanoparticles was carried out by precipitation in diethylene glycol solutions (Fe1), by cryochemical method (Fe2) and in microemulsion using Triton X-100 (Fe3), Brij-35 (Fe4) and CTAB (Fe5) as surfactants [31]. The X-ray diffraction revealed that the nanoparticles obtained were characterized by a cubic spinel crystal structure (JCPDS Card Number 19-0629 [50]) (Fig. 1). Also, the diffraction patterns of samples Fe3, Fe4, and Fe5 showed an additional phase of goethite FeOOH [50]. As can be estimated from the results of the TEM study (Fig. 1), synthesized nanoparticles are nanosized and characterized by weak agglomeration. The mean diameter of NPs varies from 6 to 15 nm depending on the method of synthesis.

Magnetic measurements for synthesized nanoparticles were performed at the temperatures 300 and 10 K. As shown in Fig. 2, in both cases, magnetization tends to saturation at the high fields. Magnetic parameters are collected in Table 2.

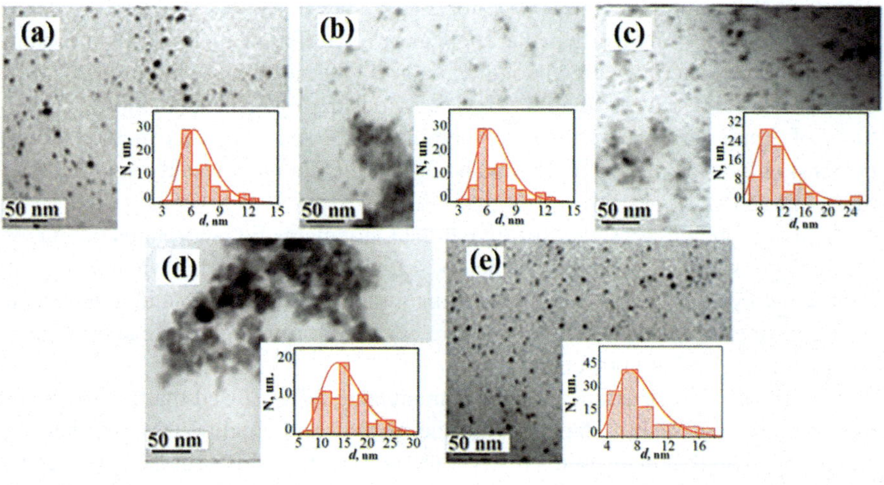

Fig. 1 TEM images of nanoparticles synthesized by various methods: (**a**) Fe1; (**b**) Fe2; (**c**) Fe3; (**d**) Fe4; (**e**) Fe5. Insets show the diagrams of size distribution for corresponding ensembles of nanoparticles

Table 1 Characterization of different techniques employed in this work to synthesize MNPs with the parameters tailored for MHT

Synthesis method	MNPs with a spinel structural type		MNPs with a perovskite structural type	
	Advantages	Disadvantages	Advantages	Disadvantages
Cryochemical method	A possibility to synthesize a large amount of MNPs. The particles are weakly agglomerated and nanosized. Relatively high heating efficiency under the action of AMF	Problems with tuning the NPs size. Fe_3O_4 particles are prone to oxidize into γ-Fe_2O_3 ones	Not employed in this work	
Synthesis in reverse microemulsions	A possibility to synthesize weakly agglomerated crystalline MNPs. Particle size can be controlled through changing surfactants	Problems with a synthesis of a large amount of MNPs. Small particles often display degraded magnetic and heating characteristics	Particle size can be controlled through changing surfactants	Amorphous NPs formed at the initial stage of the synthesis should be subjected to additional heat treatment that results in partial NPs agglomeration. Problems with a synthesis of a large amount of MNPs
Precipitation in nonaqueous solutions	A possibility to synthesize weakly agglomerated crystalline NPs of small sizes	Problems with a synthesis of a large amount of MNPs. Small particles often display degraded magnetic and heating characteristics	–	Amorphous particles formed at the initial stage of the synthesis should be subjected to additional heat treatment that results in partial NP agglomeration. Problems with a synthesis of a large amount of MNPs

(continued)

Table 1 (continued)

Synthesis method	MNPs with a spinel structural type		MNPs with a perovskite structural type	
	Advantages	Disadvantages	Advantages	Disadvantages
Solgel method	Not employed in this work.		A possibility to synthesize a large amount of MNPs. The formation of the crystalline structure occurs in one stage (without the formation of intermediate phases). At the final stage, the NP agglomeration can be controlled by changing the pH of the medium	Amorphous particles formed at the initial stage of synthesis should be subjected to additional heat treatment that results in the increase of NPs sizes and their partial agglomeration (formation of "bridges" between the particles)

Analysis of the curves showed that the synthesized nanoparticles exhibit behavior close to superparamagnetic [31] and are characterized by weak magnetic hysteresis at both temperatures (see inserts in Fig. 2) and have low coercivity (<2.5 kA/m).

Figure 3 shows plots of magnetic fluid heating temperature (T_{fluid}) versus residence time (τ) in the applied alternating magnetic field. As can be seen, the magnetic fluid based on Fe2 nanoparticles is characterized by a significantly higher rate and temperature of heating than Fe1, Fe3, Fe4, and Fe5 samples. Calculated values of specific loss power (SLP) show that depending on the method of synthesis we can receive nanoparticles characterized by SLP value from 0.2 to 34 W/g. Such behavior of magnetic losses can be related to many factors, particularly to the difference in NPs size (from 6 to 15 nm), variation in magnetic anisotropy energy, etc. [51, 52].

4.2 Spinel Ferrite Nanoparticles with Core/Shell Architecture

Core/shell architecture has acquired increasing interest due to the possibility of combining different materials and fabricating nanostructures with improved characteristics [13, 53]. In addition to varying size, shape, and composition, tuning of magnetic properties through the interface coupling of different magnetic materials becomes a prevailing strategy, introducing a new variable for the rational material design and property control in fundamental science and technological applications [54, 55].

Fig. 2 Hysteresis loops at 300 K and 10 K for Fe_3O_4 nanoparticles synthesized by various methods

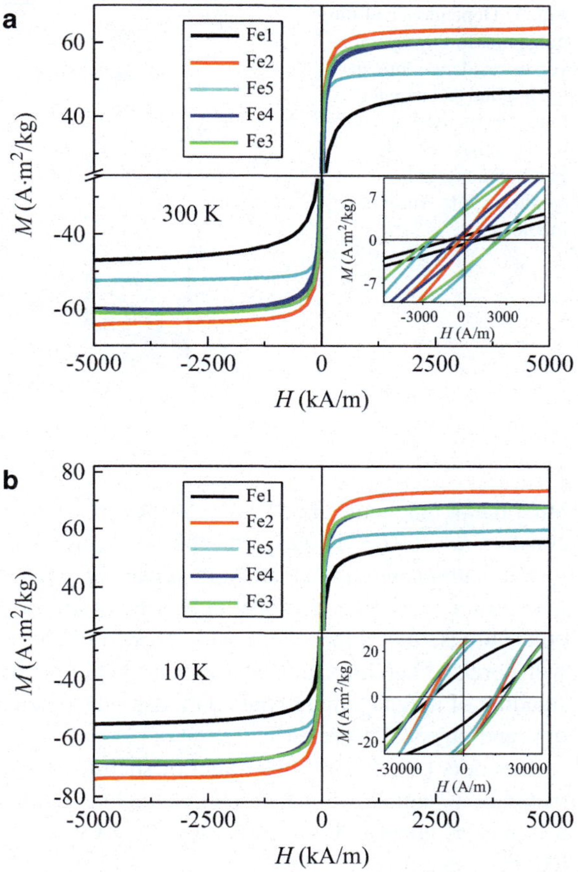

Table 2 The properties of Fe_3O_4 nanoparticles synthesized by various methods

Sample	d_{TEM}, nm	H_c, A/m		M_s, Am²/kg		T_b, K	SLP, W/g(Fe_3O_4)
		300 K	*10 K*	*300 K*	*10 K*		
Fe1	6.9±1.9	955	14800	47.0	55.3	92	0.19
Fe2	11.0 ±3.2	318	15760	64.4	74.4	400	33.80
Fe3	11.2±4.1	2550	19660	60.1	67.8	400	25.8
Fe4	15.6±4.8	636	18900	59.8	67.5	400	19.74
Fe5	8.5±3.3	2300	14400	52.2	59.9	400	16.10

Note: d_{TEM} mean size of the nanoparticles obtained from TEM images, *SLP* specific loss power, M_s saturation magnetization, T_b blocking temperature, H_c coercivity

As one of the most important and widely utilized magnetic materials, the spinel ferrite system MFe_2O_4 (M = Fe, Co, Mn, etc.) consists of both magnetically hard and soft materials. For example, cobalt ferrite ($CoFe_2O_4$) is magnetically hard with a large magnetocrystalline anisotropy constant $K > 10^6$ erg/cm³ [12, 56]. On the

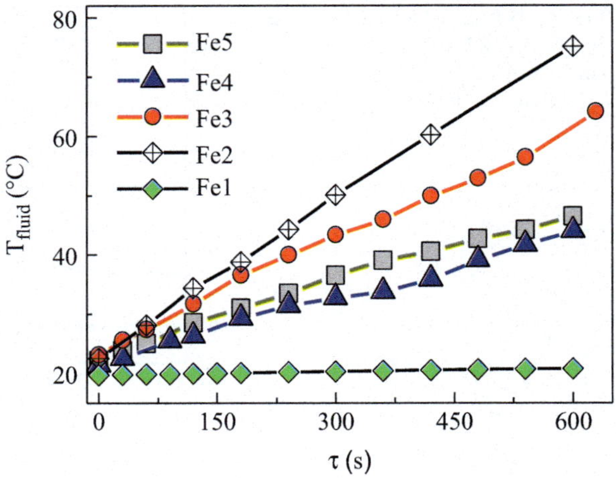

Fig. 3 Dependence of the magnetic fluid temperature versus residence time in the applied AC magnetic field. Magnetic fluids were prepared based on the Fe_3O_4 nanoparticles synthesized by various methods

other hand, magnetite (Fe_3O_4) is a ferrite with a much smaller magnetic anisotropy constant $K \sim (10^4 \div 10^5)$ erg/cm^3 [57, 58]. Due to the same crystallographic structure and almost negligible lattice mismatch among these spinel ferrites, it should be markedly controllable to epitaxially grow a uniformed shell over a core. Among other things, such kind of the well-defined bimagnetic spinel ferrite nanocrystals with core/shell architecture can provide a better platform for the fundamental understanding of magnetism and the relationship between crystalline structure, morphology, and physical properties.

In works [14, 59, 60], the effect of shell thickness on DC magnetic parameters (saturation magnetization, coercivity, blocking temperature) and AC losses (specific loss power, intrinsic loss power) has been studied for $Fe_3O_4/CoFe_2O_4$ core-/shell-like magnetic nanoparticles with a fixed diameter of core ~6.3 nm and an effective thickness of shell up to 2.5 nm. Also, the comparison of the properties of $Fe_3O_4/CoFe_2O_4$ core/shell nanoparticles and the mechanical mixture of the Fe_3O_4 and $CoFe_2O_4$ nanoparticles taken in appropriate proportions was done, which emphasized the role of interface phenomena at the Fe_3O_4–$CoFe_2O_4$ boundary in the formation of magnetic properties core/shell nanoparticles and discussed the way to tune and optimize the parameters of compositional nanoparticles for their use in various technological and biomedical applications.

The synthesis of core/shell structures has remained a complex approach, since it should enable a high crystalline quality of both core and shell, guarantee their good epitaxy, and minimize intermixing processes at the core–shell interface. For this reason, the results of earlier works carried out in our and other groups were taken into account while fabricating $Fe_3O_4/CoFe_2O_4$ core-/shell-like nanoparticles [31, 59, 61–63]. In our previous works, the combination of X-ray diffraction and ^{57}Fe Mössbauer spectroscopy studies made it possible to conclude that among several MNPs synthesis methods, only the method of coprecipitation from diethylene glycol (DEG) solutions allowed fabrication of magnetite MNPs with the smallest amount of maghemite and goethite phases [31, 59]. Therefore, it is the method of coprecipitation from DEG solutions that was chosen to fabricate the nanoparticles in the present work.

Broad studies were carried out to make sure that synthesized nanoparticles have the core/shell architecture and are not the mechanical mixture of Fe_3O_4 and $CoFe_2O_4$. Considering that individual Fe_3O_4 and $CoFe_2O_4$ MNPs have the same crystalline structure with close lattice parameters, the shell cannot be reliably distinguished from the core by the contrast of the TEM image. Therefore, before this work, a number of special experiments were carried out to check if the composite Fe_3O_4/$CoFe_2O_4$ MNPs have a core-/shell-like structure [31, 59]. The experiments included a comparative analysis of the most intensive (311) XRD peak collected from separate Fe_3O_4 and $CoFe_2O_4$ MNPs, a mechanical mixture composed of these compounds, and composite Fe_3O_4/$CoFe_2O_4$ particles. Besides, corresponding TEM and ^{57}Fe Mössbauer investigations were carried out on the individual and composite MNPs. As described in detail in Ref. [60], the results confirmed the formation of a core/shell-like structure rather than a mechanical mixture. At the same time, as follows from Refs. [59, 61, 63], the particles may contain quite a noticeable transitional layer between the core and the shell, which may be composed of $Co_{1-x}Fe_{2+x}O_4$ phases with a gradient of Co concentration from the surface to the core.

Results of XRD study of the synthesized nanostructures were analyzed in detail in Ref. [14]. All synthesized samples have a cubic spinel structure (JCPDS 19-0629) with no traces of impurity.

As can be estimated from the results of TEM investigations, the size of the Fe_3O_4 NPs is ~ 6.3 nm and grows with the increase of effective thickness (t_{shell}) of the $CoFe_2O_4$ shell according to the calculations (Fig. 4).

Studies of field and temperature dependences of the magnetization of Fe_3O_4/$CoFe_2O_4$ core/shell nanoparticles with different shell thicknesses have shown that the addition of the shell leads to a sharp increase in coercive force as well as in the blocking temperature [14, 60]. Effective anisotropy constants K_{eff} were calculated based on a simple approach of coexisting superparamagnetic and blocked magnetic nanoparticles [12, 13, 64]. It was shown that shell addition and subsequent increase

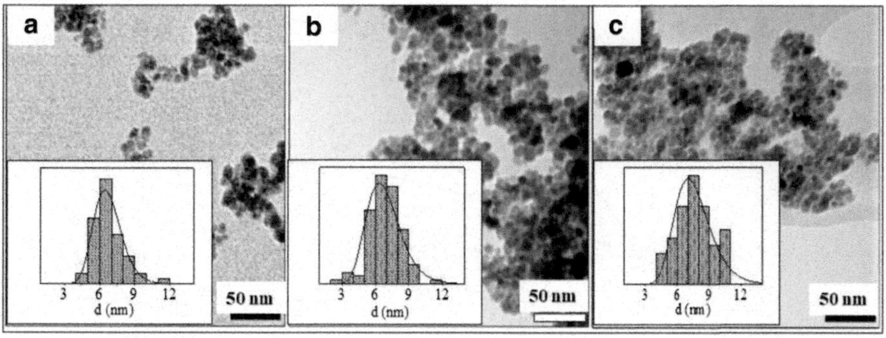

Fig. 4 TEM images of Fe_3O_4/$CoFe_2O_4$ nanoparticles with $tCoFe_2O_4$ = 0 nm (**a**); 0.05 nm (**b**); and 1 nm (**c**). Insets show the diagrams of size distribution for corresponding ensembles of nanoparticles (the units of abscissa axes are nanometers)

in its thickness lead to the strong increase of K_{eff} with a simultaneous decrease in magnetization. It is noteworthy that qualitatively similar behavior was observed in $MnFe_2O_4/CoFe_2O_4$ and $ZnFe_2O_4/CoFe_2O_4$ core/shell nanoparticles [60].

To obtain the AC magnetic heating characteristics of the synthesized MNPs, the time dependence of heat generation was studied under AMF with fixed amplitude $H_{max} = 100$ Oe and frequency $f = 300$ kHz [18, 65]. The plots of magnetic fluid temperature (T_{fluid}) versus heating time in external AMF (τ) for the fluids based on synthesized MNPs are shown in Fig. 5. Based on these data, the values of SLP were determined and compared with those calculated from the area of magnetic hysteresis loops. Figure 6 compares both sets of SLP vs t_{shell} dependences. Open squares show the data determined from experimental T_{fluid} vs t dependences. Filled triangles mark SLP values calculated as

$$SLP = \mu_0 f A_{loop} / \rho, \tag{1}$$

where $\mu_0 = 4\pi\ 10^{-7}$ H/m and ρ is the MNPs material mass density [18, 66]. It is seen that general trends in both sets of SLP vs t_{shell} dependences comply with each other. The discrepancies observed between experimental and calculated SLP values can originate from a nonnegligible effect of frequency change on coercivity and hysteresis loop shape [12, 13, 64].

As seen in Fig. 6, the addition of 1.0 nm $CoFe_2O_4$ shell to the Fe_3O_4 core strongly enhances the heating efficiency of the obtained composite MNPs. Thus, core/shell-like architecture can serve as an efficient tool to govern both the DC and AC behavior of magnetic nanoparticles.

Magnetic and calorimetric properties of the core/shell nanoparticles based on other spinel ferrites ($MnFe_2O_4/CoFe_2O_4$, $ZnFe_2O_4/CoFe_2O_4$, etc.) also were studied by our and other research groups [67–69]. The results show that the addition of the shell has a significant impact on the saturation magnetization, effective anisotropy

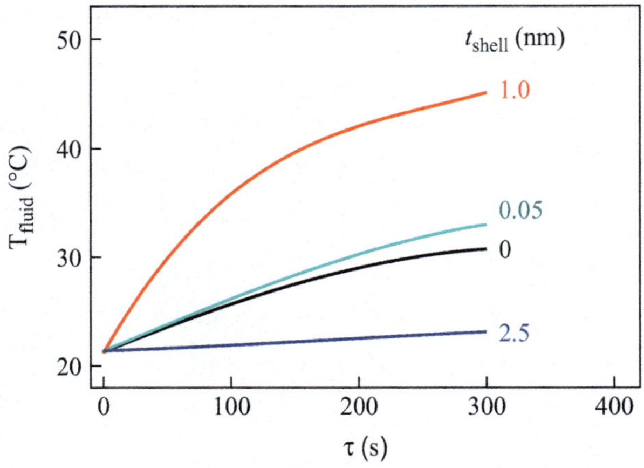

Fig. 5 Fluid temperature T_{fluid} versus heating time τ dependences for magnetic fluids based on Fe/Co(t_{shell}) MNPs

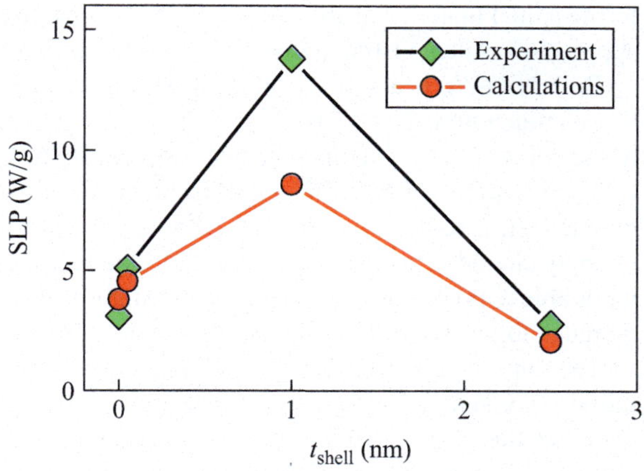

Fig. 6 Experimentally determined (open squares) and calculated (filled triangles) SLP values as functions of shell thickness for Fe/Co(t_{shell}) MNPs

constants, temperature behavior of compositional nanoparticles, and their efficiency of heating in the applied AC magnetic field.

Major trends in the heating abilities as a function of the core size, the nature, and the thickness of the shell were outlined in Ref. [70] based on the systematic fundamental study of $CoFe_2O_4/MnFe_2O_4$ and $CoFe_2O_4$/spinel iron oxide core–shell MNPs confirmed by experiments conducted on the water-based ferrofluids. At the same time, the authors stress that all the findings seem to depict a more complex scenario behind the heating abilities of bimagnetic core–shell systems than simple relationships with single magnetic or microstructural parameter(s). It follows from all the available data that the coating of MNPs with the shells leads to the transformation of the parameters of both the *initial nanoparticle* and *its shell* and, as a result, it makes the particles acquire *new characteristics that are typical neither for the core nor for the shell.*

Obtained results pave the way to tune and optimize open the parameters of spinel ferrite nanoparticles with core/shell-like architecture for their use in various technological and biomedical applications.

5 Materials with Automatic Preset Temperature Control when Used in MHT

5.1 Nickel–Zinc Spinel Nanoparticles: Magnetic Characterization and Prospects of the Use in Self-Controlled Magnetic Hyperthermia

It was noted above that for the use of materials in self-controlled MHT, the Curie temperature of these materials must be slightly above the therapeutic range of hyperthermia. One of the effective methods for controlling the Curie temperature in ferrite spinel materials is the use of solid solutions. In [15], the effectiveness of this approach was demonstrated using nickel–zinc spinels as an example.

Bulk $NiFe_2O_4$ is a soft ferrimagnet with mass magnetization ~56 A·m^2/kg and Curie temperature T_C≈658 K [67, 71]. On the contrary, bulk $ZnFe_2O_4$ is an antiferromagnet with a Neel temperature of ~9.5 K [16, 67]. Heavily Zn-substituted (x > 0.5) bulk samples of $Ni_{1-x}Zn_xFe_2O_4$ display reduced values of saturation magnetization and Curie temperature [4, 15]. Although the properties of these ferrites can be strongly altered at the nanoscale, it is expected that Zn substitution for Ni will make it possible to preset the Curie temperature slightly above the hyperthermia therapeutic range to reach the self-controlled heating regime.

The important issues were: (1) to find the regularities of the effect of zinc substitution on magnetic parameters and heating characteristics of nickel–zinc spinel nanoparticles; (2) determine the concentration region where Curie temperature is close to the range of hyperthermia therapy (~330 K); (3) experimentally verify the idea of controllability of the MNP thermal response to the action of AC magnetic field; (4) weigh up the prospects of the use of synthesized MNPs in self-controlled magnetic nanohyperthermia.

Nanopowders of $Ni_{1-x}Zn_xFe_2O_4$ (x = 0 – 1) with mean grain size of ~50 nm (see inset in Fig. 7) were synthesized via a method of hydroxide precipitation. Details of the samples preparation are described in Refs. [72, 73]

Figure 7 presents concentration dependences of spontaneous magnetization at 140 K and 300 K. Nonmonotonous character of $M_s(x)$ dependences implies that the substitution process is accompanied by the processes of redistribution of cation ions between different sites of spinel structure [73]

Bulk $NiFe_2O_4$ is a well-known inverse spinel with Ni^{2+} ions on B sites and Fe^{3+} ions distributed equally among A and B sites, where A and B indicate tetrahedral and octahedral sites of the spinel structure, respectively [74]. Morr and Haneda have

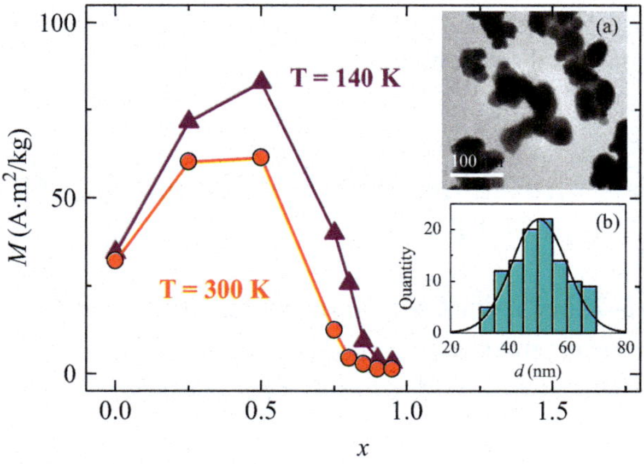

Fig. 7 Dependences of mass magnetization on zinc content for $Ni_{1-x}Zn_xFe_2O_4$ nanoparticles. Inset shows (**a**) a fragment of a microphotograph of $Ni_{0.25}Zn_{0.75}Fe_2O_4$ MNPs and (**b**) the diagram of MNPs size distribution

shown that $NiFe_2O_4$ in the ultrafine form exhibits noncollinear spin structure and that the magnetic moment at low temperatures is appreciably lower than the value for the bulk material [67]. They have proposed a model wherein the particle consists of a core with the collinear spin arrangement and a surface layer with the magnetic moment inclined to the direction of core magnetization. Our results agree with this picture: M_s of $NiFe_2O_4$ MNPs (~35 A·m²/kg) is by far smaller than that of the bulk sample (~56 A·m²/kg).

Bulk $ZnFe_2O_4$ has a normal spinel structure, in which all Zn^{2+} ions occupy tetrahedral positions (A sites) and all Fe^{3+} ions—octahedral positions (B sites). Zinc ferrite is paramagnetic at room temperature, and the transition to a magnetically ordered (antiferromagnetic) state occurs only at temperatures of about 9.5 K [67, 75]. The magnetic properties change significantly when the particle size decreases to a few tens of nanometers: At this scale, a cation inversion occurs; i.e., a part of zinc ions occupy the B sites and part of iron ions—the A sites [16, 75, 76]. Thus, there is a transformation of structure from normal to mixed spinel which disturbs a balance in the system of competing magnetic interactions and, as a rule, leads to the appearance of resultant spontaneous magnetization. The degree of cation inversion depends on several factors, including the particle size [53]. The contribution from the subsurface region which is usually characterized by a noncollinear magnetic ordering depends on the particle size as well [72, 75].

As seen from Fig. 7, the substitution of Zn for Ni results in an initial rise of spontaneous magnetization followed by a rapid decrease as x exceeds 0.75. As x crosses 0.85, the spontaneous magnetization levels off at $M_s \approx 2$ A·m²/kg. Similar trends were observed in works [77] and ascribed to the incomplete ordering of Ni^{2+} and Fe^{3+} ions between the octahedral and tetrahedral sites in the spinel.

To obtain the AC magnetic heating characteristics of the synthesized MNPs, the time dependence of heat generation was studied under AMF with fixed amplitude $\mu_0 H_{max}=10$ mT and frequency $f=300$ kHz.

The plots of magnetic fluid temperature versus residence time in external AMF for the fluids based on synthesized MNPs are shown in Fig. 8. For the samples with

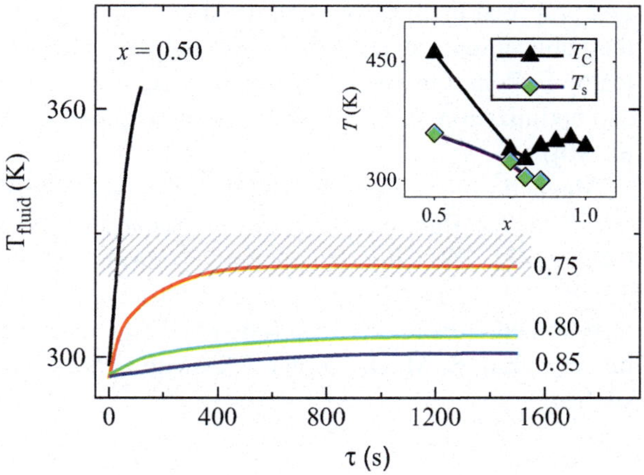

Fig. 8 Fluid temperature T_{fluid} versus heating time τ dependences for magnetic fluids based on $Ni_{1-x}Zn_xFe_2O_4$ nanoparticles. The temperature region, which is suitable for hyperthermia treatment, is hatched. Inset shows concentration dependences of Curie temperature (T_C) and maximal temperature of the sample heating under the action of AMF (T_s)

$x \leq 0.5$, T_{fluid} rapidly grows and in less than 1 min approaches the temperature of fluid boiling. This makes it impossible for an accurate determination of the temperature at which T_{fluid} achieves saturation. What is more, as follows from the data of work [19, 78–85], the precision of calorimetric measurements gets worsened in the vicinity of the fluid boiling point and the results of measurements become unreliable. For this reason, further analysis will mainly be focused on the samples with a higher zinc content ($x > 0.5$), and in some cases, the data for the sample with x=0.5 will be presented for reference.

It is important that after an abrupt initial temperature rise, the $T_{\text{fluid}}(\tau)$ curves reach saturation at a certain T_s value. Inset of Fig. 8 compares the values of T_s and Curie temperature T_C of the nanoparticles. It is observed that these values are quite close for the samples with x near 0.75, although for other samples there is a noticeable difference between T_s and T_C (for the rest of the samples, the difference exceeds a few tens of Kelvins).

The observed discrepancy between T_s and T_C for the samples with $x > 0.8$ can be easily understood since these samples display negligible heating efficiency which is not enough to appropriately heat the samples. Among other factors affecting a degree of proximity of T_s to T_C, it should noted that T_s is sensitive to the features of heat exchange with the environment, magnetization value, dispersion of magnetic parameters, etc.

The hatched region in Fig. 8 shows the temperature range which is suitable for hyperthermia therapy. It is seen that MNPs with $x \approx 0.75$ display T_s value within this region, and thus, *they are promising heat mediators for self-controlled magnetic nanohyperthermia.*

5.2 (La, Sr)MnO_3 Nanoparticles as Promising Heat Mediators for Self-Controlled Magnetic Hyperthermia

The idea of using manganite nanoparticles to implement self-controlled MHT was first proposed in 2002 [86], but research in this direction was intensified only after the publication of experimental results by a group of Czech and French scientists led by E. Pollert in 2006 [87]. Today, there is an active scientific work on the study and optimization of the properties of these materials (see, for example, [21, 48, 81, 88–90]).

Magnetic state of $La_{1-x}Sr_xMnO_3$ manganite strongly depends on strontium content x [78, 79]. Bulk samples with Sr concentration $0.15 \leq x \leq 0.60$ are ferromagnetic [38, 39, 48]. The Curie temperature T_C displays a maximum at $x = 0.33$ ($T_{C\max}$ = 370 K) and is quite sharply reduced as x deviates from 0.3. Since the development of heat mediators for self-controlled MHT requires NPs with T_C close to 320 K, in this work $La_{1-x}Sr_xMnO_3$ nanoparticles with a chemical composition of $x = 0.22$ were studied.

A study of the properties of LSMO MNPs synthesized by deposition from DEG, deposition from the reverse microemulsions and the solgel method [39] showed that the amorphous nonmagnetic precursor is formed immediately after synthesis in all three methods. A single-phase crystalline structure begins to form in one stage at a temperature of 600 °C and is completely formed at 800 °C. Manganite nanoparticles synthesized by three different methods have average sizes of 20–40 nm with a narrow size distribution [18].

In this work, the deposition from the reverse microemulsions was used for synthesizing LSMO MNPs. X-ray diffraction studies confirmed the single-phase structure of the samples [88, 89, 91, 92], and TEM measurements showed that the average size of nanoparticles is about 30 nm (see inserts in Fig. 9).

Figure 9 shows the temperature dependences of mass magnetizations M_{fc} and M_{zfc}, obtained in the magnetic field $\mu_0 H = 2$ mT. The $M_{fc}(T)$ and $M_{zfc}(T)$ curves coincide at temperatures higher than 320 K but diverge in the low-temperature region. $M_{fc}(T)$ is a declining function of temperature, while $M_{zfc}(T)$ is a curve with a maximum. The quasistatic (i.e., low frequency, with $10^{-4} < f < 10^{-2}$ c^{-1}) blocking temperature defined as a temperature at which $M_{zfc}(T)$ achieves a maximum is $T_b = 285$ K.

In Fig. 10, hysteresis loops $M(H)$ obtained at $T = 295$ K and 10 K are presented. Both curves have a tendency to saturation in strong fields ($\mu_0 H > 300$ mT). The values of saturation magnetization, defined as magnetization in a field of 1 T, are as follows: $M_s|_{T=10K} \approx 81$ A·m²/kg and $M_s|_{T=295K} \approx 47$ A·m²/kg. The M vs. H curves are characterized by coercivities $\mu_0 H_c \approx 22.7$ and 1.2 mT at $T = 10$ K and 295 K, respectively (see inset of Fig. 10).

To obtain the AC magnetic heating characteristics of the synthesized LSMO MNPs, the time dependence of heat generation was studied under AMF with a fixed amplitude H_{amp}, varying in the range from 5.6 to 10.5 mT in different experiments, and frequency $f = 300$ kHz. The plots of magnetic fluid temperature versus residence time in an AMF, $T_{fluid}(\tau)$, for the fluids based on the synthesized LSMO MNPs

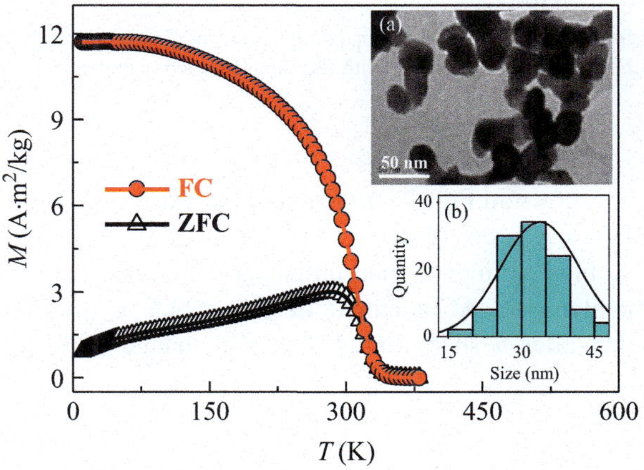

Fig. 9 Temperature dependences of mass magnetizations M_{fc} and M_{zfc} for LSMO MNPs. The inset shows (**a**) the microphotograph of LSMO MNPs and (**b**) their size distribution diagram

Fig. 10 Magnetization loops $M(H)$ at $T = 295$ K and 10 K. The inset shows the same dependences in the region of weak fields

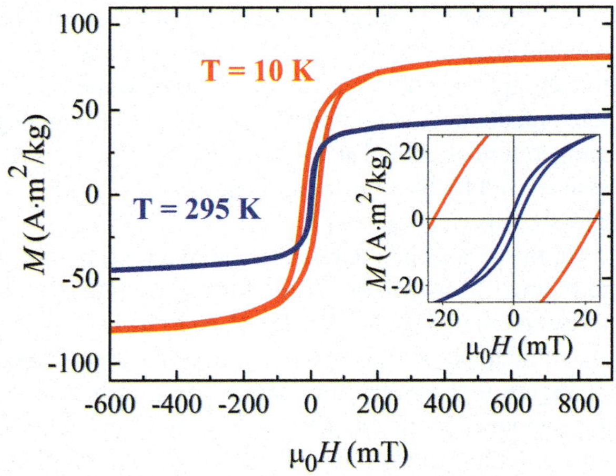

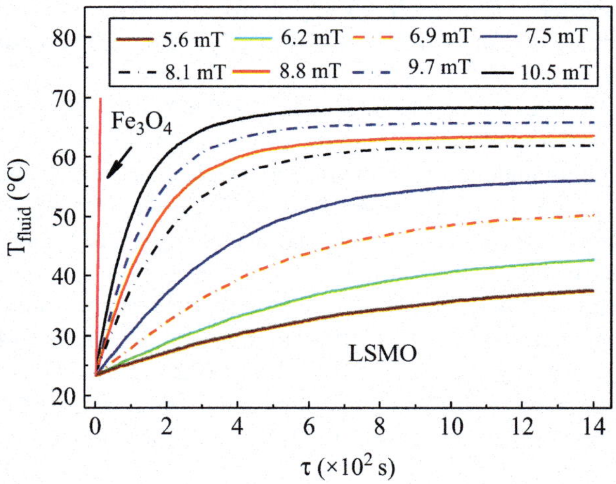

Fig. 11 Time dependences of the fluid temperature $T_{\text{fluid}}(\tau)$ for the fluids based on LSMO MNPs in an AMF with different amplitude values: $\mu_0 H_{\text{amp}} = 5.6; 6.2; 6.9; 7.5; 8.1; 8.8; 9.7; 10.5$ mT and frequency $f = 300$ kHz. For comparison, the $T_{\text{fluid}}(\tau)$ dependence is shown for the fluid based on Fe_3O_4 MNPs in an AMF with the same frequency and $\mu_0 H_{\text{amp}} = 10.5$ mT

are shown in Fig. 11. Magnetic fluid warms up due to the magnetic losses in AMF $H(t) = H_{\text{amp}} \cos(2\pi f t)$.

In experimental measurements, the initial temperature of magnetic fluid before applying the AC magnetic field was 23.5 °C. It is seen from the curves of Fig. 12 that after a sharp initial rise, the change of T_{fluid} with τ becomes slower and

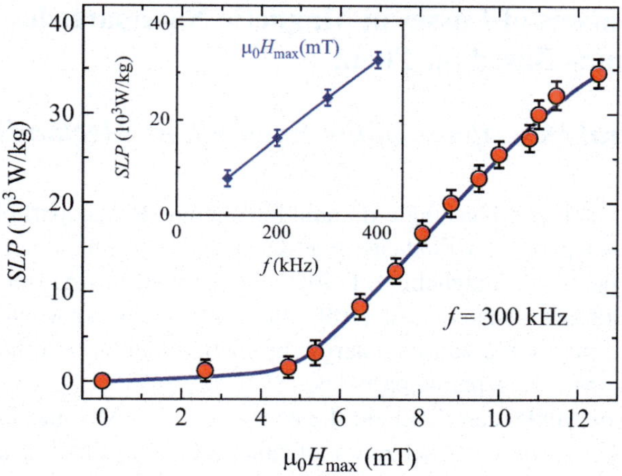

Fig. 12 SLP vs. H_{max} dependence measured at $f = 300$ kHz. The inset shows SLP vs. f dependence obtained at $\mu_0 H_{max} = 10$ mT

eventually these dependences tend to saturate. For comparison, the $T_{fluid}(\tau)$ dependence is added for the fluid based on Fe_3O_4 MNPs (~12 nm in size) subjected to the AMF with the same frequency and $H_{amp} = 10.5$ mT. The temperature of the Fe_3O_4 MNP-based fluid rapidly grows and immediately approaches the temperature of boiling.

With the increase in H_{amp}, the saturation temperature T_s rises, but eventually, it approaches the Curie temperature of the LSMO nanopowder T_C ~ 70 °C [9, 93–96]. This means that the idea of self-controlled heating where the heating efficiency sharply reduces as the fluid temperature approaches T_C of LSMO MNPs is valid. It should be noted that relatively small values of H_{amp} (< 10.5 mT) are not enough to warm up the magnetic fluid to the T_C of LSMO nanopowder, and in these cases, the thermal equilibrium occurs at lower temperatures, i.e., at T_s being lower than 70 °C.

The initial slope of each curve plotted in Fig. 12 provides information about specific loss power [97, 98]. The SLP values, which depend on the parameters of the applied AC field, were calculated for a range of magnetic field amplitudes $\mu_0 H_{max}$ and frequencies f. The details of the SLP calculation are described in Refs. [93, 99, 100].

Figure 12 shows SLP vs. H_{max} dependence obtained at $f = 300$ kHz. SLP is negligibly small at weak magnetic fields ($\mu_0 H_{max} < 5$ mT), rapidly rises in the region from ~5 to ~10 mT, and then displays a tendency to saturation. In the region from 6 to 9 mT, the SLP(H_{max}) dependence is almost linear.

It was reported in Refs. [20, 93, 97, 101] that the obtained value is high enough for the use of the MNPs as HT heating mediators in the treatment of cancer diseases. The mechanisms responsible for high heating efficiency of LSMO MNPs will be discussed below.

6 Mechanisms of Losses in Magnetic Nanoparticles and Fluids Based on Them

6.1 General Description of the Energy Loss Mechanisms

If a magnetic fluid based on single-domain ferromagnetic nanoparticles is subjected to AMF action, a part of the field energy is dissipated by the nanoparticles and transformed into heat. For magnetic HT, low-concentrated suspensions of MNPs in dielectric fluids are usually used [100]. In this case, interparticle interaction is assumed to be negligible and the energy dissipation can be described as a sum of processes occurring in separate particles.

The array of MNPs can dissipate the energy of the AMF into the environment due to the processes of remagnetization of nanoparticles or due to the rotation of the nanoparticle as a whole. The last-mentioned process is called Brown relaxation (BR losses). Remagnetization losses depend on the relationship between the characteristic parameters of MNP (saturation magnetization M_s, the effective magnetic anisotropy constant K_{eff}, the coercive force H_C, the volume of the MNP V and the relaxation time of the MNPs system τ and external parameters, the frequency f of the external magnetic field, as well as the temperature T) and require separate consideration.

1. ***BR losses***. In fluid suspensions, the relatively high anisotropy barrier may be overcome by rotation of the whole particle under the influence of the AMF. In this case, frictional losses due to viscosity η of the carrier liquid arise. The resulting relaxation time is given by [99, 100, 102]

$$\tau_B = \frac{4\pi\eta R_h^{\ 3}}{k_B T}, \tag{2}$$

where R_h is the hydrodynamically effective radius, which may differ from the geometrical one [103]. The specific losses power caused by this mechanism is calculated by the formula (3) provided below.

2. ***Remagnetization losses.*** The behavior of single-domain MNPs depends on the energy balance between anisotropy ($\varepsilon_a = K_{\mathrm{eff}}\cdot V$) and thermal fluctuation ($\varepsilon_T \approx k_B\cdot T$) contributions (here, k_B is the Boltzmann constant). In the limit $\varepsilon_T >> \varepsilon_a$, the magnetic moment of a particle can be considered as freely fluctuating. This results in the phenomenon of superparamagnetism, where the ensemble of particles behaves like a paramagnet consisting of particles with very large magnetic moments. In the opposite limit, when $\varepsilon_T << \varepsilon_a$, the particle magnetic moment is blocked on a time scale given by the experiment and lies parallel or antiparallel to the easy magnetization axis. The transition to the latter state occurs at the so-called blocking temperature T_b, below which the particle moments appear frozen on the time scale of the measurement τ_m. The T_b value depends on the measurement duration and can be determined from the temperature dependences of magnetization: The zero-field-cooled (ZFC) magnetization M_{zfc} measured in a weak magnetic field achieves a maximum at T_b [66]. The ZFC magnetization curve is

usually measured in quasistatic measurements with increasing temperature in a weak magnetic field after cooling the specimen in the absence of a magnetic field down to $T << T_b$. The measurement duration in such a case is of the order of hundreds of seconds.

For two cases, particularly, when the nanoparticle is (1) in the superparamagnetic state or (2) in the blocked state, analytical approaches have been developed to estimate the energy losses due to remagnetization. To describe energy losses in the first case (superparamagnetic behavior), we talk about Neel relaxation (NR losses: also called Neel–Arrhenius or Neel–Brown relaxation). In the second case (blocked state), the term hysteresis losses or remagnetization based on the Stoner–Wohlfarth model (SW losses) is used [9]. In the region of transition from the equilibrium superparamagnetic to the blocked state, only numerical simulation is correct [66, 98, 100, 101, 104].

2.1. NR losses. If thermally activated fluctuations of the magnetic moment direction are fast compared to the measurement duration ($\tau < \tau_m$), the behavior of the ensemble of MNPs can be described by a linear response theory (LRT) with relaxation time taken in Neel–Brown form [18]. In this case, specific power dissipated by an ensemble of MNPs is described by [9]:

$$P_{\mathrm{NR}} = f\mu_0 \int_0^{1/f} \tilde{M} dH = f\pi\mu_0\chi'' H_{\max}{}^2 = \frac{2\mu_0\pi^2 f^2\tau}{1+(2\pi f\tau)^2}\chi_0 H_{\max}{}^2. \qquad (3)$$

It is seen that P_{NR} displays a quadratic dependence on AMF amplitude $H_{\max}$. As a function of frequency, it is quadratic at low values of f ($2\pi f\tau << 1$), deviates from quadratic law as f increases, and tends to saturation at high frequencies ($2\pi f\tau >> 1$), where the saturated value reads:

$$P_{\mathrm{NR}}^{f\to\infty} = \frac{\mu_0\chi_0 H_{\max}{}^2}{2\tau}. \qquad (4)$$

Heating efficiency under such conditions ($2\pi f\tau >> 1$) does not depend on frequency.

In a general case, both the Brown relaxation process and the Neel relaxation process may be present, but the faster one is dominant.

2.2. SW losses. If thermally activated fluctuations of the magnetic moment direction are not fast in comparison with the measurement duration ($\tau > \tau_m$), the system is in the blocked state and the dependence of power losses on the magnetic field is of threshold character.

For the case of weak magnetic fields ($H_{\max} << H_a$), the reaction of the system to the external magnetic field will be very weak, remagnetization practically would not occur (only a minor hysteresis loop will be observed), and, thus, the power of dissipation will be negligible. If the amplitude of the magnetic field equals or exceeds the anisotropy field, the shape and area of the hysteresis loop will approach, but would not exceed a theoretical limit which is described by the Stoner–Wohlfarth model [66, 104].

If an ensemble of MNPs is put into an AMF, the amount of heat dissipated in MNPs during one cycle of the magnetic field is equal to the area of hysteresis loop S. The dissipated energy per second per unit volume of MNPs is then given by:

$$P_{\mathrm{SW}} = Sf. \tag{5}$$

For the MNP in a blocked state and with its easy axis aligned along the magnetic field direction, the hysteresis loop is a rectangle with temperature-dependent H_C. At low temperatures ($T \rightarrow 0$), the temperature-activated repopulation of energy minima is suppressed, and remagnetization can only occur when the energy barrier is fully removed by the magnetic field. This occurs when $H_{\max}$ reaches anisotropy field H_a [18]. Thus, for the Stoner–Wohlfarth model $H_c \rightarrow H_a$ as $T \rightarrow 0$. For an optimal case where $H_{\max} = H_c$:

$$P_{\mathrm{SW}} = 4\mu_0 H_{\max} M_s f. \tag{6}$$

To estimate P_{SW} in all other cases, it is convenient to use the formula:

$$P_{\mathrm{SW}} = 4\alpha\mu_0 H_{\max} M_s f, \tag{7}$$

where α is a dimensionless parameter which equals the ratio between the areas of actual hysteresis loop and ideal rectangle loop, with the latter having $\alpha = 1$. For the ensemble of MNPs with randomly oriented anisotropy axes and optimally chosen relation between H_c and $H_{\max}$ ($H_c = 0.81H_{\max}$), $\alpha = 0.39$ [66, 104]. When $H_{\max}$ is less than a field at which magnetization saturation occurs, α is a function of $H_{\max}$.

As a result, the upper limit of P_{SW} for the ensemble of MNPs with randomly oriented anisotropy axes is:

$$P_{\mathrm{SW}}^{\mathrm{limit}} = 4 \cdot \alpha \cdot \mu_0 H_{\max} M_s f \,|_{\alpha=0.39}\,. \tag{8}$$

The values of α observed in the experiment do not exceed 0.3 [20, 71, 80, 98, 104].

6.2 Low- and High-Frequency Measurements

A usual way to model remagnetization processes and calculate SLP values is to use the MNP parameters extracted from LF measurements [18]. It should be noted, however, that such parameters as blocking temperature T_b and coercivity H_c are frequency dependent. This means that not only SLP values calculated with the use of magnetostatic parameters could be incorrect, but also the mechanisms of AC losses could become different as measurement conditions change from magnetostatic to HF (~100 kHz) ones.

It was noted above that blocking temperature T_b is the temperature at which a condition [18]

$$\tau_m = \tau\big|_{T=T_b}. \tag{9}$$

is satisfied. At this temperature, the rate of thermally activated processes becomes insufficient to get the system into equilibrium, and MNP magnetic moment remains blocked in a metastable state.

Formula (9) makes it possible to estimate the dependence of blocking temperature on measurement time. Assuming that a time scale of measurements is $1/f$, T_b can be roughly estimated as

$$T_b \approx K_{\text{eff}} V / \left(k_B \ln\left(\frac{1}{f\tau_0} \right) \right). \tag{10}$$

If MNP magnetic parameters, in particular K_{eff}, were temperature independent, the change in the AMF frequency from ~0.01 Hz (magnetostatic measurements) to ~10^5 Hz (the frequencies at which usually HT treatment is carried out) would result in the increase of T_b by (2–3) times. Thus, for the case where the T_b^{hf} value exceeds ~150 K, the T_b^{lf} value can exceed room temperature. This means that the mechanism responsible for AC losses in the vicinity of room temperature will be SW mechanism, rather than NR one, as follows from LF measurements.

Figure 13 illustrates $M_{zfc}^{lf}(T)$ and $M_{zfc}^{hf}(T)$ dependences obtained from LF and HF measurements, respectively. For each of these curves, magnetization reaches a maximum at temperature which equals the blocking temperature. Since $T_b^{lf} \ll T_b^{hf}$, the temperature region within which SW mechanism is responsible for the losses is by far wider in the latter case (HF measurements) than in the former one (LF measurements).

The scheme in Fig. 13 is given under the assumption that $T_b^{hf} \ll T_C$. For the case, where T_b^{hf} is close to T_C, the relation between T_b^{lf} and T_b^{hf} will qualitatively be the same, but quantitative difference between them will change. Naturally, T_b cannot exceed T_C.

Thus, approaches based on modeling of remagnetization processes at high frequencies (~ 100 kHz) using magnetic parameters of MNPs obtained from magnetostatic measurements are not always correct. Because the blocking temperature and coercive force depend on the frequency, the high-frequency remagnetization processes can be qualitatively different from the low-frequency ones.

6.3 Energy Losses in (La, Sr)MnO_3 Nanoparticles

As was mentioned above, the approaches based on the use of quasistatic measurement results are not always correct for describing processes occurring in alternating fields with frequencies used in MHT (~ 100 kHz). Let's consider the situation on the example of the properties of the LSMO MNPs described in the previous section.

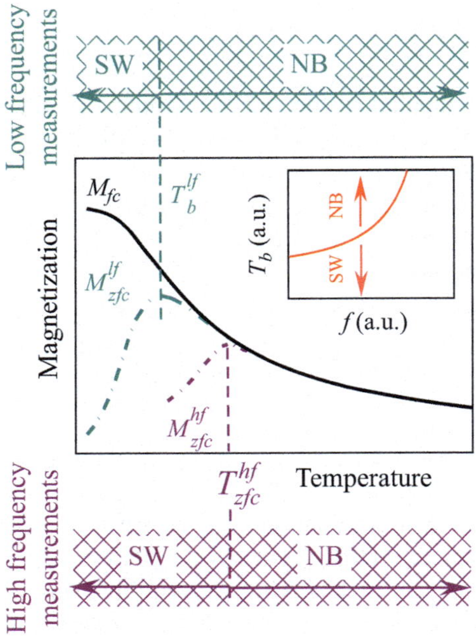

Fig. 13 Illustration of the behavior of magnetization resulted from LF and HF measurements. The inset presents the frequency dependence of blocking temperature T_b. M_{fc}—field-cooled magnetization, M_{zfc}^{lf} and M_{zfc}^{hf}—the zero-field-cooled magnetization values resulted from LF and HF measurements, respectively. T_b^{lf} and T_b^{hf}—the values of blocking temperatures resulted from LF and HF measurements, respectively. Top and bottom diagrams schematically illustrate temperature regions, where either SW or NR mechanism of losses dominates for LF and HF cases, respectively

As seen in Fig. 9, the blocking temperature of the LSMO MNPs is lower than room temperature, so in the range of room temperature and above, it should be expected that the NR mechanism is dominant. But the data in Fig. 12 suggest another. First, the dependence of SLP on AMF amplitude is of threshold character: SLP becomes noticeable only when H_{max} exceeds a certain value. Second, the SLP is almost a linear function of AMF frequency (see the inset of Fig. 12). Such kinds of SLP(H_{max}) and SLP(f) dependences are characteristic only for SW losses. As follows from formulae (2), (3) and (4), NR and BR losses display quadratic dependence on AMF amplitude. What concerns frequency dependence of SLP, NR and BR losses either display quadratic SLP(f) dependence or are frequency independent. Thus, the character of SLP(H_{max}) and SLP(f) dependences indicates that SW mechanism is responsible for LSMO MNPs heating.

The calculations performed in [21, 105–107] show that for the LSMO MNPs under investigation only SW mechanism can provide sufficiently high energy losses which correspond to experimentally obtained value of SLP ($\mu_0 H_{max}$ = 10 mT) ≈ 25×10^3 W/kg; the other kinds of AC losses are so weak that they cannot be responsible for LSMO MNP heating. It also follows from the calculations that SLP can be increased by about an order of magnitude by means of optimization of the parameters of LSMO MNPs (weakening agglomeration, narrowing size distribution, and attaining uniformity of geometrical and magnetic parameters) and corresponding adjustment of AMF parameters.

The obtained data indicate the directions of further optimization of nanoparticle parameters for use in MHT [39].

7 Biomedical Investigations

It was mentioned above that NPs of lanthanum–strontium manganite are promising for application in self-controlled magnetic HT. However, the issues concerning their toxicity and biocompatibility have been investigated only episodically to this time [6, 107].

7.1 Development of Magnetic Fluids

To investigate the biomedical features of magnetic LSMO NPs as the heat mediators for magnetic hyperthermia, it is necessary to develop the magnetic fluid for the possibility to inject such NPs into the tumor. Magnetic fluid (MF) is the suspension of nanosized ferromagnetic or ferrimagnetic particles, stabilized in the polar or nonpolar medium using surfactants or polymers.

To perform the biomedical investigations of manganite NPs, particles synthesized via the solgel method (NPs synthesized in such way had one of the highest SLP values [108]) were dispersed in 0.1% aqueous agarose solution by ultrasound for obtaining the magnetic fluid with the concentration of magnetic phase of 50 mg/ml [109].

7.2 *Performing the Biomedical Investigations of Magnetic Fluid Based on Manganite NPs* in vitro *and* in vivo

At present, it is known about the active investigations of Fe_3O_4 NPs in medicine due to their biocompatibility [22]. Therefore, to estimate the prospects of LSMO NPs in this area, it is important to investigate their biomedical characteristics and compare them with known Fe_3O_4 NPs. The first stage of biomedical investigations of magnetic fluid based on the synthesized LSMO NPs included a number of *in vitro* investigations [110].

Comparison of magnetic fluids based on Fe_3O_4 and LSMO NPs was performed on the ST and MCF-7S cell lines using four characteristics: *cytotoxicity,*

Table 3 Comparison of cytological properties of Fe_3O_4 and LSMO NPs

Characteristic	Fe_3O_4	LSMO
Cytotoxicity	Maximum permissible concentration	
	1.1 mg/ml	5 mg/ml
Genotoxicity	Possess a small genotoxic effect in the concentration of 0.9 mg/ml	Does not possess any genotoxic effect
Antioxidant activity	Does not display antioxidant activity	Displays antioxidant activity in the concentration range of 0.65–5 mg/ml
Antiviral activity	Possesses antiviral activity	Possesses antiviral activity

genotoxicity, antioxidant, and antivirus activity. The main results of performed studies are summarized in Table 3.

Investigations of *cytotoxicity* on ST and MCF-7S cell lines showed that LSMO NPs had a constant maximum permissible concentration in contrast to Fe_3O_4 and it was higher (5.5 mg/ml) than the concentration of Fe_3O_4 particles (0.1–1.1 mg/ml).

Genotoxicity on NPs was investigated for magnetic fluids with the concentrations of Fe_3O_4 and LSMO, which were limited by the maximal nontoxic concentrations, established in the previous investigation of cytotoxicity. Performed studies showed that the magnetic fluid based on LSMO NPs *did not possess a genotoxic effect* on the ST and MCF-7S cell lines when Fe_3O_4 NPs in the concentration of 0.9 mg/ml displayed small genotoxic effect both on the ST and MCF-7S cell lines.

It was also established that both types of NPs possessed small *antiviral activity* in low concentrations. However, for LSMO NPs, this concentration was higher and equal 0.25–1.25 mg/ml.

As it is known, the growth of oncological tumors relates to chronic inflammation as well as with the presence of a high level of reactive oxygen species [22]. Therefore, investigations of the effect of magnetic NPs on the decreasing *the oxidative stress* were performed, and the presence of possible antioxidant properties was studied. Investigations were carried out on the ST cell line cultures, and they showed that Fe_3O_4 magnetite NPs could protect the cells from the oxidative stress in the concentrations of 1.1–1.4 mg/ml. However, as it was found out previously, such doses were already toxic for the cells. At the same time, LSMO manganite NPs possessed antioxidant properties in the concentrations of 0.65–5 mg/ml, which were nontoxic. Obtained data are very important since the main attention usually pays to the oxidative damage of magnetic NPs and not to their antioxidant activity.

The complex of performed cytological investigations described in [111] has shown that low cytotoxicity and genotoxicity of LSMO NPs make them more attractive for biological application compared to Fe_3O_4 NPs, which are characterized by higher toxicity. Moreover, LSMO NPs possess additional antioxidant activity in the quite wide concentration range that can be used in the different schemes of prevention and treatment.

At the same time, complex *in vitro* and *in vivo* investigations showed that LSMO NPs had low toxicity during the intratumoral injection and *did not cause the defection of the histological structure and functions of the main organs*. Therefore, they can be used for performing further investigations as the heat mediators for magnetic hyperthermia.

Investigations of the toxicity of LSMO manganite NPs *in vivo* and their effect on the functioning animal organs were performed by authors [95, 97].

Acute toxicity of LSMO NPs was investigated on white mice of C57B1/6 line with the bodyweight of 20–22 g (each experimental group included six animals). The criteria of the toxicity were the death of animals in each group (in %), which was calculated according to the general methods that determined LD_{10}, LD_{50}, LD_{100}, etc. These parameters are defined as doses, where index means the percentage of dead animals, and maximum permissible dose, which does not lead to the death of any animal and does not cause the falling of the bodyweight more than 10%.

LSMO manganite NPs were injected into animals in such concentrations: 50, 100, 200, 300, and 400 mg per kg of body weight intraperitoneally and once in physiological solution. The duration of the investigations was 14 days. The loss of the bodyweight began at the dose of 200 mg/kg, but it was negligible and recovered just on the fourth day after the injection of NPs. The most significant weight loss was registered just on the first day after injection of NPs at the concentration of 300 mg/kg; however, animals gradually increased their weight on the second and third days, and the bodyweight of animals began to recover and even increase on the fourth day. The dose of 400 mg/kg caused the significant falling the body weight, and animals began to increase their weight gradually only on the seventh day after injection of NPs.

Obtained results indicate that maximum permissible dose for LSMO manganite equals 300 mg/kg of body weight and LD_{50}–1800 mg/kg of body weight.

Histological investigations were performed on the white rats with the Guerin carcinoma and on the mice of the C57B1/6 line with the Luice carcinoma of lungs. Morphological studies were carried out after 48 h of the injection of the nanocomposite. Any changes were not observed in the histological structure of the tumor (Guerin carcinoma in rats and Luice carcinoma in mice) as well as in normal organs such as liver, kidney, lungs, heart, and spleen after the injection of LSMO NPs in the dose of 300 mg/kg.

Investigations also showed that any toxic effect on the functions of the liver and kidneys was not present after the injection of LSMO nanocomposite into the tumor.

Complex *in vitro* and *in vivo* studies showed that LSMO NPs had low toxicity and did not cause the defection of the histological structure and functions of the main organs during the intratumoral injection. Therefore, they can be used for performing further investigations as the heat mediators for magnetic hyperthermia.

Further, in vivo investigations of the magnetic fluid based on the LSMO NPs as the heat mediators for magnetic hyperthermia were performed on the tumors of Guerin carcinoma and Volker carcinosarcomas [112]. Tumors were inculcated to the left and right thighs of the white rat, and the volume of the tumor was waited to reach the value of ~1 cm^3. After that, magnetic fluid was injected into the tumor in the right thigh, and this tumor was subjected to the effect of an alternating magnetic field for 30 min (treatment was performed 1–2 times with the break of 1–2 days). The magnetic fluid was not injected into the tumor of the left thigh. It was used for comparison.

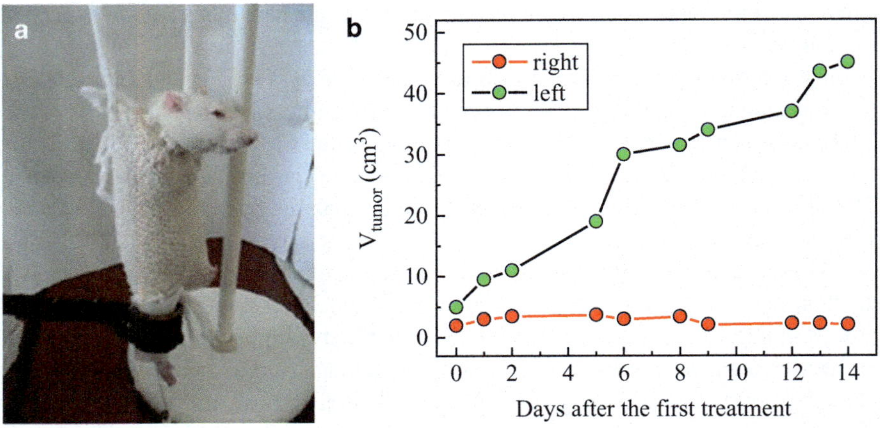

Fig. 14 (**a**) Experiments performing the magnetic hyperthermia on the previously inculcated tumor to the right thigh of the white rat; right thigh is placed into the coil of the generator of the alternating magnetic field. (**b**) Growth of the tumor of Guerin carcinoma in the right thigh after the injection of the magnetic fluid and the effect of the magnetic field (red curve) and nontreated tumor in the left thigh (black curve)

Such an approach led to the delay of the growth of the tumor in the right thigh; i.e., the volume of the tumor remained almost unchanged in contrast to the volume of the nontreated tumor (left thigh). The temperature in the tumor of the right thigh equaled in the average of 45–46.5 °C after the 30th minute of the heating.

The dependence of the efficiency of magnetic hyperthermia on the number of heating sessions was also established. In a series of experiments, heating was repeated for 3–4 times with the break of 2–3 days that led to the more significant decreasing the growth of tumor (Fig. 14): Volume of the tumor became 2.7 cm^3 on the 13th day after the first treatment, and volume of the nontreated tumors reached 47 cm^3. Increasing the volume of tumors was scarcely observed after the fourfold processing of the tumor with the magnetic fluid by the alternating magnetic field, and, in some cases, the regression of the tumor up to 35% was registered.

Performed investigations showed that the application of magnetic NPs of LSMO manganite in combination with the effect of the alternating magnetic field allows decreasing and even finishing the growth of oncological tumors after some time, and in the number of cases, it leads to the complete death of the tumor. This implies that LSMO NPs are promising materials for applications in magnetic hyperthermia.

8 Unresolved Issues and Prospects

Despite the recent progress in heat generating MNPs, there remain challenges to be addressed for the practical applications of magnetothermal effects of MNPs in biological systems. For example, the MNPs should be more precisely tailored to be biocompatible in physiological environments, ensuring the biosafety. For drug release control or hyperthermia applications, the MNPs are needed to incorporate surface coating to have a prolonged blood half-life for the enhanced tissue-specific delivery [113–115]. For precise stimulation and regulation of cellular activities, fine controllability of local temperatures with the nanoscale spatial resolution is a critical factor, and thus, new methods to measure absolute temperature and thermal gradient in the close vicinity of MNP surfaces are required. To address these issues, notable efforts have been made and should be further confirmed in terms of surface chemistries (e.g., stealth coating of PEG or zwitterions and conjugation of targeting molecules), nanobiosystem interaction studies (e.g., pharmacokinetics, biodistribution, and cellular uptake), and nanoscale temperature measuring techniques based on various chemical strategies (e.g., heat-responsive azo compound and fluorescent polymers) [113–115].

In principle, MNPs upon the incorporation of functional molecules (e.g., antibodies, chemotherapy drugs, siRNAs, aptamers, and cell-penetrating peptides) can be a versatile platform capable of the noninvasive, remote regulation of cellular activities with molecular-level specificity and without penetration depth limits. Based on these characteristic features, the nanoscale heating of MNPs holds great potential to extend the target from the single-cell level to the whole body for the better understanding and precise control of biological systems.

References

1. Cancer Tomorrow (n.d.). https://gco.iarc.fr/tomorrow/home. Accessed 9 May 2020.
2. K.W. Baumann, J.M. Baust, K.K. Snyder, J.G. Baust, R.G. van Buskirk, Characterization of pancreatic cancer cell thermal response to heat ablation or cryoablation. Technol. Cancer Res. Treat. **16**, 393–405 (2017). https://doi.org/10.1177/1533034616655658
3. K.F. Chu, D.E. Dupuy, Thermal ablation of tumors: Biological mechanisms and advances in therapy. Nat. Rev. Cancer **14**, 199–208 (2014). https://doi.org/10.1038/nrc3672
4. A.G. Belous, S.O. Solopan, Y.D. Stupin, B.S. Khomenko, L.L. Kovalenko, O.P. Fedorchuk, *Method of Synthesis of Nickel-Manganese-Zinc Spinel Ferrites for Microwave Applications*. Ukrainian Patent, No 123709 (2018)
5. A.G. Belous, E.D. Solovyova, S.O. Solopan, O.v. Yelenich, L.N. Bubnovskaya, S.P. Osinsky, Properties and potential applications of ferromagnetic nanostructures in medicine and microwave engineering. Solid State Phenom. **230**, 95–100 (2015). https://doi.org/10.4028/www.scientific.net/SSP.230.95
6. A.G. Belous, S.O. Solopan, O.V. Yelenich, A.I. Tovstolytkin, T.V. Kolodiazhnyi, S.P. Osinsky, L.N. Bubnovskaya, Nanoparticles of spinel and perovskite ferromagnets and prospects for their application in medicine. AIP Conf. Proc. **1627**, 13–18 (2014). https://doi.org/10.1063/1.4901650
7. A.G. Belous, A.I. Tovstolytkin, S.A. Solopan, Y.Y. Shlapa, O.P. Fedorchuk, Nanosized oxide magnetics: synthesis, properties, application. Ukr. Chem. J. **83**, 3–24 (2017)
8. D.S. Nikam, S.V. Jadhav, V.M. Khot, M.R. Phadatare, S.H. Pawar, Study of AC magnetic heating characteristics of $Co_{0.5}Zn_{0.5}Fe_2O_4$ nanoparticles for magnetic hyperthermia therapy. J. Magn. Magn. Mater. **349**, 208–213 (2014). https://doi.org/10.1016/j.jmmm.2013.08.039
9. M. Veverka, K. Závěta, O. Kaman, P. Veverka, K. Knížek, E. Pollert, M. Burian, P. Kašpar, Magnetic heating by silica-coated Co-Zn ferrite particles. J. Phys. D. Appl. Phys. **47**, 065503 (2014). https://doi.org/10.1088/0022-3727/47/6/065503

10. Z. Shaterabadi, G. Nabiyouni, M. Soleymani, Physics responsible for heating efficiency and self-controlled temperature rise of magnetic nanoparticles in magnetic hyperthermia therapy. Prog. Biophys. Mol. Biol. **133**, 9–19 (2018). https://doi.org/10.1016/j.pbiomolbio.2017.10.001
11. A. López-Ortega, M. Estrader, G. Salazar-Alvarez, A.G. Roca, J. Nogués, Applications of exchange coupled bi-magnetic hard/soft and soft/hard magnetic core/shell nanoparticles. Phys. Rep. **553**, 1–32 (2015). https://doi.org/10.1016/j.physrep.2014.09.007
12. Q. Song, Z.J. Zhang, Controlled synthesis and magnetic properties of bimagnetic spinel ferrite $CoFe_2O_4$ and $MnFe_2O_4$ nanocrystals with core-shell architecture. J. Am. Chem. Soc. **134**, 10182–10190 (2012). https://doi.org/10.1021/ja302856z
13. Q. Zhang, I. Castellanos-Rubio, R. Munshi, I. Orue, B. Pelaz, K.I. Gries, W.J. Parak, P. del Pino, A. Pralle, Model driven optimization of magnetic anisotropy of exchange-coupled core-shell ferrite nanoparticles for maximal hysteretic loss. Chem. Mater. **27**, 7380–7387 (2015). https://doi.org/10.1021/acs.chemmater.5b03261
14. D. Polishchuk, N. Nedelko, S. Solopan, A. Ślawska-Waniewska, V. Zamorskyi, A. Tovstolytkin, A. Belous, Profound interfacial effects in $CoFe_2O_4/Fe_3O_4$ and $Fe_3O_4/CoFe_2O_4$ core/shell nanoparticles. Nanoscale Res. Lett. **13**, 67 (2018). https://doi.org/10.1186/s11671-018-2481-x
15. A.I. Tovstolytkin, M.M. Kulyk, V.M. Kalita, S.M. Ryabchenko, V.O. Zamorskyi, O.P. Fedorchuk, S.O. Solopan, A.G. Belous, Nickel-zinc spinel nanoferrites: Magnetic characterization and prospects of the use in self-controlled magnetic hyperthermia. J. Magn. Magn. Mater. **473**, 422–427 (2019). https://doi.org/10.1016/j.jmmm.2018.10.075
16. I. Sharifi, H. Shokrollahi, S. Amiri, Ferrite-based magnetic nanofluids used in hyperthermia applications. J. Magn. Magn. Mater. **324**, 903–915 (2012). https://doi.org/10.1016/j.jmmm.2011.10.017
17. Soshin Chikazumi, *Physics of Ferromagnetism, 1997*. Google books (n.d.). https://books.google.com.ua/books?id=AZVfuxXF2GsC&printsec=frontcover&hl=uk&source=gbs_ge_summary_r&cad=0#v=onepage&q&f=false. Accessed 25 Mar 2020.
18. V.M. Kalita, A.I. Tovstolytkin, S.M. Ryabchenko, O.V. Yelenich, S.O. Solopan, A.G. Belous, Mechanisms of AC losses in magnetic fluids based on substituted manganites. Phys. Chem. Chem. Phys. **17**, 18087–18097 (2015). https://doi.org/10.1039/c5cp02822a
19. Y. Shlapa, M. Kulyk, V. Kalita, T. Polek, A. Tovstolytkin, J.-M. Greneche, S. Solopan, A. Belous, Iron-doped (La,Sr)MnO_3 manganites as promising mediators of self-controlled magnetic nanohyperthermia. Nanoscale Res. Lett. **11**, 24 (2016). https://doi.org/10.1186/s11671-015-1223-6
20. E. Pollert, O. Kaman, P. Veverka, M. Veverka, M. Marysko, K. Záveta, M. Kacenka, I. Lukes, P. Jendelová, P. Kaspar, M. Burian, V. Herynek, Core-shell $La_{(1-x)}Sr_{(x)}MnO_3$ nanoparticles as colloidal mediators for magnetic fluid hyperthermia. Philos. Trans. R. Soc. **368**, 4389 (2010). https://doi.org/10.1098/rsta.2010.0123
21. A.I. Tovstolytkin, Y.M. Lytvynenko, A.V. Bodnaruk, O.V. Bondar, V.M. Kalita, S.M. Ryabchenko, Y.Y. Shlapa, S.O. Solopan, A.G. Belous, Unusual magnetic and calorimetric properties of lanthanum-strontium manganite nanoparticles. J. Magn. Magn. Mater. **498**, 166088 (2020). https://doi.org/10.1016/j.jmmm.2019.166088
22. O. Shydlovska, N. Zholobak, S. Dybkova, S. Osinsky, L. Bubnovskaya, O. Yelenich, S. Solopan, A. Belous, Synthesis and comparative characteristics of biological activities of (La, Sr)MnO_3 and Fe_3O_4 nanoparticles. Eur. J. Nanomed. **9**, 33–43 (2017). De Gruyter. https://doi.org/10.1515/ejnm-2016-0028
23. S.L. Easo, P.V. Mohanan, Dextran stabilized iron oxide nanoparticles: Synthesis, characterization and in vitro studies. Carbohydr. Polym. **92**, 726–732 (2013). https://doi.org/10.1016/j.carbpol.2012.09.098
24. R. Qiao, C. Yang, M. Gao, Superparamagnetic iron oxide nanoparticles: From preparations to in vivo MRI applications. J. Mater. Chem. **19**, 6274–6293 (2009). https://doi.org/10.1039/b902394a

25. S. Stolnik, L. Illum, S.S. Davis, Long circulating microparticulate drug carriers. Adv. Drug Deliv. Rev. **64**, 290–301 (2012). https://doi.org/10.1016/j.addr.2012.09.029
26. R.Y. Hong, B. Feng, L.L. Chen, G.H. Liu, H.Z. Li, Y. Zheng, D.G. Wei, Synthesis, characterization and MRI application of dextran-coated Fe_3O_4 magnetic nanoparticles. Biochem. Eng. J. **42**, 290–300 (2008). https://doi.org/10.1016/j.bej.2008.07.009
27. Q.A. Pankhurst, N.T.K. Thanh, S.K. Jones, J. Dobson, Progress in applications of magnetic nanoparticles in biomedicine. J. Phys. D. Appl. Phys. **42**, 224001 (2009). https://doi.org/10.1088/0022-3727/42/22/224001
28. I.I. Kaneva, D.G. Krutogin, V.G. Andreev, L.M. Letuk, *Ferrite Materials and Components of Magnetoelectronics* (Laboratornyi praktikum, MISiS, Moskow, 2005)
29. L.A. Reznitskii, A.S. Guzei, Thermodynamic properties of alkaline earth Titanates, Zirconates, and Hafnates. Russ. Chem. Rev. **47**, 99–119 (1978). https://doi.org/10.1070/rc1978v047n02abeh002213
30. A.G. Belous, Y.V. Pashkova, O.I. Vunov, et al., Effect of processing conditions on the phase transformations, structure and magnetoresistive properties of $La_{0.7}Sr_{0.3}MnO_{3\pm\gamma}$ manganites. Ukr. Chem. J. **71**, 17–23 (2005)
31. O.V. Yelenich, S.O. Solopan, T.V. Kolodiazhnyi, J.M. Greneche, A.G. Belous, Synthesis of iron oxide nanoparticles by different methods and study of their properties. Solid State Phenom. **230**, 108–113 (2015). https://doi.org/10.4028/www.scientific.net/SSP.230.108
32. S.O. Solopan, Y.D. Fateev, A.G. Belous, Features of synthesis of weakly agglomerated Fe_3O_4 nanoparticles in microemulsions. Ukr. Chem. J. **78**, 3–8 (2012)
33. O.V. Yelenich, S.O. Solopan, V.V. Trachevskii, A.G. Belous, Synthesis and properties of AFe_2O_4 (A = Mn, Fe, Co, Ni, Zn) nanoparticles produced by deposition from diethylene glycol solution. Russ. J. Inorg. Chem. **58**, 901–905 (2013)
34. O.V. Yelenich, S.A. Solopan, M. Troschenkov Yu, A.G. Belous, Synthesis of weakly agglomerated nanoparticles based Fe_3O_4 and $CoFe_2O_4$ in non-aqueous solutions. Ukr. Chem. J. **78**, 11–15 (2012)
35. O.V. Yelenich, S.O. Solopan, A.G. Belous, Polyol synthesis and properties of AFe_2O_4 nanoparticles (A = Mn, Fe, Co, Ni, Zn) with spinel structure. Solid State Phenom. **200**, 149–155 (2013). https://doi.org/10.4028/www.scientific.net/SSP.200.149
36. O. Yelenich, S. Solopan, T. Kolodiazhnyi, Y. Tykhonenko, A. Tovstolytkin, A. Belous, Magnetic properties and AC losses in AFe_2O_4 (A = Mn, Co, Ni, Zn) nanoparticles synthesized from nonaqueous solution. J. Chem. **2015**, 532198 (2015). https://doi.org/10.1155/2015/532198
37. A.V. Nagornyi, M.V. Avdeev, O.V. Yelenich, S.O. Solopan, A.G. Belous, A.V. Shulenina, V.A. Turchenko, D.V. Soloviov, L.A. Bulavin, V.L. Aksenov, Structural aspects of Fe_3O_4/$CoFe_2O_4$ magnetic nanoparticles according to X-ray and neutron scattering. J. Surf. Invest. **12**, 737–743 (2018). https://doi.org/10.1134/S102745101804033X
38. Y. Shlapa, S. Solopan, O. Yelenich, V. Trachevskii, A. Belous, Synthesis of ferromagnetic $La_{1-x}Sr_xMnO_3$ nanoparticles by precipitation from diethylene glycol solution and their properties. J. Adv. Ceram. **5**, 197–203 (2016). https://doi.org/10.1007/s40145-016-0190-4
39. Y. Shlapa, S. Solopan, A. Belous, A. Tovstolytkin, Effect of synthesis method of $La_{1-x}Sr_xMnO_3$ manganite nanoparticles on their properties. Nanoscale Res. Lett. **13**, 1–7 (2018). https://doi.org/10.1186/s11671-017-2431-z
40. A.K. Ganguli, A. Ganguly, S. Vaidya, Microemulsion-based synthesis of nanocrystalline materials. Chem. Soc. Rev. **39**, 474–485 (2010)
41. S.A. Solopan, A.G. Belous, S.P. Osinsky, A.I. Tovstolytkin, Features of synthesis of nanomaterials based on the modified lanthanum-strontium manganites for the hyperthermia of tumors. Nanosyst. Nanomater. Nanotechnol. **8**, 775–786 (2010)
42. Y.Y. Shlapa, S.A. Solopan, A.G. Belous, Magnetothermic effect in core/shell nanocomposite $(La,Sr)MnO_3/SiO_2$. Theor. Exp. Chem. **54**, 92–98 (2018). https://doi.org/10.1007/s11237-018-9551-0
43. A.G. Belous, S.A. Solopan, O.V. Yelenich, Y.Y. Shlapa, S.P. Osinsky, Synthesis and properties of ferromagnetic nanostructures and their possible application in medicine and UHF

technic, in *Fundamental Problems of Development of Novel Materials and Compounds of Chemical Production*, (2016), pp. 91–101

44. S.A. Solopan, A.V. Yelenich, A.I. Tovstolytkin, A.G. Belous, Effect of the features of synthesis of (La,Sr)MnO_3 nanoparticles on their properties. Nanosyst. Nanomater. Nanotechnol. **2**, 189–198 (2014)
45. Y.Y. Shlapa, S.O. Solopan, Y.V. Odynets, A.G. Belous, Organic-inorganic composite structures based on ferromagnetic (La,Sr)MnO_3 manganite nanoparticles. Ukr. Chem. J. **81**, 99–103 (2015)
46. S.O. Solopan, O.I. V'Yunov, A.G. Belous, T.I. Polek, A.I. Tovstolytkin, Effect of nanoparticles agglomeration on electrical properties of $La_{1-x}A_xMnO_3$ (A = Sr, Ba) nanopowder and ceramic solid solutions. Solid State Sci. **14**, 501–505 (2012). https://doi.org/10.1016/j.solidstatesciences.2012.01.030
47. O.V. Yelenich, S.O. Solopan, A.G. Belous, Synthesis of $NiFe_2O_4$ nanoparticles in non-aqueous solutions and their properties. Ukr. Chem. J. **79**, 3–8 (2013)
48. Y. Shlapa, S. Solopan, A. Belous, Nanoparticles of $La_{1-x}Sr_xMnO_3$ ($0.23 \leq x \leq 0.25$) manganite: Features of synthesis and crystallochemical properties. J. Magn. Magn. Mater. **510**, 166902 (2020)
49. Y. Shlapa, S. Solopan, A. Belous, Effect of mechano-chemical processing in the synthesis of weakly agglomerated ferromagnetic $La_{1-x}Sr_xMnO_3$ nanoparticles on their properties, in *2015 IEEE 35th International Conference on Electronics and Nanotechnology, ELNANO 2015 – Conference Proceedings*, (2015), pp. 282–286. https://doi.org/10.1109/ELNANO.2015.7146892
50. Y.T. He, S.J. Traina, Transformation of magnetite to goethite under alkaline pH conditions. Clay Miner. **42**, 13–19 (2007). https://doi.org/10.1180/claymin.2007.042.1.02
51. A.P. Khandhar, R.M. Ferguson, J.A. Simon, K.M. Krishnan, Enhancing cancer therapeutics using size-optimized magnetic fluid hyperthermia. J. Appl. Phys. **111**, 07B306 (2012). https://doi.org/10.1063/1.3671427
52. R. Hergt, W. Andrae, C.G. d'Ambly, I. Hilger, W.A. Kaiser, U. Richter, H.G. Schmidt, Physical limits of hyperthermia using magnetite fine particles. IEEE Trans. Magn. **34**, 3745–3754 (1998). https://doi.org/10.1109/20.718537
53. N.H. Cho, T.C. Cheong, J.H. Min, J.H. Wu, S.J. Lee, D. Kim, J.S. Yang, S. Kim, Y.K. Kim, S.Y. Seong, A multifunctional core-shell nanoparticle for dendritic cell-based cancer immunotherapy. Nat. Nanotechnol. **6**, 675–682 (2011). https://doi.org/10.1038/nnano.2011.149
54. S.H. Noh, W. Na, J.T. Jang, J.H. Lee, E.J. Lee, S.H. Moon, Y. Lim, J.S. Shin, J. Cheon, Nanoscale magnetism control via surface and exchange anisotropy for optimized ferrimagnetic hysteresis. Nano Lett. **12**, 3716–3721 (2012). https://doi.org/10.1021/nl301499u
55. P. Mélinon, S. Begin-Colin, J.L. Duvail, F. Gauffre, N.H. Boime, G. Ledoux, J. Plain, P. Reiss, F. Silly, B. Warot-Fonrose, Engineered inorganic core/shell nanoparticles. Phys. Rep. **543**, 163–197 (2014). https://doi.org/10.1016/j.physrep.2014.05.003
56. A.A. Sattar, H.M. El-Sayed, I. Alsuqia, Structural and magnetic properties of $CoFe_2O_4$/$NiFe_2O_4$ core/shell nanocomposite prepared by the hydrothermal method. J. Magn. Magn. Mater. **395**, 89–96 (2015). https://doi.org/10.1016/j.jmmm.2015.07.039
57. A.H. Habib, C.L. Ondeck, P. Chaudhary, M.R. Bockstaller, M.E. McHenry, Evaluation of iron-cobalt/ferrite core-shell nanoparticles for cancer thermotherapy. J. Appl. Phys. **103**, 07A307 (2008). https://doi.org/10.1063/1.2830975
58. R. Ghosh, L. Pradhan, Y.P. Devi, S.S. Meena, R. Tewari, A. Kumar, S. Sharma, N.S. Gajbhiye, R.K. Vatsa, B.N. Pandey, R.S. Ningthoujam, Induction heating studies of Fe_3O_4 magnetic nanoparticles capped with oleic acid and polyethylene glycol for hyperthermia. J. Mater. Chem. **21**, 13388–13398 (2011). https://doi.org/10.1039/c1jm10092k
59. O.V. Yelenich, S.O. Solopan, J.M. Greneche, A.G. Belous, Synthesis and properties MFe_2O_4 (M = Fe, Co) nanoparticles and core-shell structures. Solid State Sci. **46**, 19–26 (2015). https://doi.org/10.1016/j.solidstatesciences.2015.05.011

60. S.O. Solopan, N. Nedelko, S. Lewińska, A. Ślawska-Waniewska, V.O. Zamorskyi, A.I. Tovstolytkin, A.G. Belous, Core/shell architecture as an efficient tool to tune DC magnetic parameters and AC losses in spinel ferrite nanoparticles. J. Alloys Compd. **788**, 1203–1210 (2019). https://doi.org/10.1016/j.jallcom.2019.02.276
61. G. Franceschin, T. Gaudisson, N. Menguy, B.C. Dodrill, N. Yaacoub, J.-M. Grenèche, R. Valenzuela, S. Ammar, Exchange-biased $Fe_{3-x}O_4$-CoO granular composites of different morphologies prepared by seed-mediated growth in polyol: From core-shell to multicore embedded structures. Part. Part. Syst. Charact. **35**, 1800104 (2018). https://doi.org/10.1002/ppsc.201800104
62. N. Flores-Martinez, G. Franceschin, T. Gaudisson, P. Beaunier, N. Yaacoub, J.-M. Grenèche, R. Valenzuela, S. Ammar, Giant exchange-bias in polyol-made $CoFe_2O_4$-CoO core-shell like nanoparticles. Part. Part. Syst. Charact. **35**, 1800290 (2018). https://doi.org/10.1002/ppsc.201800290
63. T. Gaudisson, R. Sayed-Hassan, N. Yaacoub, G. Franceschin, S. Nowak, J.M. Grenèche, N. Menguy, P. Sainctavit, S. Ammar, On the exact crystal structure of exchange-biased Fe_3O_4-CoO nanoaggregates produced by seed-mediated growth in polyol. CrystEngComm **18**, 3799–3807 (2016). https://doi.org/10.1039/c6ce00700g
64. O. Masala, D. Hoffman, N. Sundaram, K. Page, T. Proffen, G. Lawes, R. Seshadri, Preparation of magnetic spinel ferrite core/shell nanoparticles: Soft ferrites on hard ferrites and vice versa. Solid State Sci. **8**, 1015–1022 (2006). https://doi.org/10.1016/j.solidstatesciences.2006.04.014
65. U.M. Engelmann, C. Shasha, E. Teeman, I. Slabu, K.M. Krishnan, Predicting size-dependent heating efficiency of magnetic nanoparticles from experiment and stochastic Néel-Brown Langevin simulation. J. Magn. Magn. Mater. **471**, 450–456 (2019). https://doi.org/10.1016/j.jmmm.2018.09.041
66. J. Carrey, B. Mehdaoui, M. Respaud, Simple models for dynamic hysteresis loop calculations of magnetic single-domain nanoparticles: Application to magnetic hyperthermia optimization. J. Appl. Phys. **109**, 083921 (2011). https://doi.org/10.1063/1.3551582
67. S. Chikazumi, C.D. (Chad D.) Graham, S. Chikazumi, *Physics of Ferromagnetism* (Oxford University Press, 1997). https://global.oup.com/academic/product/physics-of-ferromagnetism-9780198517764?cc=ua&lang=en&. Accessed 27 May 2019.
68. J.H. Lee, Y.M. Huh, Y.W. Jun, J.W. Seo, J.T. Jang, H.T. Song, S. Kim, E.J. Cho, H.G. Yoon, J.S. Suh, J. Cheon, Artificially engineered magnetic nanoparticles for ultra-sensitive molecular imaging. Nat. Med. **13**, 95–99 (2007). https://doi.org/10.1038/nm1467
69. C.N. Chinnasamy, A. Narayanasamy, N. Ponpandian, K. Chattopadhyay, K. Shinoda, B. Jeyadevan, K. Tohji, K. Nakatsuka, T. Furubayashi, I. Nakatani, Mixed spinel structure in nanocrystalline $NiFe_2O_4$. Phys. Rev. B Condens. Matter Mater. Phys. **63**, 184108 (2001). https://doi.org/10.1103/PhysRevB.63.184108
70. M. Sanna Angotzi, V. Mameli, C. Cara, A. Musinu, C. Sangregorio, D. Niznansky, H.L. Xin, J. Vejpravova, C. Cannas, Coupled hard-soft spinel ferrite-based core-shell nanoarchitectures: Magnetic properties and heating abilities. Nanoscale Adv. **2**, 3191–3201 (2020). https://doi.org/10.1039/d0na00134a
71. O.V. Yelenich, S.O. Solopan, T.V. Kolodiazhnyi, V.V. Dzyublyuk, A.I. Tovstolytkin, A.G. Belous, Magnetic properties and high heating efficiency of $ZnFe_2O_4$ nanoparticles. Mater. Chem. Phys. **146**, 129–135 (2014). https://doi.org/10.1016/j.matchemphys.2014.03.010
72. P. Gao, X. Hua, V. Degirmenci, D. Rooney, M. Khraisheh, R. Pollard, R.M. Bowman, E.V. Rebrov, Structural and magnetic properties of $Ni_{1-x}Zn_xFe_2O_4$ (x=0, 0.5 and 1) nanopowders prepared by sol-gel method. J. Magn. Magn. Mater. **348**, 44–50 (2013). https://doi.org/10.1016/j.jmmm.2013.07.060
73. C. Srinivas, B.V. Tirupanyam, A. Satish, V. Seshubai, D.L. Sastry, O.F. Caltun, Effect of Ni^{2+} substitution on structural and magnetic properties of Ni-Zn ferrite nanoparticles. J. Magn. Magn. Mater. **382**, 15–19 (2015). https://doi.org/10.1016/j.jmmm.2015.01.008

74. A.H. Morr, K. Haneda, Magnetic structure of small $NiFe^2O^4$ particles. J. Appl. Phys. **52**, 2496–2498 (1981). https://doi.org/10.1063/1.328979
75. B.T. Naughton, D.R. Clarke, Lattice expansion and saturation magnetization of nickel–zinc ferrite nanoparticles prepared by aqueous precipitation. J. Am. Ceram. Soc. **90**, 3541–3546 (2007). https://doi.org/10.1111/j.1551-2916.2007.01980.x
76. N. Daffé, F. Choueikani, S. Neveu, M.A. Arrio, A. Juhin, P. Ohresser, V. Dupuis, P. Sainctavit, Magnetic anisotropies and cationic distribution in $CoFe_2O_4$ nanoparticles prepared by co-precipitation route: Influence of particle size and stoichiometry. J. Magn. Magn. Mater. **460**, 243–252 (2018). https://doi.org/10.1016/j.jmmm.2018.03.041
77. R.R. Wildeboer, P. Southern, Q.A. Pankhurst, On the reliable measurement of specific absorption rates and intrinsic loss parameters in magnetic hyperthermia materials. J. Phys. D. Appl. Phys. **47**, 495003 (2014). https://doi.org/10.1088/0022-3727/47/49/495003
78. A. Urushibara, Y. Moritomo, T. Arima, A. Asamitsu, G. Kido, Y. Tokura, Insulator-metal transition and giant magnetoresistance in $La_{1-x}Sr_xMnO_3$. Phys. Rev. B **51**, 14103–14109 (1995). https://doi.org/10.1103/PhysRevB.51.14103
79. A.G. Belous, O.I. V'yunov, E.V. Pashkova, O.Z. Yanchevskii, A.I. Tovstolytkin, A.M. Pogorelyi, Effects of chemical composition and sintering temperature on the structure of $La_{1-x}Sr_xMnO_{3\pm\gamma}$ solid solutions. Inorg. Mater. **39**, 161–170 (2003). https://doi.org/10.1023/A:1022198613723
80. D.H. Manh, P.T. Phong, P.H. Nam, D.K. Tung, N.X. Phuc, I.J. Lee, Structural and magnetic study of $La_{0.7}Sr_{0.3}MnO_3$ nanoparticles and AC magnetic heating characteristics for hyperthermia applications. Phys. B Condens. Matter **444**, 94–102 (2014). https://doi.org/10.1016/j.physb.2014.03.025
81. Y. Shlapa, I. Timashkov, S. Solopan, M. Rajnak, J. Kovac, M. Molcan, M. Timko, A. Belous, Ce-doped manganites of lanthanum-strontium as promising inducers of magnetic hyperthermia, in *IEEE 39th International Conference on Electronics and Nanotechnology, ELNANO 2019*, (2019), pp. 439–433
82. I.P. Timashkov, Y.Y. Shlapa, S.A. Solopan, A.G. Belous, Synthesis and crystallochemical properties of cerium-substituted (La,Sr)MnO_3 manganite nanoparticles. Ukr. Chem. J. **85**, 17–24 (2019). Ukrainian Chemical, Institute of General and Inorganic Chemistry named after V.I. Vernadsky NAS of Ukraine. https://doi.org/10.33609/0041-6045.85.9.2019.17-24
83. Y. Shlapa, S. Solopan, A. Bodnaruk, M. Kulyk, V. Kalita, Y. Tykhonenko-Polishchuk, A. Tovstolytkin, A. Belous, Effect of synthesis temperature on structure and magnetic properties of $(La, Nd)_{0.7}Sr_{0.3}MnO_3$ nanoparticles. Nanoscale Res. Lett. **12**, 1–7 (2017). https://doi.org/10.1186/s11671-017-1884-4
84. Y. Shlapa, S. Solopan, A. Bodnaruk, M. Kulyk, V. Kalita, Y. Tykhonenko-Polishchuk, A. Tovstolytkin, V. Zinchenko, A. Belous, Lanthanum-strontium manganites for magnetic nanohyperthermia: Fine tuning of parameters by substitutions in lanthanum sublattice. J. Alloys Compd. **702**, 31–37 (2017). https://doi.org/10.1016/j.jallcom.2017.01.222
85. T.I. Polek, A.I. Tovstolytkin, M.M. Kulyk, Y.Y. Shlapa, S.O. Solopan, A.G. Belous, Crystallographic, magnetic and magnetoresistive properties of $La_{0,77}Sr_{0,23}Mn_{1-y}Fe_yO_3$ ceramics. Metalophys. Novel Technol. **38**, 479–492 (2016)
86. B. Hildebrandt, P. Wust, O. Ahlers, A. Dieing, G. Sreenivasa, T. Kerner, R. Felix, H. Riess, The cellular and molecular basis of hyperthermia. Crit. Rev. Oncol. Hematol. **43**, 33–56 (2002). https://doi.org/10.1016/S1040-8428(01)00179-2
87. S. Vasseur, E. Duguet, J. Portier, G. Goglio, S. Mornet, E. Hadová, K. Knížek, M. Maryško, P. Veverka, E. Pollert, Erratum to "Lanthanum manganese perovskite nanoparticles as possible in vivo mediators for magnetic hyperthermia". J. Magn. Magn. Mater. **307**, 330 (2006). https://doi.org/10.1016/j.jmmm.2006.06.034
88. N.K. Prasad, K. Rathinasamy, D. Panda, D. Bahadur, TC-tuned biocompatible suspension of $La_{0.73}Sr_{0.27}MnO_3$ for magnetic hyperthermia. J. Biomed. Mater. Res. B Appl. Biomater. **85B**, 409–416 (2008). https://doi.org/10.1002/jbm.b.30959

89. A.K. Pradhan, R. Bah, R.B. Konda, R. Mundle, H. Mustafa, O. Bamiduro, R.R. Rakhimov, X. Wei, D.J. Sellmyer, Synthesis and magnetic characterizations of manganite-based composite nanoparticles for biomedical applications. J. Appl. Phys. **103**, 07F704 (2008). https://doi.org/10.1063/1.2829906
90. E. Pollert, K. Knížek, M. Maryško, P. Kašpar, S. Vasseur, E. Duguet, New T_c-tuned magnetic nanoparticles for self-controlled hyperthermia. J. Magn. Magn. Mater. **316**, 122–125 (2007). https://doi.org/10.1016/j.jmmm.2007.02.031
91. A. Belous, A. Tovstolytkin, S. Solopan, Y. Shlapa, O. Fedorchuk, Synthesis, properties and applications of some magnetic oxide based nanoparticles and films. Acta Phys. Polon. **A133**, 1006 (2018). https://doi.org/10.12693/APhysPolA.133.1006
92. K. Zhang, T. Holloway, J. Pradhan, M. Bahoura, R. Bah, R.R. Rakhimov, A.K. Pradhan, R. Prabakaran, G.T. Ramesh, Synthesis and magnetic characterizations of $La_{1-x}Sr_xMnO_3$ nanoparticles for biomedical applications. J. Nanosci. Nanotechnol. **10**, 5520–5526 (2010). https://doi.org/10.1166/jnn.2010.2437
93. A. Belous, S. Solopan, A. Yelenich, L. Bubnovskaya, S. Osinsky, Nanoparticles of ferromagnetic materials and possibilities of their application in the hyperthermia of malignant tumors, in *2013 IEEE 33rd International Scientific Conference Electronics and Nanotechnology, ELNANO 2013 – Conference Proceedings*, (2013), pp. 280–284. https://doi.org/10.1109/ELNANO.2013.6552060
94. A.I. Tovstolytkin, Y.Y. Shlapa, S.O. Solopan, A.V. Bodnaruk, M.M. Kulyk, V.M. Kalita, V.O. Zamorskyi, S.M. Ryabchenko, A.G. Belous, Manganite nanoparticles as promising heat mediators for magnetic hyperthermia: Comparison of different chemical substitutions. Acta Phys. Pol. A **133**, 1017–1020 (2018)
95. L. Bubnovskaya, A. Belous, A. Solopan, A. Podoltsev, I. Kondratenko, A. Kovelskaya, T. Sergienko, S. Osinsky, Nanohyperthermia of malignant tumors. II. in vivo tumor heating with manganese perovskite nanoparticles. Exp. Oncol. **34**, 336–339 (2012)
96. L. Bubnovskaya, A. Belous, S. Solopan, A. Kovelskaya, L. Bovkun, A. Podoltsev, I. Kondtratenko, S. Osinsky, Magnetic fluid hyperthermia of rodent tumors using manganese perovskite nanoparticles. J. Nanopart. **2014**, 1–9 (2014). https://doi.org/10.1155/2014/278761
97. S. Solopan, A. Belous, A. Yelenich, L. Bubnovskaya, A. Kovelskaya, A. Podoltsev, I. Kondratenko, S. Osinsky, Nanohyperthermia of malignant tumors. I. Lanthanum-strontium manganite magnetic fluid as potential inducer of tumor hyperthermia. Exp. Oncol. **33**, 130–135 (2011)
98. O.V. Yelenich, S.O. Solopan, T.V. Kolodiazhnyi, V.V. Dzyublyuk, A.I. Tovstolytkin, A.G. Belous, Superparamagnetic behavior and AC-losses in $NiFe_2O_4$ nanoparticles. Solid State Sci. **20**, 115–119 (2013). https://doi.org/10.1016/j.solidstatesciences.2013.03.013
99. A. Jordan, R. Scholz, P. Wust, H. Fähling, R. Felix, Magnetic fluid hyperthermia (MFH): Cancer treatment with AC magnetic field induced excitation of biocompatible superparamagnetic nanoparticles. J. Magn. Magn. Mater. **201**, 413–419 (1999). https://doi.org/10.1016/S0304-8853(99)00088-8
100. R. Hergt, S. Dutz, R. Müller, M. Zeisberger, Magnetic particle hyperthermia: nanoparticle magnetism and materials development for cancer therapy. J. Phys. Condens. Matter **18**, S2919–S2934 (2006). https://doi.org/10.1088/0953-8984/18/38/S26
101. R.E. Rosensweig, Heating magnetic fluid with alternating magnetic field. J. Magn. Magn. Mater. **252**, 370–374 (2002). https://doi.org/10.1016/S0304-8853(02)00706-0
102. S. Dutz, R. Hergt, Magnetic particle hyperthermia—A promising tumour therapy? Nanotechnology **25**, 452001 (2014). https://doi.org/10.1088/0957-4484/25/45/452001
103. S. Bedanta, W. Kleemann, Supermagnetism. J. Phys. D. Appl. Phys. **42**, 013001 (2009). https://doi.org/10.1088/0022-3727/42/1/013001
104. B. Mehdaoui, A. Meffre, J. Carrey, S. Lachaize, L.-M. Lacroix, M. Gougeon, B. Chaudret, M. Respaud, Optimal size of nanoparticles for magnetic hyperthermia: A combined theoretical and experimental study. Adv. Funct. Mater. **21**, 4573–4581 (2011). https://doi.org/10.1002/adfm.201101243

105. D.M. Polishchuk, Y.O. Tykhonenko-Polishchuk, S.O. Solopan, A.V. Bodnaruk, M.M. Kulyk, A.I. Tovstolytkin, A.M. Pogorily, Features of the magnetic state of ensembles of nanoparticles of substituted manganites: Experiment and model calculations. Low Temp. Phys. **43**, 570–577 (2017)
106. A.V. Bodnaruk, V.M. Kalita, M.M. Kulyk, S.M. Ryabchenko, A.I. Tovstolytkin, S.O. Solopan, A.G. Belous, Critical behavior of ensembles of superparamagnetic nanoparticles with dispersions of magnetic parameters. J. Phys. Condens. Matter **31**, 375801 (2019). https://doi.org/10.1088/1361-648X/ab26fa
107. V.M. Kalita, D.M. Polishchuk, D.G. Kovalchuk, A.V. Bodnaruk, S.O. Solopan, A.I. Tovstolytkin, S.M. Ryabchenko, A.G. Belous, Interplay between superparamagnetic and blocked behavior in an ensemble of lanthanum-strontium manganite nanoparticles. Phys. Chem. Chem. Phys. **19**, 27015–27024 (2017). https://doi.org/10.1039/c7cp05547a
108. A.V. Nagornyi, V.I. Petrenko, M.V. Avdeev, O.V. Yelenich, S.O. Solopan, A.G. Belous, A.Y. Gruzinov, O.I. Ivankov, L.A. Bulavin, Structural aspects of magnetic fluid stabilization in aqueous agarose solutions. J. Magn. Magn. Mater. **431**, 16–19 (2017). https://doi.org/10.1016/j.jmmm.2016.10.018
109. N. Singh, G.J.S. Jenkins, R. Asadi, S.H. Doak, Potential toxicity of superparamagnetic iron oxide nanoparticles (SPION). Nano Rev. **1**, 5358 (2010). https://doi.org/10.3402/nano.v1i0.5358
110. S. Pecorini, L. Gibellini, S. de Biasi, E. Bianchini, M. Nasi, A. Cossarizza, M. Pinti, Mitochondria, oxidative stress, cancer, and aging, in *Geriatric Oncology*, ed. by M. Extermann, (Springer International Publishing, Cham, 2020), pp. 183–204. https://doi.org/10.1007/978-3-319-57415-8_80
111. L. Bubnovskaya, N. Lukyanova, A. Belous, S. Solopan, Y. Shlapa, A. Podoltsev, Some effects of nanohyperthermia on the basis of lanthanum-strontium manganites (La,Sr)MnO_3 nanoparticles as heating mediators. Acta Sci. Cancer Biol. **4**, 1–8 (2020)
112. O. Veiseh, J.W. Gunn, M. Zhang, Design and fabrication of magnetic nanoparticles for targeted drug delivery and imaging. Adv. Drug Deliv. Rev. **62**, 284–304 (2010). https://doi.org/10.1016/j.addr.2009.11.002
113. W. Lin, Introduction: Nanoparticles in medicine. Chem. Rev. **115**, 10407–10409 (2015). https://doi.org/10.1021/acs.chemrev.5b00534
114. L. Polo-Corrales, C. Rinaldi, et al., J. Appl. Phys. **111**, 07B334 (2012). https://doi.org/10.1063/1.3680532
115. A. Riedinger, P. Guardia, A. Curcio, M.A. Garcia, R. Cingolani, L. Manna, T. Pellegrino, Subnanometer local temperature probing and remotely controlled drug release based on Azo-functionalized iron oxide nanoparticles. Nano Lett. **13**, 2399–2406 (2013). https://doi.org/10.1021/nl400188q

Synthesis and Applications of Anisotropic Magnetic Iron Oxide Nanoparticles

Lucía Gutiérrez, María del Puerto Morales, and Alejandro G. Roca

1 Introduction

Magnetic iron oxide nanocrystals, i.e. magnetite (Fe_3O_4), maghemite (γ-Fe_2O_3) or any of its intermediate stoichiometries ($Fe_{3-x}O_4$), have been applied in widespread applications since ancient times due to their appealing properties [1]. Being the only iron oxide with Fe ions in two oxidation states (+2, +3), magnetite is an abundant material that can be found in the nature (rocks) together with titanomagnetite [2]. Magnetite nanocrystals can also be found inside some bacteria (magnetotactic bacteria) where it is mineralized in intracellular vesicles, generally 30–100 nm nanocubes surrounded by a membrane and displayed in chains (magnetosomes) [3]. Finally, these nanocrystals can be "artificially" synthesized by top-down methods like high-energy ball milling, crushing or grinding, [4] or "bottom-up" approaches involving colloidal methods like coprecipitation, hydro−/solvothermal and thermal decomposition, rendering better structural properties in terms of crystallinity, purity and solvent solubility [5]. Moreover, magnetite particles in the nanometre range size can be

L. Gutiérrez (✉)
Instituto de Ciencia de Materiales de Aragón, ICMA-CSIC, Universidad de Zaragoza, Zaragoza, Spain

Centro de Investigación Biomédica en Red de Bioingeniería, Biomateriales y Nanomedicina (CIBER-BBN), Madrid, Spain

Department of Analytical Chemistry, Universidad de Zaragoza, Zaragoza, Spain
e-mail: lu@unizar.es

M. P. Morales
Instituto de Ciencia de Materiales de Madrid, ICMM-CSIC, Madrid, Spain

A. G. Roca
MagneticNanostrutures Group, Institut Català de Nanociència i Nanotecnologia, Barcelona, Spain

A. G. Roca et al. (eds.), *Surfaces and Interfaces of Metal Oxide Thin Films, Multilayers, Nanoparticles and Nano-composites*,
https://doi.org/10.1007/978-3-030-74073-3_3

synthesized breaking down the limit of superparamagnetism (random alignment of the magnetic moment in the absence of a magnetic field) leading to very interesting reversible magnetic behaviour that paves the way for new applications.

Magnetite nanocrystals have been used along human history for a wide spectrum of applications. They were used as pigments for ceramics in terracotta pottery since ancient times (ancient Greece) [6] and more recently as pigment for inks, dyes or any other material that required black colour, being the favourite one as magnetite absorbed light much better than other black candidates, and thus, it was more reliable [7]. Then, other application where magnetite nanocrystals were well-established was their use in water purification processes, e.g. heavy metal removal [8]. In the latest decades of the twentieth century, their applicability boosted due to magnetite electronic and magnetic properties at the microscale and nanoscale. Magnetite exhibited a great performance for magnetic recording media, spintronics and biomedicine, in both diagnosis (contrast agents in MRI, magnetic resonance imaging) and therapy (drug delivery vehicles and nanoheaters in magnetic and optical hyperthermia) [9]. For each application, scientists have designed and optimized the material in terms of particle size, shape, aggregation state and internal crystal structure to maximize their performance (Fig. 1).

However, magnetite is not the most stable iron oxide phase and tends to oxidize to maghemite. [10] The oxidation kinetics is faster as the temperature increases, as the atmosphere is richer in oxygen, when particle size drops or the crystallinity is poorer. Magnetic properties, in terms of saturation magnetization and coercivity, decrease after this phase transformation, but not dramatically [11]. Regarding the electrical properties, magnetite is a semimetal due to the electron hoping between the Fe ions of the octahedral sublattice of its structure, while maghemite is an insulator.

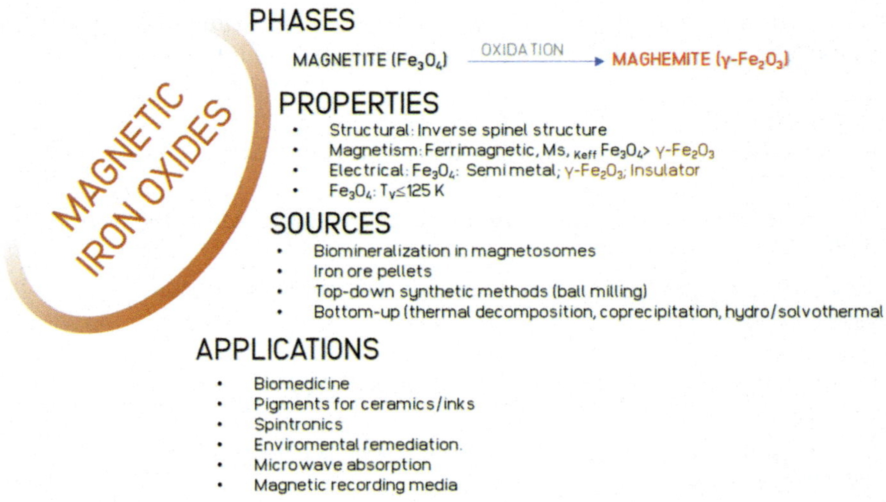

Fig. 1 Schematic description of the most interesting magnetic iron oxide phases, properties, sources and applications that will be considered in this chapter

The aim of this chapter is to present an overview on anisotropic magnetic nanoparticles and their applications in several areas. These nanostructures, with different morphology from the spherical one, have anisotropic properties that depend on the direction of the atoms within the nanostructure and a larger specific surface area. Anisotropic magnetic iron oxide nanoparticles could bring several advantages such as their higher surface-to-volume ratio, enhanced magnetic properties and selected reactivity due to the different facet exposure to the solvent. However, the synthesis of these nanostructures requires the use of different protocols or modification of the established ones. [1]The first section will be focused on the most feasible and successful routes for the preparation of magnetic nanoparticles with a specific morphology. Then, in a second part, we will report the most striking results on the use of these morphologies on diverse applications, ranging from environmental ones to most technological ones, like spintronics or biomedicine. As a summary, magnetite can be considered as an old friend that never gets old fashion and evolves adapting every time to new and more technological applications. This transformation can be understood by its multiple structural, electrical, and magnetic properties and also biodegradability. Thus, with more advanced synthetic techniques, it is possible to tailor the structural properties of magnetite nanocrystals making them an attractive material for the intended application.

2 Preparation of Anisotropic Iron Oxide Nanoparticles

For each application, the concept of "ideal" nanoparticles, i.e. the ones that exhibit the best performance, is different. Therefore, the ideal nanoparticle needs to be tailored mainly by their size and shape. In general, a synthetic route that leads to nanoparticles with an accurate control on their size and shape, low size distribution, aggregation state, high crystallinity and chemical purity is the main goal. The resulting particles would be ideal candidates for the intended applications since all of them will contribute to the desire effect. In this sense, colloidal routes developed in liquid media render nanoparticles accomplishing these properties, with low-cost production and large scalability. Moreover, the obtention of particles in liquid media enables an easier manipulation and postprocessing procedures (e.g. surface modification, deposition forming self-assembly, dispersion in different media).

Magnetite has a cubic cell with an inverse spinel structure (Fd3m) with a cell parameter of 8.394 Å [12]. Fe(II) ions are located in octahedral sites, and Fe(III) ions are located in both octahedral and tetrahedral sites (Fig. 2). The complete oxidation to maghemite keeps the same inverse spinel structure, but all Fe(II) ions are oxidized to Fe(III) and vacancies are created in octahedral positions, distributed coherently [13, 14]. After such transformation, due to the oxidation of the Fe ions, the cell parameter decreases to 8.33–8.35 Å depending on the vacancy order.

The synthesis of anisotropic, i.e. nonspherical, magnetic iron oxide nanoparticles is not trivial (Fig. 3). Thus, structures with hexagonal, orthorhombic or monoclinic crystal symmetry whose facets are perpendicular to the long axis have higher energy

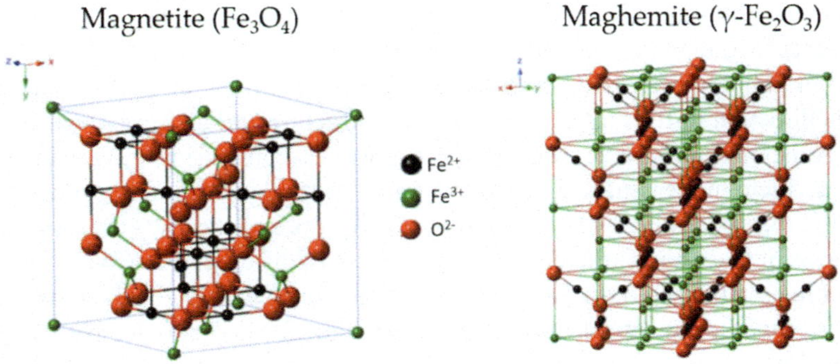

Fig. 2 Crystal structure of magnetite and maghemite. (Adapted with permission from Ref. [15]. © 2015 National Institute for Materials Science)

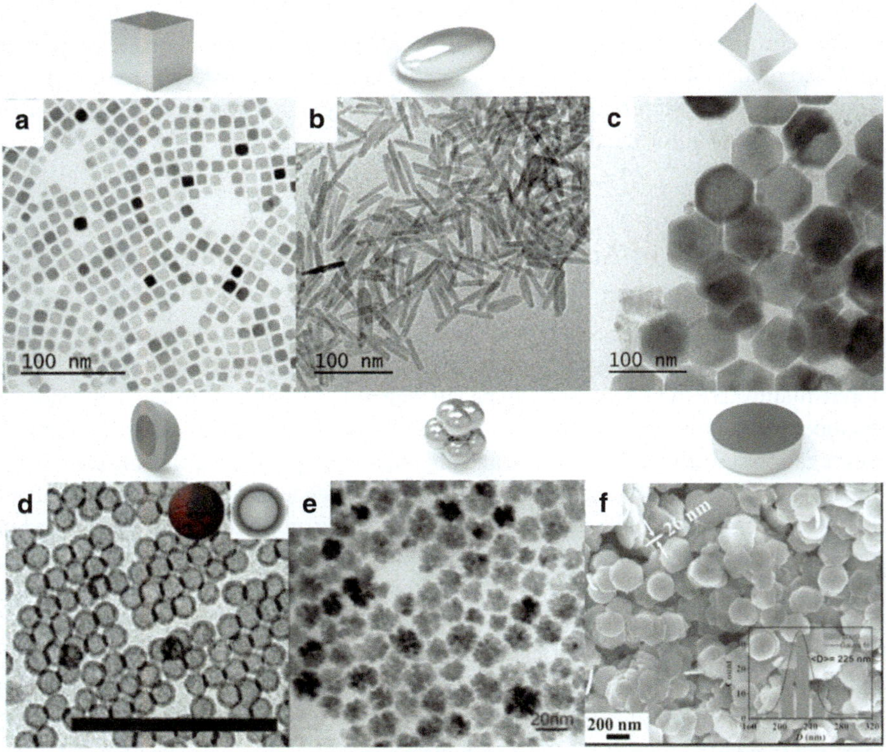

Fig. 3 TEM image of magnetic iron oxide nanoparticles with different morphology: (**a**) nanocubes, (**b**) elongated nanoparticles, (**c**) octahedrons, (**d**) hollow nanoparticles, reprinted with permission from Ref. [31]. Copyright 2007 American Chemical Society; (**e**) nanoflower, reprinted with permission from [32]. Copyright 2017 American Chemical Society. (**f**) SEM image of magnetite nanodiscs. (Adapted with permission from Ref. [33], © 2015, by Wiley)

and lead to elongated structures [16]. However, for cubic crystals, like magnetite, the isotropic growth is favoured. Then, there are two main strategies to induce the growth of anisotropic magnetic iron oxide magnetic nanostructures. The first approach is based on the use of an iron oxide/oxohydroxide as shape template. This strategy consists of two stages where first, an iron oxyhydroxide with the target shape is synthesized (i.e. goethite or akaganeite in elongated nanoparticles) and, in a second step, the oxyhydroxide is reduced to magnetite keeping the desired morphology. The second strategy consists on modifying the reaction kinetics to favour a kinetic regime. In this strategy, there is an essential role played by ligands (molecules or polymers) which bind to metal cations to form intermediate species prior nucleation or to regulate the particle growth through specific adsorption on certain facets of the particle surface. Other key parameters to control the reaction regime are the reaction temperature, the heating rate and the monomer concentration [17, 18].

Magnetite nanocubes could be synthesized in a wide range of sizes, in organic and aqueous media. Using the thermal decomposition in organic media, iron(III) acetylacetonate seems to be the best iron precursor (Fig. 3a). Well-defined ferrimagnetic magnetite nanocubes up to 170 nm can be synthesized using iron (III) acetylacetonate as precursor and oleic acid as surfactant in dibenzyl ether [19]. In this synthesis, the particle size can be modified by changing the iron precursor concentration and reflux time. The addition of a second ligand changes the reaction kinetics and thus decreases the particle size down to 20 nm. Ligands of different nature like decanoic acid, [20] tryoctylphosphine oxide [21], chloride ions [22] or sodium oleate [23] modify the decomposition kinetic and control particles growth by adhering to selective facets of the initial nuclei. Interestingly, a novel strategy using a solvent mixture composed by dibenzyl ether, 1-octadecene and 1-tetradecene with oleic acid and sodium oleate has been reported for the synthesis of more stable and reproducible particles than just using dibenzyl ether [24]. Using this synthesis approach, by increasing the temperature at which the reaction mixture is degassed, Fe_3O_4 nanocubes of 10 nm could be grown. Using an aqueous route, the hydrolysis of iron (II) sulphate in alkaline media mixed with ethanol in the presence of a mild oxidant renders magnetite nanocubes from 30 to 170 nm [25]. This synthesis has been recently scaled up to grams per hour in a continuous flow approach [26].

Magnetite octahedrons exposing the {111} facets are grown through the decomposition of iron (III) oleate in noncoordinating high-boiling point solvents in the presence of an amine or ammonium quaternary salts (Fig. 3c). For example, the decomposition of iron (III) oleate in tetracosane in the presence of oleylamine leads to 21 nm Fe_3O_4 octahedrons [27]. Using trioctylammonium bromide in conjunction with oleic acid as surfactants and squalene as solvent leads to 50 nm Fe_3O_4 octahedrons. [28, 29] Interestingly, the decomposition of iron (0) pentacarbonyl in the presence of a mixture of surfactants and oleylamine/oleic acid 10:1 leads to 50 nm octahedral particles [30].

Elongated or rod-like Fe_3O_4 nanoparticles have been traditionally synthesized using a shape template as intermediate, but nowadays they can also be prepared directly in organic media (Fig. 3b). Due to its crystal structure, hexagonal hematite

(α-Fe_2O_3) [34], orthorhombic goethite (α-Fe_2O_3) [35] or monoclinic akaganeite (β-Fe_2O_3) [36] have been ideal candidates for magnetite templating. In the case of direct synthesis of hematite, the synthesis is very slow (it takes several days) and large nanoparticles (hundreds of nm) are obtained [37]. The concentration of the iron salt, the ageing time and the presence of phosphate ions [37] and urea [38] affect their final length. Interestingly, using akaganeite for templating, the reaction time is reduced and also the particle dimensions [39]. This synthesis could be controlled by modifying the pH but also with the presence of polymers with amine groups, such as polyethylenimine [40, 41]. During the last years, a solvothermal method to grow magnetite nanorods in one step has been developed [42, 43]. This method uses iron (0) pentacarbonyl as precursor in the presence of hexadecylamine and oleic acid, using 1-octanol as solvent. The resulting magnetite rods have lengths from 25 to hundreds of nm and aspect ratios from 5 to 10.

Magnetite discs/plates need the use of an intermediate step for their final synthesis (Fig. 3f). By performing a synthesis where the growth through the long axis is inhibited, it is possible to synthesize hematite disk/plate like nanoparticles that serve as templates. Thus, the use of sodium acetate as inhibitor in the hydrolysis of iron (III) chloride in a mixture of water/ethanol is a robust synthesis route [33, 44, 45]. Then, the reduction could be performed by annealing under inert/reducing atmosphere or in organic media with oleic acid. By this route, it is possible to synthesize discs from 400 x 8 nm to structures of 40 x 40 nm.

Hollow magnetite nanoparticles are obtained through different strategies (Fig. 3d). The sodium resulting from the decomposition of sodium oleate at 380 °C is able to carve iron oxide nanocubes synthesized using iron (III) oleate as precursor [46]. Starting from Fe nanospheres, by passing or generating an oxygen flow through the reaction at high temperatures, the hollowing effect is achieved due to the inward diffusion of oxygen and outward diffusion of the iron cations [31, 47].

Flower-like magnetite nanoparticles, i.e. clusters of magnetite nanoparticles of a controlled size, are mainly synthesized by the polyol method using ethylene glycol as solvent (Fig. 3e). For the cluster formation it is critical the addition of N-methyl diethanolamine, which is used to control the aggregation of the primary cores (5–20 nm) in larger particles of around 30–50 nm [48, 49].

3 Applications

The control of the nanoparticle morphology could bring different benefits such as the enhancement of the magnetic properties due to the shape anisotropy and specific surface area increase in comparison to the spherical morphology. In addition, each morphology exposes different facets of the nanoparticle which determine the final activity of the nanoparticles [50]. All these tuneable properties postulate the anisotropic magnetic iron oxide nanoparticles for widespread applications.

3.1 *Biomedicine*

Magnetic iron oxide nanoparticles have been widely studied in the biomedical field due to their appealing magnetic properties which can be finely controlled by their size, shape and aggregation state [5, 9]. These particles are biocompatible and biodegradable, since iron is a nutrient and can be metabolized by the organisms [51]. Indeed, iron oxide nanoparticles are the only magnetic material approved by the Food and Drug Administration (FDA) for biomedical applications [52]. Although iron oxide nanoparticles can be biofunctionalized with proteins for specific cell targeting, remarkably, due to their magnetic character, this type of particles can be driven to and accumulated at the desired organ/tissue by magnetic actuation (active targeting) [53]. Magnetic iron oxide nanoparticles can be active agents in both diagnosis and therapy.

Diagnosis

In diagnosis, magnetic iron oxide nanoparticles can be used as T_2 contrast agents in magnetic resonance imaging. This imaging technique is characterized by a high spatial resolution and high soft-tissue contrast. Iron oxide magnetic nanoparticles have a high magnetic moment that produces inhomogeneities in the local magnetic field of the water protons leading to a reduction of the MRI signal intensity on T_2 (transverse relaxation time) weighted images [52]. There were different formulations based on superparamagnetic iron oxide nanoparticles approved by the Food and Drug Administration, but most of them (Feridex, Endorem or Resovist) were removed from the market due to the lack of clinical use [54]. Only Ferumoxytol remains in market due to its low hydrodynamic size (< 40 nm) and its long circulating half-life (15 hours) [55]. To study the capacity of magnetic nanoparticles as contrast agents, one of the variables to study is the longitudinal or transversal relaxivity (r_1 or r_2, respectively), which is a direct measurement of how the magnetic nanoparticles decrease the relaxation time of the water protons due to the perturbation of their magnetic local field.

In general, faceted magnetic iron oxide nanoparticles present better magnetic properties [36, 56] and are more suitable for MRI than their spherical counterparts. Moreover, it seems that a higher specific surface area favours larger r_2 values (Fig. 4). Therefore, **elongated nanoparticles** are particularly a good morphology to be used as MRI contrast agents. Magnetite rods of 25 and 50 nm length (both with 5 nm in diameter) have r_2 values of 670 and 905 $mM^{-1}\cdot s^{-1}$, respectively [57]. Thicker magnetite nanorods (70x12 nm), that possess a transversal relaxivity of 608 $mM^{-1}\cdot s^{-1}$ and nanorods (30x4 nm) with a r_2 of 312 $mM^{-1}\cdot s^{-1}$ [58], also exhibit better performances than the commercial MRI contrast agents (Ferumoxytol; $r_2 = 68\ mM^{-1}\cdot s^{-1}$).

The exchange interactions given by the clustering of the maghemite particles when they are displayed in **flower-shaped morphology** also brings benefits to their relaxometric properties. 30 nm clusters made of primary units of 10 nm exhibit a r_2

Fig. 4 Magnetic iron oxide morphologies that have shown a better performance for MRI and their fundamental properties

of 365 $mM^{-1}·s^{-1}$, near 2 times higher than 10 nm single-core particles made by the same method (polyol). This improvement is due to the cooperative behaviour of the magnetic moment, which increases the intensity of the local magnetic field.

Single 23 nm **magnetite nanocubes** reach a transversal relaxivity value of 398 $mM^{-1}·s^{-1}$ [59]. However, it should be noted that the configuration of the nanocubes in solution affects their relaxometric properties as when they are clustered in 200 nm beads, their r_2 decays down to 161 $mM^{-1}·s^{-1}$. **Magnetite octopods** of around 30 nm present a considerable high transversal relaxivity value of 679 $mM^{-1}·s^{-1}$ (measured at 7 T) much larger than their spherical equivalent (126 $mM^{-1}·s^{-1}$) [60].

During the last decade, the research on the use the iron oxides as **T_1 contrast agents** (positive contrast) has been intensified [61, 62]. Magnetic iron oxide nanoparticles with a relatively small size and/or a large surface area increase the exposure of Fe ions to water interface and are ideal candidates for T_1 contrast agents (Fig. 4). Due to its small size, particles tend to be "more paramagnetic" than superparamagnetic exposing their five unpaired Fe^{3+}/Fe^{2+} electrons [62]. Moreover, their longer circulation time and better biocompatibility than gadolinium complexes (the main commercial T_1 contrast agent) postulates iron oxides as a promising material for this application [63]. Also, ultrathin iron oxide nanowhiskers (20x2 nm) have shown interesting properties as T_1 contrast agent with a high r_1 (6.3 $mM^{-1}·s^{-1}$) and low r_2 (11.15 $mM^{-1}·s^{-1}$) [64], slightly better than 3 nm spherical ultrasmall iron oxide nanoparticles (r_1 = 5.2 $mM^{-1}·s^{-1}$; r_2 = 10.4 $mM^{-1}·s^{-1}$) [63].

3.2 *Therapy*

One of the main problems of cancer and other diseases is the difficulty of achieving enough drug concentration at the action site to produce the desired therapeutic effect. Magnetic iron oxide nanoparticles, through two different approaches, magnetothermia and photothermia, are valuable tools for the design of new and/or more efficient therapies (Fig. 5).

In **magnetic hyperthermia**, magnetic iron oxide nanoparticles can release heat under the application of an alternating magnetic field (50–700 kHz) [65–67]. Since the generated heat affects the tumour locally, it would cause minimal damage to healthy tissues. Therefore, magnetic hyperthermia is an alternative to reduce the side effects of chemotherapy. There are several mechanisms (susceptibility loss, hysteresis losses and viscous heating) by which the magnetic nanoparticles can dissipate heat [68–70]. The particles structural and colloidal properties (i.e. particle size, shape, aggregation state, interactions), the media (viscosity) and the AC field (frequency and amplitude) will determine which of the aforementioned mechanisms is the main one to produce heat. The capability of magnetic nanoparticles to release heat is evaluated by the specific absorption rate (SAR), which can be defined as the power dissipated per unit mass [71].

In the latest years, an intense research has been performed regarding the properties of **magnetite nanocubes for magnetic hyperthermia**. The comparison of nanocubes (19.5 nm) and spheres with the same volume (24 nm) leads to SAR values favourable to the nanocubes. [72] However, comparing heating properties of 53 nm spheres and 43 nm cubes, it was found better heating properties for the spherical ones because the cubes were forming strong and isometric aggregates [73]. Cubes of different sizes (13–38 nm) have been studied finding the best SAR for 19 nm particles 2277 $W \cdot g^{-1}$ (700 kHz, 300 Oe) [74]. Interestingly, the configuration of possible aggregates deeply affects the SAR values. 3D aggregates of 38 nm nanocubes have a SAR of less than 300 $W \cdot g^{-1}$; however, when they were isolated, the SAR increased up to 1400 $W \cdot g^{-1}$ [75]. Nevertheless, it has been demonstrated that the formation of chain-like aggregates aligned in the magnetic field boosts the SAR value by five times compared to 3D aggregates [73]. The measurement of SAR evaluates the possible performance of the nanoparticles; however, when hyperthermia is performed in vivo, it suffers from low heating efficiency due to the high viscosity of the tumour site and particle aggregation upon intracellular internalization

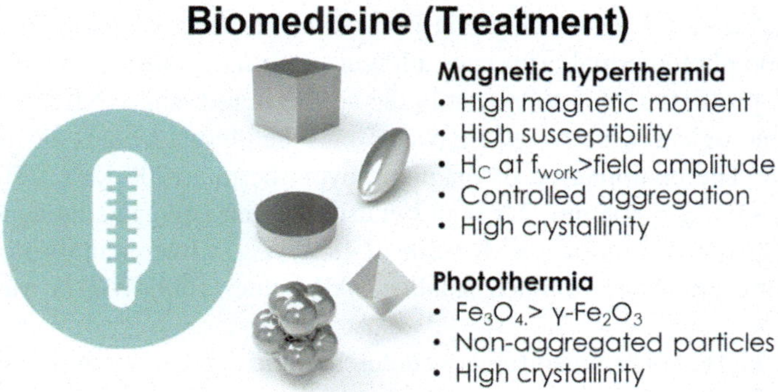

Fig. 5 Magnetic iron oxide morphologies that have shown a better performance for hyperthermia and their fundamental properties

[76]. Thus, the confinement of magnetic nanoparticles in cavities of hundreds of nanometres would enable a certain degree of nanoparticle rotation and minimize aggregation effects, so the heating power would be better preserved intracellularly [77].

Octahedral magnetite nanoparticles have also an optimum morphology to work as nanoheaters in magnetic hyperthermia. Small octahedrons of 6 and 12 nm exhibit SAR values of 163 and 275 $W \cdot g^{-1}$ (247 kHz, 310 Oe) [78]., 40 nm octahedrons reach up to 2483 $W \cdot g^{-1}$ (358 kHz, 800 Oe) and finally 98 nm octahedrons, around the monodomain limit of magnetite, have been reported to reach 2629 $W \cdot g^{-1}$ [79].

Elongated magnetite nanoparticles display moderate heating release properties when compared to other morphologies. The drawback is the high field amplitude needed to obtain a proper heating release due to the high switching field needed to reverse the magnetic moment. Magnetite nanorods grown from the reduction of akaganeite with dimensions of 45x10 nm exhibit a SAR of 1072 $W \cdot g^{-1}$ (390 kHz, 415 Oe), whereas longer rods (400x40 nm) displayed lower SAR values. High crystalline rods made by solvothermal approach exhibit SAR values of 1300 $W \cdot g^{-1}$ (310 kHz, 800 Oe). In contrast, smaller rods of 41x7nm only reached 540 $W \cdot g^{-1}$ [43].

Magnetite nanoplates of certain dimensions have the advantage of the formation of magnetic vortex configurations where spins are arranged circularly and oriented parallel to the field [80]. As a consequence, these particles display negligible coercivity and remanent magnetization which results in extraordinary SAR values, up to 4400 $W \cdot g^{-1}$ for nanoplates of 225 nm in diameter and 26 nm in thickness [81]. However, for nanodiscs without vortex configuration (150–200x10–15 nm), only values of 245 $W \cdot g^{-1}$ (180 kHz, 12 Oe) have been described [44].

Superparamagnetic γ-Fe_2O_3 nanoflowers of 50 nm containing spherical 11 nm particles grown by the polyol method can reach a SAR value of 1790 $W \cdot g^{-1}$, much larger than SAR of single 11 nm (48 $W \cdot g^{-1}$) [49]. Controlled aggregation seems to boost the increase of SAR; however, a compromise between colloidal stability and heating performance should be achieved. In addition to that, nanoparticle cell affinity needs to be maximized through conjugation with biological targeting molecules to increase nanoparticle concentration inside the cells [82].

During the last 5 years, magnetic iron oxides, especially magnetite, have also shown good photothermal properties allowing their use in **photothermal therapy**. Magnetite presents some absorption in the near-infrared range (NIR), especially in the first biological window where water or other biomolecules do not present absorption. The combination of magnetic hyperthermia (900 kHz, 250 Oe) with phototherapy (λ_{las} = 808 nm, 2.5 $W \cdot cm^{-2}$) using 20 nm aggregated magnetite nanocubes resulted in a combined SAR value of 4850 $W \cdot g^{-1}$. Interestingly, by doing just magnetic hyperthermia, a much smaller SAR value (1000 $W \cdot g^{-1}$) was achieved [83–85].

Interestingly, some anisotropic structures could lead to **magnetomechanical actuation** by applying an AC magnetic field of low frequency, just to promote the rotation of the particles. It has been reported that Fe_3O_4 nanorods (200x50 nm) [86] induce cancer cell death on HeLa cells through mechanical oscillations under the exposure of a 35 kHz field, associated with the Brownian motion of the nanorods [87]. Nanodiscs could also induce cell death through magnetomechanical effects. Nanodiscs made of Permalloy coated with gold (1000x60 nm) were able to exert mechanical forces to N10 glioma cells leading to apoptosis. The AC field of low frequency (60 Hz) induced a shift of the vortex structure, creating an oscillation which was then transmitted to a mechanical force applied to the cell [88]. This strategy opens a new pathway where magnetic iron oxide nanodiscs could be applied.

3.3 *Magnetic Recording Media*

In the second half of the twentieth century, magnetic recording media were based on acicular iron oxide magnetic nanoparticles (γ-Fe_2O_3) with a length size between 100 and 500 nm and aspect ratios from 6 to 10. They were used as magnetic recording material because of their chemical stability, high magnetic moment aligned with the largest dimension of the particle and high Curie temperature (590–675 °C) to keep moderate coercivity (not too high to allow successful writing and not too small to resist changes or degradation of signals) (Fig. 6) [89, 90]. Elongated morphology was preferred as shape anisotropy is the main contributor to the effective anisotropy, being up to two orders of magnitude higher than the magnetocrystalline one for particles with axial ratio larger than 6. On the other side, the microstructure of the particles, including defects, pores and inhomogeneities, were shown to decrease the effective anisotropy [91, 92]. These problems were inherent to the synthetic routes employed for the growth the magnetic iron oxides using lepidocrocite (γ-FeOOH) or goethite (α-FeOOH) as shape templates, as the subsequent dehydration was the main cause of pore and defects formation [93, 94]. An alternative route was based on the hydrothermal treatment of $Fe(OH)_3$ which leads to hematite (α-Fe_2O_3) bypassing the formation of iron oxyhydroxides and the formation of pores [34, 95]. In the last decade of the twentieth century, iron oxides, as well as chromium dioxide particles, were replaced by metal particulate materials, which possess twice the magnetization of iron oxides and higher coercivity [96–98].

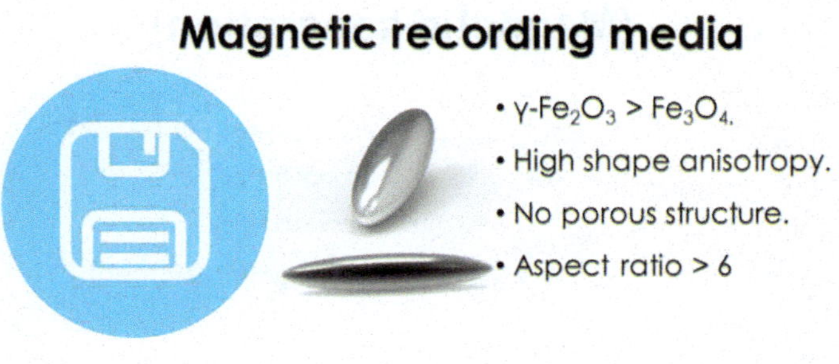

Fig. 6 Magnetic iron oxide morphology tested for magnetic recording media and their fundamental properties

3.4 Water Treatment

One of the major challenges in the environmental field is water treatment for the removal of heavy metal ions like As, Pb, Hg, Cr, Cd and Ni, which cause harmful effects on humans and animals [8]. Moreover, organic pollutants such as detergents and pesticides, all of them from human source, can also be found in water and represent another type of environmental threats. They mainly consist in chlorinated and nonchlorinated aliphatic and aromatic molecules [99]. Traditional methods for water treatment include centrifugation, coagulation, filtration and reverse osmosis. The use of **iron oxide nanoparticles for water remediation** as absorbents of metal ions presents several advantages such as their relatively low cost, the efficient recovery of the material by a simple magnetic separation procedure and also its high specific surface area (Fig. 7) [100]. It has been demonstrated that the absorption capacity of As(III) and As(V) increases from 0.6 to 59.5 mg As·g^{-1} as the magnetite particle size decreases from 300 to 11 nm. These differences were explained by the distinct exposed surface area and the different reactivity of the surfaces (decrease of tetrahedral site occupancy) [101–103]. The main mechanism related to the absorption of pollutants at the particles surface is the complexation and/or electrostatic interaction, followed by ion exchange between the iron surface and the toxic ions [104, 105]. The performance of the iron oxide nanoparticles depends on the crystalline structure (Fe_3O_4 or γ-Fe_2O_3), the particle concentration, the surface coating and the presence of interfering ions (especially nitrate and phosphate ions) [106, 107].

Due to its porous 3D morphology and high specific surface area, **micron-sized nanoflowers** have shown great potential in the absorption of As(V) and Cr(VI). Depending on the phase, values for the absorption of As(V) of 4.65 and 4.75 mg·g^{-1} have been found for Fe_3O_4 and γ-Fe_2O_3, respectively, while smaller differences have been found for the absorption of Cr(VI) (4.38 and 3.86 mg·g^{-1} for Fe_3O_4 and γ-Fe_2O_3) [108]. These nanoflowers have been also tested for the absorption of Orange II reaching absorption values around 43.5 mg·g^{-1}. **Hollow iron oxide nanocubes** with a length of 7 nm and a shell thickness of around 2 nm have shown a powerful capacity for the absorption of As(V) (326 mg·g^{-1}) and As(III) (190 mg·g^{-1})

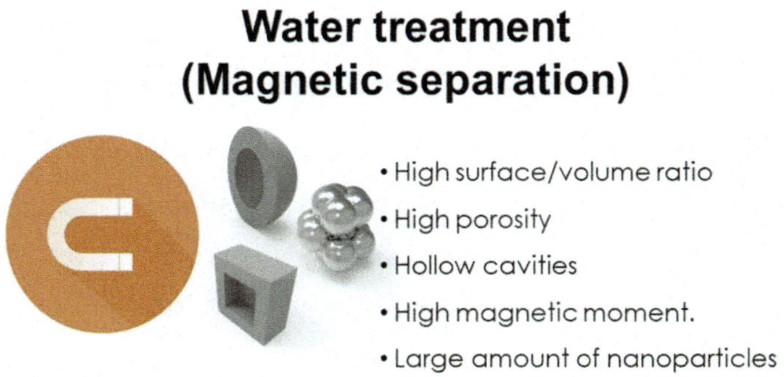

Fig. 7 Magnetic iron oxide morphologies that have shown a better performance for water treatment and their fundamental properties

[109]. **Hollow spheres** between 200 and 300 nm in diameter and 20 nm shell thickness were tested for the absorption and removal of Neutral Red dye reaching values up to 90 $mg\cdot g^{-1}$ [110].

Despite these encouraging results, magnetic nanoparticles have important limitations in their use for water remediation, mainly related to the large quantities required for this application and the difficulties to fulfil the regulations for obtaining drinking water [8, 111].

3.5 Spintronics

Magnetite is a high-valuable material for application in spintronics because it is classified as a half-metallic ferromagnet (bandgap of 0.1–0.15 eV) with high values of spin polarization but also favourable Curie temperature (T_C = 850 K) [112]. Density functional theory (DFT) has predicted a value of −100% spin polarization at the Fermi level for the bulk material [113].

Recently, if has been observed that **8 nm magnetite octahedrons** capped with oleylamine exhibit interesting results in tunnelling magnetoresistance (TMR). [114] The oleylamine surface layer between the particles acts as insulating barrier providing multiple tunnel injection junctions where intergranular tunnelling is possible. TMR at room temperature reaches 38% increasing up to 69% at 180 K, much higher than those measured for spherical 8 nm particles coated with oleylamine (24% at room temperature and 41% at 180 K). The better performance of octahedral particles is a consequence of their better stoichiometry at the surface, presenting less defects that increase the spin polarization of the material and so the TMR. Moreover, the fact that (111) facets are exposed facilitates the strong coupling of oleylamine molecules on them (Fig. 8). **Magnetite nanorods** (75 × 9 nm, aspect ratio 8.3) were also tested as potential material for spintronics, and it was measured a TMR of 14% at room temperature when they were randomly oriented [115]. However, once they were oriented, they could increase its TMR up to 19% and to 31% at 130 K (close to the Verwey transition) because the relative magnetization is maximized due to the parallel orientation of the nanorods. The study of the particle dimensions with the

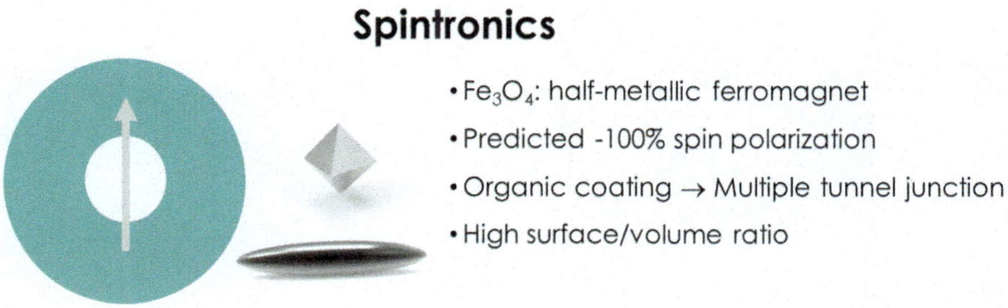

Fig. 8 Magnetic iron oxide morphologies that have shown a better performance for spintronics and their fundamental properties

TMR leads to a TMR decreases with the nanorod dimension, mainly for the diameter. Thus, the spin polarization of the 75 x 9 nm magnetite rods was 46%.

3.6 Microwave Absorption

In recent years, there has been an upsurge of electronic devices and communications (i.e. smartphones, satellite broadcasting, local area network (LAN), radars, etc.) operating in the gigahertz range (GHz) whose main drawback is the electromagnetic noise leading to poorer communications. Moreover, microwave radiation induces a considerable amount of health threats to organisms causing the breakdown of DNA strands, increasing heart rate, weakening the immune response and inducing cancer [116]. In this regard, **iron oxide magnetic nanoparticles** have shown high microwave absorption and low reflection over a broad frequency range. Tuning the particles microwave absorption properties is possible by controlling their size and shape and consequently their magnetic properties, that is high saturation magnetization values and low coercivity that favour larger permeability (Fig. 9) [117]. The development of Fe_3O_4 composites with large anisotropy may exceed the Snoeck limit, having high permittivity and reaching better MW absorption properties [118, 119].

Fe_3O_4 nanodiscs with a thickness of 30 nm have shown an increase of the permittivity with the diameter (from 80 to 500 nm) and resonance frequency [120]. However, the permeability shows an opposite trend (2.0 for the 80 nm and 1.8 for 500 nm discs at 0.1 GHz) due to the increase in shape anisotropy as the diameter grows. The smallest nanodiscs (30x 80 nm) exhibit a high microwave absorption as a consequence of its lower permittivity (RL < -10 dB) with a wide frequency bandwidth of about 2–18 GHz, with a sheet thickness of around 2.1 mm. In the same way, nanorods of Fe_3O_4/Fe/SiO_2 of 1 μm x 80 nm present properties as microwave absorbers since the reflection loss below −10 dB is up to 6.96 GHz with a sheet thickness of 2 mm [121]. Porous magnetite **flowers** with a diameter of 2.5 μm

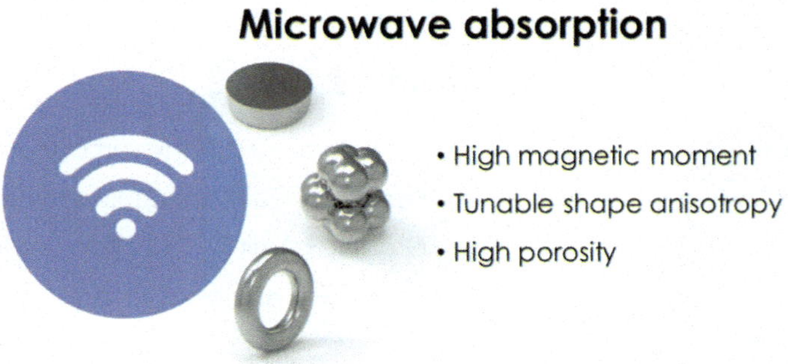

Fig. 9 Magnetic iron oxide morphologies that have shown a better performance for microwave absorption and their fundamental properties

composed of cores of 50 nm present a bandwidth with a RL < -10 dB of 3.8 GHz with a minimum of −28.31 dB at 13.26 GHz. Magnetite **nanorings** coated by carbon of 160 x 70 nm exhibit a strong reflection loss value of 61.54 dB at 16.9 GHz with a thickness of 1.5 mm and a low filling ratio of 25%. Authors attribute this attenuation ability to the eddy current loss enhanced by combination of confinement vortex and train-driven vortex [122].

It can be concluded that the performance of Fe_3O_4 for microwave absorption may require a high anisotropic material that also contains certain porosity in its structure increasing the interfacial area throughout the composite [123].

3.7 *Li-Ion Batteries*

The use of Fe_3O_4 as material for the design of Li-ion batteries (LIBs) is due to its high theoretical capacity (900–1000 $mA{\cdot}h{\cdot}g^{-1}$), low cost, environmental-friendly and high rate performance [124, 125]. The current generation of LIBs are based on electrode materials in which Li^+ ions are stored by insertion between structural layers during charging and extracted from the layers during discharging without significant structural change, which gives excellent cycling performance. The electrochemical reaction between Li and magnetite occurs as follows:

$$Fe_3O_4 + 8Li^+ + 8e^- \rightarrow 3Fe^0 + 4Li_2O$$

Fe_3O_4 presents several drawbacks for their application as LIBs such as stress due to volume changes after lithium ions charge and discharge and the high surface area of nanomaterials which causes aggregation of the iron oxide nanoparticles and electrolyte decomposition forming a thick solid electrolyte interphase on the electrode surface [126, 127]. Fortunately, the use of high porous and/or hollow nanostructures with high surface area has shown good properties as LIBs materials because they shorten the diffusion pathway of Li ions and electrons and increase the electrochemical reaction area leading to an improvement penetration of the electrolyte and accommodating the strain caused by the lithium inclusion/removal (Fig. 10).

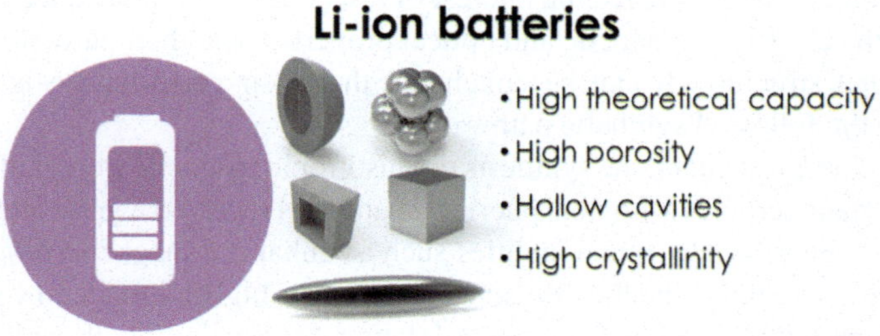

Fig. 10 Magnetic iron oxide morphologies that have shown a better performance for Li-ion batteries and their fundamental properties

For example, high porous **Fe_3O_4 nanorods** with lengths within 400 and 700 nm and diameters between 20 and 80 nm coming from the reduction of goethite nanorods exhibit a high reversible capacity of 843.5 mA·h·g-1 after the 50th cycle at 0.1 C (10 h discharge time) due to the high dense porous structure and the elongated morphology [128]. A composite of **Fe_3O_4 nanoflowers** in carbon with a diameter of 2 μm and sheets of 50 nm in thickness exhibits high capacities up to 1030 mA·h·g^{-1} at a density rate of 0.2 C (5 h discharge time) up to 150 cycles [129]. Noteworthy, it has been observed a different electrochemical activity in the facets exposed to the media by the Fe_3O_4 nanostructures. Thus, (220) facets which can be found in nanoprisms are more electroactive than (111) which are characteristic of octahedrons. [130] Moreover, comparing nanorods, nanoplates and spherical particles, the nanorods present a higher electrochemical activity and larger working capacities [42].

Hollow γ-Fe_2O_3 nanoparticles with a void around 15 nm and a shell thickness near 4 nm can host lithium ions serving as cathode material reaching a capacity of 219 mA·h·g^{-1} with a Coulombic efficiency of 99.7% [131]. In the case of smaller hollow particles (5.7 nm in void with less than 2 nm in shell thickness) capacities up to 132 mA·h·g^{-1} were displayed, being the high concentration of vacancies and chemical stability of γ-Fe_2O_3 during the voltage window the reasons for these high capacity values. On the other hand, **16 nm Fe_3O_4 nanocubes** synthesized by a L-serine-assisted solvothermal approach presents a high specific capacitance of 695 mA·h·g^{-1} at 0.2 C and an excellent Coulombic efficiency of above 95% after the 11th cycle [132]. **Rhombic dodecahedral Fe_3O_4 nanocrystals** (NCs) exposing {110} facets exhibit a high initial discharge capacity of 1147 mA·h·g^{-1} at 0.2 C and a good cycle performance (362 mA·h·g^{-1} at 0.2 C after 100 cycles and 191 mA·h·g^{-1} at 1 C up to 130 cycles) [133].

4 Conclusions and Perspectives

Since ancient times, magnetic iron oxide nanoparticles have been present in people's daily life being remarkable the feasibility of this material to adapt thank to their appealing properties. The possibility of synthesizing these particles using versatile synthetic routes boosted the interest of researchers to establish a clear relation between the electrical, magnetic and optical properties and their particle size, shape and internal structure. It can be concluded that their performance is intimately related to the followed synthetic pathway.

During the latest years, the synthesis of anisotropic magnetic iron oxide nanoparticles, i.e. nanoparticles with nonspherical shape, has received a great attention. The anisometry leads to new functionalities such as enhanced magnetic properties due to the increase in the magnetic effective anisotropy, a higher surface to volume ratio and different particle facets exposed modulating the reactivity with molecules, particles or other entities. These properties are very advantageous in widespread applications such as biomedicine as theranostic agents (MRI and magnetic/optical hyperthermia) and others such as magnetic recording media, environmental remediation, Li-ion batteries, spintronics or microwave absorption.

This chapter summarizes the most reproducible and successful synthetic routes that lead to iron oxide magnetic nanoparticles with nonspherical morphology and also their applications related to the actual societal challenges. Most of the morphologies considered can be synthesized by thermal decomposition and solvothermal synthesis. Although thermal decomposition produces monodisperse and highly crystalline nanoparticles, this strategy also uses organic solvents which are not environmentally friendly, and for some applications, a further step to transfer them to water is needed. On the other side, solvothermal synthesis produces nanoparticles stable in aqueous media, with less sophisticated protocols and more environmentally friendly solvents. However, this synthesis route is more difficult of scaling up and requires longer time and in many cases a final step for the reduction of the resulting nanoparticles, from hematite to magnetite. Summing up, among the different synthesis routes described here, factors like time, cost, energy consumption, environmental issues and scalability should be considered when translating these processes to the industry.

The applications reviewed in this chapter are very broad, from environmental remediation and biomedicine to spintronics and energy (Li-ion batteries). Most of the research on magnetic nanoparticles for these applications (except magnetic recording media) has been carried out with spherical particles, whose properties depend on particle size, aggregation, crystallinity and coating. However, during the latest decades of the twentieth century, iron oxide and iron metal particles with elongated morphology were the ideal structure for magnetic recording media due to the enhancement of the anisotropy constant. In the latest years, the literature regarding anisometric iron oxides nanoparticles has boosted, demonstrating the interest in them with other areas respect to their spherical equivalents. For example, the use of magnetic nanocubes and nanodiscs in magnetic hyperthermia leads to an increment in the SAR values, also showing a great performance in in vitro studies. Interestingly, the performance of these particles as heating agents can be increased if magnetic hyperthermia is combined with photothermal therapy. On the other hand, nanostructures with high specific surface area that exhibits porosity are very advantageous for Li-ion batteries because they shorten the pathway of Li ions and electrons leading to an improvement penetration of the electrolyte.

In general, the excellent properties of magnetite and maghemite confer these materials of a unique adaptability setting this material far away of being "old fashion" for practical applications. In the case of magnetite, these properties are based on its structure of inverse spinel with an interesting electron hoping between Fe^{2+} and Fe^{3+} ions at octahedral sites. This material has been subject of interest in diverse fields such as magnetism, optics and electronics. Interestingly, most of its properties abruptly change when crossing the Verwey transition. Enabling more degrees of freedom by shaping the morphology of the nanoparticles different from the spherical one has brought advantageous benefits in different applications. For most of the applications, only a few morphologies have been tested yet. There are still morphologies with enormous potential such as plates, rings and nanotubes due to their magnetic structure (vortex) or specific surface area that have not been tested properly in a variety of applications. Moreover, for some morphologies, there is room to improve the synthetic routes. These routes should be more robust, reproducible and scalable in order to make monodispersed particles with tuneable properties. Moreover, for the demanding application, there should exist more protocols about how to process the material to exhibit the best performance without losing any of its inherent properties.

Acknowledgements This work was supported by the European Commission Framework Programme 7 (NanoMag project, No. 604448). We acknowledge the Spanish Ministry of Economy and Competitiveness through MAT2016-77391, MAT2017-88148-R, RTI2018-095495-J-I00 and PGC2018-096016-B-I00 grants, and Consejo Superior de Investigaciones Científicas through PIE201960E062 project. L.G. acknowledges financial support from the Ramón y Cajal subprogramme (RYC-2014-15512). A.G.R. thanks the Generalitat de Catalunya through 2017-SGR-292 project and the financial support of Ramon Areces Foundation through project CIVP19A5922. The ICN2 is funded by the CERCA programme/Generalitat de Catalunya. The ICN2 is supported by the Severo Ochoa Centres of Excellence programme, funded by the Spanish Research Agency (AEI, grant no. SEV-2017-0706).

References

1. A.G. Roca, L. Gutiérrez, H. Gavilán, et al., Design strategies for shape-controlled magnetic iron oxide nanoparticles. Adv. Drug Deliv. Rev. **138**, 68–104 (2019). https://doi.org/10.1016/j.addr.2018.12.008
2. D.J. Dunlop, Ö. Özdemir, *Rock Magnetism: Fundamentals and Frontiers* (Cambridge University Press, Cambridge, 1997)
3. D. Faivre, D. Schüler, Magnetotactic Bacteria and Magnetosomes. Chem. Rev. **108**, 4875–4898 (2008). https://doi.org/10.1021/cr078258w
4. G. Priyadarshana, N. Kottegoda, A. Senaratne, et al., Synthesis of magnetite nanoparticles by top-down approach from a high purity ore. J. Nanomater. **2015**, 1–8 (2015). https://doi.org/10.1155/2015/317312
5. A.G. Roca, R. Costo, A.F. Rebolledo, et al., Progress in the preparation of magnetic nanoparticles for applications in biomedicine. J Phys D-Applied Phys **42**, 224002 (2009). https://doi.org/10.1088/0022-3727/42/22/224002
6. J. Van Klinken, Magnetization of ancient ceramics. Archaeometry **43**, 49–57 (2001). https://doi.org/10.1111/1475-4754.00004
7. K.J. Sreeram, R. Indumathy, A. Rajaram, et al., Template synthesis of highly crystalline and monodisperse iron oxide pigments of nanosize. Mater. Res. Bull. **41**, 1875–1881 (2006). https://doi.org/10.1016/j.materresbull.2006.03.017
8. K. Simeonidis, S. Mourdikoudis, E. Kaprara, et al., Inorganic engineered nanoparticles in drinking water treatment: A critical review. Environ Sci Water Res Technol **2**, 43–70 (2016). https://doi.org/10.1039/C5EW00152H
9. M. Colombo, S. Carregal-Romero, M.F. Casula, et al., Biological applications of magnetic nanoparticles. Chem. Soc. Rev. **41**, 4306 (2012). https://doi.org/10.1039/c2cs15337h
10. A.G. Roca, J.F. Marco, M.P. Morales, C.J. Serna, Effect of nature and particle size on properties of uniform magnetite and Maghemite nanoparticles. J. Phys. Chem. C **111**, 18577–18584 (2007). https://doi.org/10.1021/jp075133m
11. B.D. Cullity, C.D. Graham, *Introduction to Magnetic Materials* (Wiley, Hoboken, 2008)

12. M.E. Fleet, The structure of magnetite. Acta Crystallogr Sect B Struct Crystallogr Cryst Chem **37**, 917–920 (1981). https://doi.org/10.1107/S0567740881004597
13. M.P. Morales, C. de Julián, J.M. González, C.J. Serna, The effect of the distribution of vacancies on the magnetic properties of γ – Fe2O3 particles. J. Mater. Res. **9**, 135–141 (1994). https://doi.org/10.1557/JMR.1994.0135
14. K.V.P.M. Shafi, A. Ulman, A. Dyal, et al., Magnetic enhancement of γ-Fe_2O_3 nanoparticles by Sonochemical coating. Chem. Mater. **14**, 1778 (2002). https://doi.org/10.1021/cm011535+
15. W. Wu, Z. Wu, T. Yu, et al., Recent progress on magnetic iron oxide nanoparticles: Synthesis, surface functional strategies and biomedical applications. Sci. Technol. Adv. Mater. **16**, 023501 (2015). https://doi.org/10.1088/1468-6996/16/2/023501
16. S.G. Kwon, T. Hyeon, Colloidal chemical synthesis and formation kinetics of uniformly sized Nanocrystals of metals, oxides, and Chalcogenides. Acc. Chem. Res. **41**, 1696–1709 (2008). https://doi.org/10.1021/ar8000537
17. L. Manna, E.C. Scher, A.P. Alivisatos, Synthesis of soluble and processable rod-, arrow-, teardrop-, and tetrapod-shaped CdSe nanocrystals. J. Am. Chem. Soc. **122**, 12700–12706 (2000). https://doi.org/10.1021/ja003055+
18. Z.A. Peng, X. Peng, Mechanisms of the shape evolution of CdSe Nanocrystals. J. Am. Chem. Soc. **123**, 1389–1395 (2001). https://doi.org/10.1021/ja0027766
19. D. Kim, N. Lee, M. Park, et al., Synthesis of uniform Ferrimagnetic magnetite Nanocubes. J. Am. Chem. Soc. **131**, 454–455 (2009). https://doi.org/10.1021/ja8086906
20. P. Guardia, J. Pérez-Juste, A. Labarta, et al., Heating rate influence on the synthesis of iron oxide nanoparticles : The case of decanoic acid. Chem. Commun. **46**, 6108–6110 (2010). https://doi.org/10.1039/c0cc01179g
21. G. Gao, X. Liu, R. Shi, et al., Shape-controlled synthesis and magnetic properties of Monodisperse Fe 3 O 4 Nanocubes. Cryst. Growth Des. **10**, 2888–2894 (2010). https://doi.org/10.1021/cg900920q
22. Z. Xu, C. Shen, Y. Tian, et al., Organic phase synthesis of monodisperse iron oxide nanocrystals using iron chloride as precursor. Nanoscale **2**, 1027 (2010). https://doi.org/10.1039/b9nr00400a
23. M.V. Kovalenko, M.I. Bodnarchuk, R.T. Lechner, et al., Fatty acid salts as stabilizers in size- and shape-controlled Nanocrystal synthesis: The case of inverse spinel Iron oxide. J. Am. Chem. Soc. **129**, 6352–6353 (2007). https://doi.org/10.1021/ja0692478
24. J. Muro-Cruces, A.G. Roca, A. López-Ortega, et al., Precise size control of the growth of Fe 3 O 4 Nanocubes over a wide size range using a rationally designed one-pot synthesis. ACS Nano **13**, 7716–7728 (2019). https://doi.org/10.1021/acsnano.9b01281
25. M. Andrés Vergés, R. Costo, A.G. Roca, et al., Uniform and water stable magnetite nanoparticles with diameters around the monodomain–multidomain limit. J. Phys. D. Appl. Phys. **41**, 134003 (2008). https://doi.org/10.1088/0022-3727/41/13/134003
26. T. Asimakidou, A. Makridis, S. Veintemillas-Verdaguer, et al., Continuous production of magnetic iron oxide nanocrystals by oxidative precipitation. Chem. Eng. J. **393**, 124593 (2020). https://doi.org/10.1016/j.cej.2020.124593
27. L. Zhang, J. Wu, H. Liao, et al., Octahedral Fe_3O_4 nanoparticles and their assembled structures. Chem Commun 4378. (2009). https://doi.org/10.1039/b906636e
28. Shavel A, Rodri B, Pacifico J, et al (2009) Shape Control in Iron Oxide Nanocrystal Synthesis , Induced by Trioctylammonium Ions Shape Control in Iron Oxide Nanocrystal Synthesis , Induced by Trioctylammonium Ions. 5843–5849. https://doi.org/10.1021/cm803201p
29. A. Shavel, L.M. Liz-Marzán, Shape control of iron oxide nanoparticles. Phys. Chem. Chem. Phys. **11**, 3762–3763 (2009). https://doi.org/10.1039/b822733k
30. J. Cheon, N.-J. Kang, S.-M. Lee, et al., Shape evolution of single-crystalline Iron oxide Nanocrystals. J. Am. Chem. Soc. **126**, 1950–1951 (2004). https://doi.org/10.1021/ja038722o
31. A. Cabot, V.F. Puntes, E. Shevchenko, et al., Vacancy coalescence during oxidation of Iron nanoparticles. J. Am. Chem. Soc. **129**, 10358–10360 (2007). https://doi.org/10.1021/ja072574a

32. G. Hemery, A.C. Keyes, E. Garaio, et al., Tuning sizes, morphologies, and magnetic properties of Monocore versus multicore Iron oxide nanoparticles through the controlled addition of water in the Polyol synthesis. Inorg. Chem. **56**, 8232–8243 (2017). https://doi.org/10.1021/acs.inorgchem.7b00956
33. Y. Yang, X. Liu, Y. Lv, et al., Orientation mediated enhancement on magnetic hyperthermia of Fe 3 O 4 Nanodisc. Adv. Funct. Mater. **25**, 812–820 (2015). https://doi.org/10.1002/adfm.201402764
34. M.P. Morales, C. Pecharroman, T. Gonzalez-Carreño, C.J. Serna, Structural characteristics of uniform γ-Fe2O3 particles with different axial (length/width) ratios. J. Solid State Chem. **108**, 158–163 (1994)
35. A.F. Rebolledo, O. Bomatí-Miguel, J.F. Marco, P. Tartaj, A facile synthetic route for the preparation of superparamagnetic iron oxide nanorods and nanorices with tunable surface functionality. Adv. Mater. **20**, 1760–1765 (2008). https://doi.org/10.1002/adma.200701782
36. M.A. Blesa, M. Mijalchik, M. Villegas, G. Rigotti, Transformation of akaganeite into magnetite in aqueous hydrazine suspensions. React Solids **2**, 85–94 (1986). https://doi.org/10.1016/0168-7336(86)80066-3
37. M.P. Morales, T. González-Carreño, C.J. Serna, The formation of alpha-Fe2O 3 monodispersed particles in solution. J. Mater. Res. **7**, 2538–2545 (1992). https://doi.org/10.1557/JMR.1992.2538
38. M. Ocaña, M.P. Morales, C.J. Serna, Homogeneous precipitation of uniform α-Fe_2O_3 particles from Iron salts solutions in the presence of urea. J. Colloid Interface Sci. **212**, 317–323 (1999). https://doi.org/10.1006/jcis.1998.6042
39. O. Bomati-Miguel, A.F. Rebolledo, P. Tartaj, Controlled formation of porous magnetic nanorods via a liquid/liquid solvothermal method. Chem Commun (Camb), 4168–4170 (2008). https://doi.org/10.1039/b805239e
40. S. Lentijo Mozo, E. Zuddas, A. Casu, A. Falqui, Synthesizing Iron oxide nanostructures: The Polyethylenenemine (PEI) role. Crystals **7**, 22 (2017). https://doi.org/10.3390/cryst7010022
41. M. Mohapatra, S. Anand, Synthesis and applications of nano-structured iron oxides / hydroxides – A review. Int. J. Eng. Sci. Technol. **2**, 127–146 (2010). https://doi.org/10.4314/ijest.v2i8.63846
42. H. Sun, B. Chen, X. Jiao, et al., Solvothermal synthesis of Tunable Electroactive magnetite Nanorods by controlling the side reaction. J. Phys. Chem. C **116**, 5476–5481 (2012). https://doi.org/10.1021/jp211986a
43. R. Das, J. Alonso, Z. Nemati Porshokouh, et al., Tunable high aspect ratio Iron oxide Nanorods for enhanced hyperthermia. J. Phys. Chem. C **120**, 10086–10093 (2016). https://doi.org/10.1021/acs.jpcc.6b02006
44. X.-L. Liu, Y. Yang, J.-P. Wu, et al., Novel magnetic vortex nanorings/nanodiscs: Synthesis and theranostic applications. Chinese Phys B **24**, 127505 (2015). https://doi.org/10.1088/1674-1056/24/12/127505
45. Y. Huang, D. Ding, M. Zhu, et al., Facile synthesis of α-Fe 2 O 3 nanodisk with superior photocatalytic performance and mechanism insight. Sci. Technol. Adv. Mater. **16**, 014801 (2015). https://doi.org/10.1088/1468-6996/16/1/014801
46. K. An, T. Hyeon, Synthesis and biomedical applications of hollow nanostructures. Nano Today **4**, 359–373 (2009). https://doi.org/10.1016/j.nantod.2009.06.013
47. A. Cabot, A.P. Alivisatos, V.F. Puntes, et al., Magnetic domains and surface effects in hollow maghemite nanoparticles. Phys. Rev. B **79**, 094419 (2009). https://doi.org/10.1103/PhysRevB.79.094419
48. L. Lartigue, P. Hugounenq, D. Alloyeau, et al., Cooperative Organization in Iron Oxide Multi-Core Nanoparticles Potentiates Their Efficiency as heating mediators and MRI contrast agents. ACS Nano **6**, 10935–10949 (2012). https://doi.org/10.1021/nn304477s
49. P. Hugounenq, M. Levy, D. Alloyeau, et al., Iron oxide Monocrystalline Nanoflowers for highly efficient magnetic hyperthermia. J. Phys. Chem. C **116**, 15702–15712 (2012). https://doi.org/10.1021/jp3025478

50. J. Liu, Z. Wu, Q. Tian, et al., Shape-controlled iron oxide nanocrystals: Synthesis, magnetic properties and energy conversion applications. CrystEngComm **18**, 6303–6326 (2016). https://doi.org/10.1039/C6CE01307D
51. R. Crichton, *Inorganic Biochemistry of Iron Metabolism* (Wiley, Chichester, 2001)
52. A.S. Thakor, J.V. Jokerst, P. Ghanouni, et al., Clinically approved nanoparticle imaging agents. J. Nucl. Med. **57**, 1833–1837 (2016). https://doi.org/10.2967/jnumed.116.181362
53. M. Angelakeris, Magnetic nanoparticles: A multifunctional vehicle for modern theranostics. Biochim Biophys Acta - Gen Subj **1861**, 1642–1651 (2017). https://doi.org/10.1016/j.bbagen.2017.02.022
54. R. Weissleder, M. Nahrendorf, M.J. Pittet, Imaging macrophages with nanoparticles. Nat. Mater. **13**, 125–138 (2014). https://doi.org/10.1038/nmat3780
55. G.B. Toth, C.G. Varallyay, A. Horvath, et al., Current and potential imaging applications of ferumoxytol for magnetic resonance imaging. Kidney Int. **92**, 47–66 (2017). https://doi.org/10.1016/j.kint.2016.12.037
56. E.D. Smolensky, H.-Y.E. Park, Y. Zhou, et al., Scaling laws at the nanosize: The effect of particle size and shape on the magnetism and relaxivity of iron oxide nanoparticle contrast agents. J. Mater. Chem. B **1**, 2818 (2013). https://doi.org/10.1039/c3tb00369h
57. A. Orza, H. Wu, Y. Xu, et al., One-step facile synthesis of highly magnetic and surface functionalized Iron oxide Nanorods for biomarker-targeted applications. ACS Appl. Mater. Interfaces **9**, 20719–20727 (2017). https://doi.org/10.1021/acsami.7b02575
58. J. Mohapatra, A. Mitra, H. Tyagi, et al., Iron oxide nanorods as high-performance magnetic resonance imaging contrast agents. Nanoscale **7**, 9174–9184 (2015). https://doi.org/10.1039/C5NR00055F
59. M.E. Materia, P. Guardia, A. Sathya, et al., Mesoscale assemblies of Iron oxide Nanocubes as heat mediators and image contrast agents. Langmuir **31**, 808–816 (2015). https://doi.org/10.1021/la503930s
60. Z. Zhao, Z. Zhou, J. Bao, et al., Octapod iron oxide nanoparticles as high-performance T2 contrast agents for magnetic resonance imaging. Nat. Commun. **4**, 2266 (2013). https://doi.org/10.1038/ncomms3266
61. Y. Bao, J.A. Sherwood, Z. Sun, Magnetic iron oxide nanoparticles as T1 contrast agents for magnetic resonance imaging. J. Mater. Chem. C **6**, 1280–1290 (2018). https://doi.org/10.1039/C7TC05854C
62. I. Fernández-Barahona, M. Muñoz-Hernando, J. Ruiz-Cabello, et al., Iron oxide nanoparticles: An alternative for positive contrast in magnetic resonance imaging. Inorganics **8**, 28 (2020). https://doi.org/10.3390/inorganics8040028
63. H. Wei, O.T. Bruns, M.G. Kaul, et al., Exceedingly small iron oxide nanoparticles as positive MRI contrast agents. Proc. Natl. Acad. Sci. **114**, 2325–2330 (2017). https://doi.org/10.1073/pnas.1620145114
64. T. Macher, J. Totenhagen, J. Sherwood, et al., Ultrathin Iron oxide Nanowhiskers as positive contrast agents for magnetic resonance imaging. Adv. Funct. Mater. **25**, 490–494 (2015). https://doi.org/10.1002/adfm.201403436
65. W. Yue, A.H. Hill, A. Harrison, W. Zhou, Mesoporous single-crystal Co3O4 templated by cage-containing mesoporous silica. Chem. Commun. (Camb.) **1**, 2518–2520 (2007). https://doi.org/10.1039/b700185a
66. Y. Yue, X. Wang, Nanoscale thermal probing. Nano Rev **3**, 11586 (2012). https://doi.org/10.3402/nano.v3i0.11586
67. Z. Zhou, D. Huang, J. Bao, et al., A synergistically enhanced T1-T2 dual-modal contrast agent. Adv. Mater. **24**, 6223–6228 (2012). https://doi.org/10.1002/adma.201203169
68. G. Vallejo-Fernandez, O. Whear, A.G. Roca, et al., Mechanisms of hyperthermia in magnetic nanoparticles. J. Phys. D. Appl. Phys. **46**, 312001 (2013). https://doi.org/10.1088/0022-3727/46/31/312001
69. S. Dutz, R. Hergt, Magnetic nanoparticle heating and heat transfer on a microscale: Basic principles, realities and physical limitations of hyperthermia for tumour therapy. Int. J. Hyperth. **29**, 790–800 (2013). https://doi.org/10.3109/02656736.2013.822993

70. A.E. Deatsch, B.A. Evans, Heating efficiency in magnetic nanoparticle hyperthermia. J. Magn. Magn. Mater. **354**, 163–172 (2014). https://doi.org/10.1016/j.jmmm.2013.11.006
71. M. Ma, Y. Wu, J. Zhou, et al., Size dependence of specific power absorption of Fe3O4 particles in AC magnetic field. J. Magn. Magn. Mater. **268**, 33–39 (2004). https://doi.org/10.1016/S0304-8853(03)00426-8
72. Z. Nemati, R. Das, J. Alonso, et al., Iron oxide Nanospheres and Nanocubes for magnetic hyperthermia therapy: A comparative study. J. Electron. Mater. **46**, 3764–3769 (2017). https://doi.org/10.1007/s11664-017-5347-6
73. D. Serantes, K. Simeonidis, M. Angelakeris, et al., Multiplying magnetic hyperthermia response by nanoparticle assembling. J. Phys. Chem. C **118**, 5927–5934 (2014). https://doi.org/10.1021/jp410717m
74. P. Guardia, R. Di Corato, L. Lartigue, et al., Water-soluble Iron oxide Nanocubes with high values of specific absorption rate for Cancer cell hyperthermia treatment. ACS Nano **6**, 3080–3091 (2012). https://doi.org/10.1021/nn2048137
75. P. Guardia, A. Riedinger, S. Nitti, et al., One pot synthesis of monodisperse water soluble iron oxide nanocrystals with high values of the specific absorption rate. J. Mater. Chem. B **2**, 4426 (2014). https://doi.org/10.1039/c4tb00061g
76. K. Mahmoudi, A. Bouras, D. Bozec, et al., Magnetic hyperthermia therapy for the treatment of glioblastoma: A review of the therapy's history, efficacy and application in humans. Int. J. Hyperth. **34**, 1316–1328 (2018). https://doi.org/10.1080/02656736.2018.1430867
77. M.V. Zyuzin, M. Cassani, M.J. Barthel, et al., Confining Iron oxide Nanocubes inside submicrometric cavities as a key strategy to preserve magnetic heat losses in an intracellular environment. ACS Appl. Mater. Interfaces **11**, 41957–41971 (2019). https://doi.org/10.1021/acsami.9b15501
78. J. Mohapatra, A. Mitra, M. Aslam, D. Bahadur, Octahedral-shaped Fe3O4 nanoparticles with enhanced specific absorption rate and R2 Relaxivity. IEEE Trans. Magn. **51**, 1–3 (2015). https://doi.org/10.1109/TMAG.2015.2439213
79. Y. Lv, Y. Yang, J. Fang, et al., Size dependent magnetic hyperthermia of octahedral Fe 3 O 4 nanoparticles. RSC Adv. **5**, 76764–76771 (2015). https://doi.org/10.1039/C5RA12558H
80. A. Wachowiak, Direct observation of internal spin structure of magnetic vortex cores. Science (80-) **298**, 577–580 (2002). https://doi.org/10.1126/science.1075302
81. M. Ma, Y. Zhang, Z. Guo, N. Gu, Facile synthesis of ultrathin magnetic iron oxide nanoplates by Schikorr reaction. Nanoscale Res. Lett. **8**, 16 (2013). https://doi.org/10.1186/1556-276X-8-16
82. S. Del Sol-Fernández, Y. Portilla-Tundidor, L. Gutiérrez, et al., Flower-like Mn-doped magnetic nanoparticles functionalized with α v β 3 -integrin-ligand to efficiently induce intracellular heat after alternating magnetic field exposition, triggering Glioma cell death. ACS Appl. Mater. Interfaces **11**, 26648–26663 (2019). https://doi.org/10.1021/acsami.9b08318
83. J. Estelrich, M. Busquets, Iron oxide nanoparticles in Photothermal therapy. Molecules **23**, 1567 (2018). https://doi.org/10.3390/molecules23071567
84. A. Espinosa, R. Di Corato, J. Kolosnjaj-Tabi, et al., Duality of Iron oxide nanoparticles in Cancer therapy: Amplification of heating efficiency by magnetic hyperthermia and Photothermal bimodal treatment. ACS Nano **10**, 2436–2446 (2016). https://doi.org/10.1021/acsnano.5b07249
85. A. Espinosa, J. Kolosnjaj-Tabi, A. Abou-Hassan, et al., Magnetic (hyper)Thermia or Photothermia? Progressive comparison of Iron oxide and gold nanoparticles heating in water, in cells, and in vivo. Adv. Funct. Mater. **28**, 1803660 (2018). https://doi.org/10.1002/adfm.201803660
86. D. Cheng, X. Li, G. Zhang, H. Shi, Morphological effect of oscillating magnetic nanoparticles in killing tumor cells. Nanoscale Res. Lett. **9**, 195 (2014). https://doi.org/10.1186/1556-276X-9-195
87. X. Yao, K. Sabyrov, T. Klein, et al., Evaluation of magnetic heating of asymmetric magnetite particles. J. Magn. Magn. Mater. **381**, 21–27 (2015). https://doi.org/10.1016/j.jmmm.2014.12.035

88. D.-H. Kim, E.A. Rozhkova, I.V. Ulasov, et al., Biofunctionalized magnetic-vortex microdiscs for targeted cancer-cell destruction. Nat. Mater. **9**, 165–171 (2010). https://doi.org/10.1038/nmat2591
89. M.P. Sharrock, Recent advances in metal particulate recording media: Toward the ultimate particle. IEEE Trans. Magn. **36**, 2420–2425 (2000). https://doi.org/10.1109/20.908453
90. M.P. Sharrock, Particulate magnetic recording media: A review. IEEE Trans. Magn. **25**, 4374–4389 (1989). https://doi.org/10.1109/20.45317
91. A. Berkowitz, Some materials considerations in particulate media. IEEE Trans. Magn. **22**, 466–471 (1986). https://doi.org/10.1109/TMAG.1986.1064462
92. A.E. Berkowitz, R.P. Goehner, E.L. Hall, P.J. Flanders, Microstructure, relaxation, and print-through in γ-Fe 2 O 3 particles. J. Appl. Phys. **57**, 3928–3930 (1985). https://doi.org/10.1063/1.334919
93. M. Stachen, M.P. Morales, M. Ocaña, C.J. Serna, Effect of precursor impurities on the magnetic properties of uniform γ-Fe2O3 ellipsoidal particles. Phys. Chem. Chem. Phys. **1**, 4465–4471 (1999). https://doi.org/10.1039/a904606b
94. H. Naono, R. Fujiwara, Micropore formation due to thermal decomposition of acicular microcrystals of α-FeOOH. J. Colloid Interface Sci. **73**, 406–415 (1980). https://doi.org/10.1016/0021-9797(80)90086-7
95. A. Corradi, S. Andress, J. French, et al., Magnetic properties of new (NP) hydrothermal particles. IEEE Trans. Magn. **20**, 33–38 (1984). https://doi.org/10.1109/TMAG.1984.1063007
96. M.P. Morales, S.A. Walton, L.S. Prichard, et al., Characterisation of advanced metal particle recording media pigments. J. Magn. Magn. Mater. **190**, 357–370 (1998). https://doi.org/10.1016/S0304-8853(98)00249-2
97. N.O. Nuñez, M.P. Morales, P. Tartaj, C.J. Serna, Preparation of high acicular and uniform goethite particles by a modified-carbonate route. J. Mater. Chem. **10**, 2561–2565 (2000). https://doi.org/10.1039/b005014h
98. R. Mendoza-Reséndez, M.P. Morales, C.J. Serna, Reduction mechanism of uniform iron oxide nanoparticles to metal used as recording media. Mater. Sci. Eng. C **23**, 1139–1142 (2003). https://doi.org/10.1016/j.msec.2003.09.126
99. T. Pradeep, Anshup, Noble metal nanoparticles for water purification: A critical review. Thin Solid Films **517**, 6441–6478 (2009). https://doi.org/10.1016/j.tsf.2009.03.195
100. A.S. Helal, E. Mazario, A. Mayoral, et al., Highly efficient and selective extraction of uranium from aqueous solution using a magnetic device: Succinyl-β-cyclodextrin-APTES@ maghemite nanoparticles. Environ. Sci. Nano **5**, 158–168 (2018). https://doi.org/10.1039/C7EN00902J
101. M. Auffan, J. Rose, J.-Y. Bottero, et al., Towards a definition of inorganic nanoparticles from an environmental, health and safety perspective. Nat. Nanotechnol. **4**, 634–641 (2009). https://doi.org/10.1038/nnano.2009.242
102. M. Auffan, J. Rose, O. Proux, et al., Enhanced adsorption of arsenic onto Maghemites nanoparticles: As(III) as a probe of the surface structure and heterogeneity. Langmuir **24**, 3215–3222 (2008). https://doi.org/10.1021/la702998x
103. S. Brice-Profeta, M.-A. Arrio, E. Tronc, et al., Magnetic order in - nanoparticles: A XMCD study. J. Magn. Magn. Mater. **288**, 354–365 (2005). https://doi.org/10.1016/j.jmmm.2004.09.120
104. G.A. Waychunas, C.S. Kim, J.F. Banfield, Nanoparticulate Iron oxide minerals in soils and sediments: Unique properties and contaminant scavenging mechanisms. J. Nanopart. Res. **7**, 409–433 (2005). https://doi.org/10.1007/s11051-005-6931-x
105. M.S. Onyango, Y. Kojima, H. Matsuda, A. Ochieng, Adsorption kinetics of arsenic removal from groundwater by Iron-modified zeolite. J Chem Engineering Japan **36**, 1516–1522 (2003). https://doi.org/10.1252/jcej.36.1516
106. S.R. Chowdhury, E.K. Yanful, Arsenic and chromium removal by mixed magnetite–maghemite nanoparticles and the effect of phosphate on removal. J. Environ. Manag. **91**, 2238–2247 (2010). https://doi.org/10.1016/j.jenvman.2010.06.003

107. A. Khodabakhshi, M.M. Amin, M. Mozaffari, Synthesis of magnetite nanoparticles and evaluation of its efficiency for arsenic removal from simulated industrial waste water. Iran J Environ Heal Sci Eng **8**, 189–200 (2011)
108. L.-S. Zhong, J.-S. Hu, H.-P. Liang, et al., Self-assembled 3D flowerlike Iron oxide nanostructures and their application in water treatment. Adv. Mater. **18**, 2426–2431 (2006). https://doi.org/10.1002/adma.200600504
109. L. Balcells, C. Martínez-Boubeta, J. Cisneros-Fernández, et al., One-step route to Iron oxide hollow Nanocuboids by cluster condensation: Implementation in water remediation technology. ACS Appl. Mater. Interfaces **8**, 28599–28606 (2016). https://doi.org/10.1021/acsami.6b08709
110. M. Iram, C. Guo, Y. Guan, et al., Adsorption and magnetic removal of neutral red dye from aqueous solution using Fe3O4 hollow nanospheres. J. Hazard. Mater. **181**, 1039–1050 (2010). https://doi.org/10.1016/j.jhazmat.2010.05.119
111. C. Martinez-Boubeta, K. Simeonidis, *Magnetic Nanoparticles for Water Purification* (Nanoscale Materials in Water Purification. Elsevier, In, 2019), pp. 521–552
112. G.S. Parkinson, U. Diebold, J. Tang, L. Malkinski, Tailoring the Interface properties of magnetite for Spintronics. Adv Magn Mater, 61–88 (2012). https://doi.org/10.5772/39101
113. Z. Zhang, S. Satpathy, Electron states, magnetism, and the Verwey transition in magnetite. Phys. Rev. B **44**, 13319–13331 (1991). https://doi.org/10.1103/PhysRevB.44.13319
114. A. Mitra, B. Barick, J. Mohapatra, et al., Large tunneling magnetoresistance in octahedral Fe_3O_4 nanoparticles. AIP Adv. **6**, 055007 (2016). https://doi.org/10.1063/1.4948798
115. A. Mitra, J. Mohapatra, H. Sharma, et al., Controlled synthesis and enhanced tunnelling magnetoresistance in oriented Fe3O4 nanorod assemblies. J. Phys. D. Appl. Phys. **51**, 085002 (2018). https://doi.org/10.1088/1361-6463/aaa697
116. G. Sun, B. Dong, M. Cao, et al., Hierarchical dendrite-like magnetic materials of Fe 3 O 4 , γ-Fe 2 O 3 , and Fe with high performance of microwave absorption. Chem. Mater. **23**, 1587–1593 (2011). https://doi.org/10.1021/cm103441u
117. C.-L. Zhu, M.-L. Zhang, Y.-J. Qiao, et al., Fe 3 O 4 /TiO 2 Core/Shell nanotubes: Synthesis and magnetic and electromagnetic wave absorption characteristics. J. Phys. Chem. C **114**, 16229–16235 (2010). https://doi.org/10.1021/jp104445m
118. R.K. Walser, W. Win, P.M. Valanju, Shape-optimized ferromagnetic particles with maximum theoretical microwave susceptibility. IEEE Trans. Magn. **34**, 1390–1392 (1998). https://doi.org/10.1109/20.706558
119. L.L. Adebayo, H. Soleimani, N. Yahya, et al., Recent advances in the development OF Fe3O4-BASED microwave absorbing materials. Ceram. Int. **46**, 1249–1268 (2020). https://doi.org/10.1016/j.ceramint.2019.09.209
120. Y. Yang, M. Li, Y. Wu, et al., Size-dependent microwave absorption properties of Fe 3 O 4 nanodiscs. RSC Adv. **6**, 25444–25448 (2016). https://doi.org/10.1039/C5RA28035D
121. Y.J. Chen, P. Gao, C.L. Zhu, et al., Synthesis, magnetic and electromagnetic wave absorption properties of porous Fe_3O_4/Fe/SiO2 core/shell nanorods. J. Appl. Phys. **106**, 054303 (2009). https://doi.org/10.1063/1.3204958
122. X. Wang, F. Pan, Z. Xiang, et al., Magnetic vortex core-shell Fe3O4@C nanorings with enhanced microwave absorption performance. Carbon N Y **157**, 130–139 (2020). https://doi.org/10.1016/j.carbon.2019.10.030
123. X. Li, B. Zhang, C. Ju, et al., Morphology-controlled synthesis and electromagnetic properties of porous Fe 3 O 4 nanostructures from Iron Alkoxide precursors. J. Phys. Chem. C **115**, 12350–12357 (2011). https://doi.org/10.1021/jp203147q
124. X. Liu, Z. Zhong, Y. Tang, B. Liang, Review on the synthesis and applications of Fe_3O_4 Nanomaterials. J. Nanomater. **2013**, 1–7 (2013). https://doi.org/10.1155/2013/902538
125. J. Liu, Z. Wu, Q. Tian, et al., Shape-controlled iron oxide nanocrystals: Synthesis, magnetic properties and energy conversion applications. CrystEngComm **18**, 6303–6326 (2014). https://doi.org/10.1039/c6ce01307d

126. O. Delmer, P. Balaya, L. Kienle, J. Maier, Enhanced potential of amorphous electrode materials: Case study of RuO_2. Adv. Mater. **20**, 501–505 (2008). https://doi.org/10.1002/adma.200701349
127. J. Ma, X. Guo, Y. Yan, et al., FeO x -based materials for electrochemical energy storage. Adv. Sci. **5**, 1700986 (2018). https://doi.org/10.1002/advs.201700986
128. Z. Xiao, Y. Xia, Z. Ren, et al., Facile synthesis of single-crystalline mesoporous α-Fe2O3 and Fe3O4 nanorods as anode materials for lithium-ion batteries. J. Mater. Chem. **22**, 20566 (2012). https://doi.org/10.1039/c2jm34083f
129. S. Jin, H. Deng, D. Long, et al., Facile synthesis of hierarchically structured Fe_3O_4/carbon micro-flowers and their application to lithium-ion battery anodes. J. Power Sources **196**, 3887–3893 (2011). https://doi.org/10.1016/j.jpowsour.2010.12.078
130. Y. Zeng, R. Hao, B. Xing, et al., One-pot synthesis of Fe_3O_4 nanoprisms with controlled electrochemical properties. Chem. Commun. **46**, 3920 (2010). https://doi.org/10.1039/c0cc00246a
131. B. Koo, H. Xiong, M.D. Slater, et al., Hollow Iron oxide nanoparticles for application in Lithium ion batteries. Nano Lett. **12**, 2429–2435 (2012). https://doi.org/10.1021/nl3004286
132. H. Cao, R. Liang, D. Qian, et al., Serine-assisted synthesis of Superparamagnetic Fe3O4 Nanocubes for Lithium ion batteries. J. Phys. Chem. C **115**, 24688–24695 (2011). https://doi.org/10.1021/jp2096482
133. X.-L. Cheng, J.-S. Jiang, D.-M. Jiang, Z.-J. Zhao, Synthesis of rhombic dodecahedral Fe 3 O 4 Nanocrystals with exposed high-energy {110} facets and their peroxidase-like activity and Lithium storage properties. J. Phys. Chem. C **118**, 12588–12598 (2014). https://doi.org/10.1021/jp412661e

Oxide Semiconductor Plasmonics for Infrared Applications

Hiroaki Matsui

1 Introduction

Plasmonic studies of oxide semiconductors, such as In_2O_3 and ZnO, have gained increasing attention given the discoveries of new optical properties with varied potential applications. Oxide semiconductors with metallic properties obtained by incorporation of impurity dopants show surface plasmon resonances (SPRs) in the infrared (IR) range [1–5]. Unlike traditional metals such as Ag and Au, SPRs can be widely tuned by control of the free carriers (electrons or holes) [6–8], thereby providing new possibilities for the optical manipulation of light. Additionally, plasmonic studies can break new ground in the areas of oxide semiconductors. A characteristic property found by these studies is that the optical nature of oxide semiconductors shows a low-loss plasmonic material in near-IR wavelengths compared with traditional metals. In particular, the band structures on oxide semiconductors comprise *s*- and *p*-orbitals [9]. Band structures related to *d* orbitals such as those exhibited by the noble metals do not exist in the bandgap [10]. Therefore, SPRs based on oxide semiconductors provide new possibilities for alternative plasmonic applications in the IR range.

Oxide semiconductor-based SPRs are excited on surfaces of different nanostructures [11–13]. Above all, nanoparticle (NP) shapes of ZnO: Al (ZAO) and In_2O_3: Sn (ITO) show localized-type surface plasmon resonances (LSPRs) due to the collective excitations of carriers into NPs [14, 15], which induce localized electric fields (*E*-fields) on the NP surfaces. To date, studies related to LSPRs have been focused

H. Matsui (✉)
Department of Bioengineering, The University of Tokyo, Tokyo, Japan

Department of Electrical Engineering and Information Systems, The University of Tokyo, Tokyo, Japan
e-mail: hiroaki@ee.t.u-tokyo.ac.jp

A. G. Roca et al. (eds.), *Surfaces and Interfaces of Metal Oxide Thin Films, Multilayers, Nanoparticles and Nano-composites*,
https://doi.org/10.1007/978-3-030-74073-3_4

on metal NPs for optical applications of waveguides and biochemical sensing [16–18]. Recently, ZAO and ITO NPs have launched as alternative plasmonic nanomaterials in the IR range by controlling the material characteristics. For example, assembled films of ITO NPs (ITO NP films) have shown increased photoluminescence and light absorption in the near-IR range [19, 20], which are related to strong *E*-fields formed on the NP film surfaces [21]. Plasmon coupling at an interparticle gap plays an important role in surface-enhanced fluorescence and Raman spectroscopy [22] and has demonstrated using metal NPs in the visible range. Therefore, the use of oxide semiconductor NPs extends the interesting surface-enhanced optical characteristics to longer wavelengths in the IR range.

One of the aims of this work is to develop and utilize the plasmonic behaviors of oxide semiconductor NPs in thermal shielding applications. Solar energy comprises ultraviolet (5%), visible (40%), and IR rays (55%) [23]. At present, IR-reflecting film sheets in the near-IR and IR ranges are needed to prevent heat transfer from sunlight. Above all, the control of solar thermal shielding through windows affected by visible light creates a comfortable environment inside automobiles and buildings. Silica aerogels [24, 25], photonic crystals [26, 27], and metal–oxide multilayers including ITO-TiN (Au)-ITO and TiO_2-Cu-TiO_2 [28–30] have been reported as conventional technologies for use in solar thermal shielding applications. Recently, oxide semiconductors have received much attention in the area of thermal shielding as a means of preventing solar and radiant heat in the near- and mid-IR range, respectively [31]. The composites and films consisting of oxide semiconductor NPs show optical properties in the IR range that are dominated by absorbance [30, 32–34]. However, present thermal shielding applications have mainly focused on cutting IR radiation not by absorption, but through reflection. To date, no paper has reported on the reflective performance of doped oxide semiconductor NPs. Plasmonic applications exhibiting thermal shielding ability have not been previously studied in detail. In this chapter, we introduce the plasmonic properties necessary for satisfying recent industry demands for a material with thermal shielding ability using ITO NP films as a concrete example.

This chapter is organized as follows. In Sect. 2, we give a description of the structural and optical properties of ITO NPs through local structural analyses. In Sect. 3, we report on plasmonic responses of ITO NPs from theoretical and experimental approaches. Elucidation of the optical processes of plasmonic excitations in ITO NPs can provide valuable information for oxide semiconductor plasmonics. Section 4 is devoted to a detailed discussion of thermal shielding using ITO NP films. In particular, we describe plasmonic responses related to the three-dimensional *E*-field coupling along the out-of-plane and in-plane directions, which is important in facilitating selective high reflections in the IR range. This mechanism becomes important in the harnessing of IR optical properties for the fabrication of solar thermal shielding. In Sect. 5, we detail some electromagnetic responses of assembled ITO NP films in the microwave region. Finally, concluding remakes are given in Sect. 6.

2 Structural and Optical Properties of ITO NPs

2.1 Structural Analyses

ITO NPs were fabricated using a metal–organic decomposition technique [35]. The metal precursor complexes of $(C_9H_{22}CO_2)_3In$ and $(C_9H_{22}CO_2)_4Sn$ were selected as starting materials. Indium and tin carboxylates with a chemical ratio of 95:5 were heated in a flask supported by a heating mantle to 350 °C and then maintained at this temperature for 4 h. The mixture was then gradually cooled to room temperature which afforded a pale blue suspension. ITO NPs with a pale blue color were obtained after repeated centrifugation and washing with processes using ethanol and final dispersal in toluene.

Figure 1a shows X-ray diffraction (XRD) patterns of ITO NPs with Sn content ($[Sn]_{XFS}$) of 0 and 5%. The Sn content of ITO NPs was determined using X-ray florescent spectroscopy (XFS). XRD results revealed that the NPs exhibited broad diffraction peaks of a colloidal sample with a crystal structure [36]. Broadening of the line widths of the XRD patterns can be attributed to structural imperfections such as defects and strains. Transmittance electron microscopy (TEM)

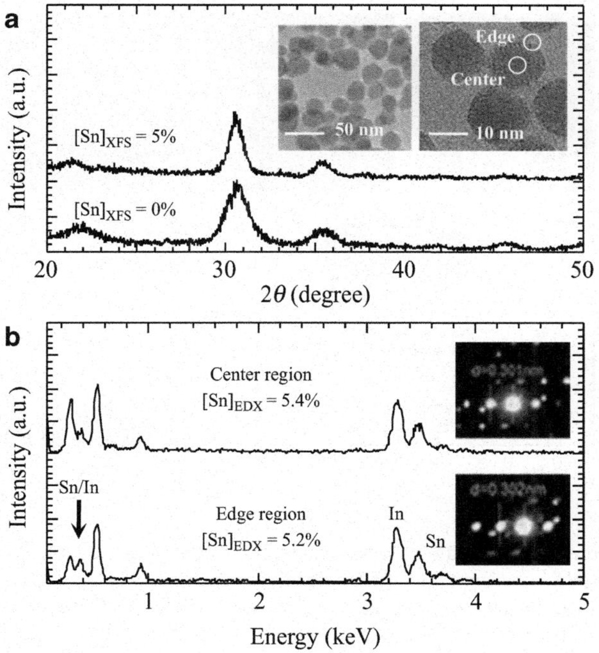

Fig. 1 (**a**) XRD patterns of ITO NPs with Sn contents of 0% and 5%. Inset shows low- (left) and high-resolution (right) TEM images of an ITO NP with a Sn content of 5%. (**b**) EDX spectra at the center and edge regions of the NP. Inset represents ED patterns of the (222) plane at the center and edge. (Fig. 1 from [20]). (Copyright ©2014 American Institute of Physics)

measurements of the NPs showed that the NPs had isolated particle features due to organic ligands ($C_{10}H_{22}O_2$) terminated on the NP surfaces [inset of Fig. 1a]. Electron diffraction (ED) analyses revealed that the lattice interval (d_{222}) along the [222] direction at the center region (d_{222} = 0.301 nm) was close to that at the edge region of the NP (d_{222} = 0.302 nm) [inset of Fig. 1b]. The experimental results of energy-dispersive X-ray microscopy showed that the Sn concentration [Sn] at the center region ([Sn] = 5.4%) was similar to that at the edge region ([Sn] = 5.2%), indicating that Sn atoms in the NP were spatially homogeneous [Fig. 1b].

We further examined the local structures of the NPs by scanning-TEM (STEM) coupled with electron energy-loss spectroscopy (STEM-EELS) [30]. STEM-EELS can directly detect a plasmonic response in a single NP. High-angle annular dark field (HAADF) images in Figs. 2a–c clear show that indium and Sn components in the NP were distributed homogeneously. The EELS spectra at the edge and vacuum regions on the STEM-acquired particle image showed a slight spectral difference in energy-loss regions from 1.0 to 0.5 eV [Fig. 2d]. In Fig. 2e, a differential EELS spectrum showed a maximum peak at 0.7 eV that was similar to the light absorption in the near-IR range, which was evidence of a LSPR excitation on the NP surface as result of spatially homogeneous doping of Sn atoms in the NP.

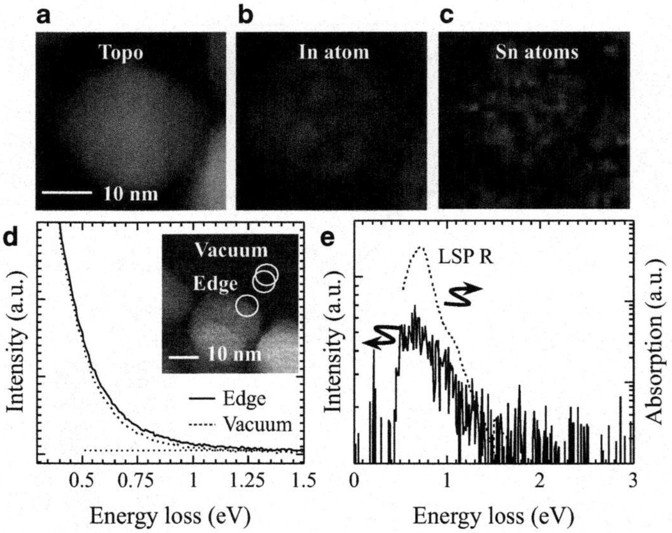

Fig. 2 (**a**) Aberration-corrected STEM image, and HAADF images of In (**b**) and Sn atoms (**c**) in the ITO NP with a Sn concentration of 5%. (**d**) EELS spectra taken in the electron probe position at vacuum (black lime) and edge (red line) regions, and a STEM-acquired particle image (inset). (**e**) Differential EELS spectrum (black line), and optical spectrum of ITO NP taken in the near-IR (green line). (Fig. 3 from [35]). (Copyright ©2014 American Institute of Physics)

2.2 Optical Properties

Optical absorptions of ITO NPs with different electron density (n_e) were examined [Fig. 3] and indicated that the systematic change in the absorption spectra is clearly related to the Sn content. Absorption measurements were performed using a Fourier transform infrared (FTIR) spectrometer. n_e values were estimated from the absorption spectra using theoretical calculations. The following equation was used to derive absorption intensity (A) from the experimental data [20]:

$$A = 4\pi kR^3 Im\left\{\frac{\varepsilon_m(\omega) - \varepsilon_d}{\varepsilon_m(\omega) + 2\varepsilon_d}\right\} \tag{1}$$

where $k = 2\pi(\varepsilon_d)^{1/2}\omega/c$ with c representing the speed of light, ε_d indicates the host dielectric constants of toluene, $\varepsilon_m(\omega)$ is the particle dielectric function, and R is the particle radius. Furthermore, $\varepsilon_m(\omega)$ employed the free-carrier Drude term with a frequency-dependent damping constant, $\Gamma(\omega)$ [20]:

$$\varepsilon_m(\omega) = 1 - \frac{\omega_p^2}{\omega(\omega + i\Gamma)} \tag{2}$$

The plasma frequency (ω_p) is given by $\omega_p^2 = ne / \varepsilon_\infty \varepsilon_0 m^*$, where ε_∞ is the high-frequency dielectric constant, ε_0 is the vacuum permittivity, and m^* is the effective electron mass. Fitted absorptions were used with parameter values of $\varepsilon_d = 2.03(n = 1.426$ refractive index of the solvent), $\varepsilon_\infty = 3.8$, and $m^* = 0.3\ m_0$ to estimate $\varepsilon_p(\omega)$. The term $\Gamma(\omega)$ based on electron–impurity scattering can be described by the following relation [21]:

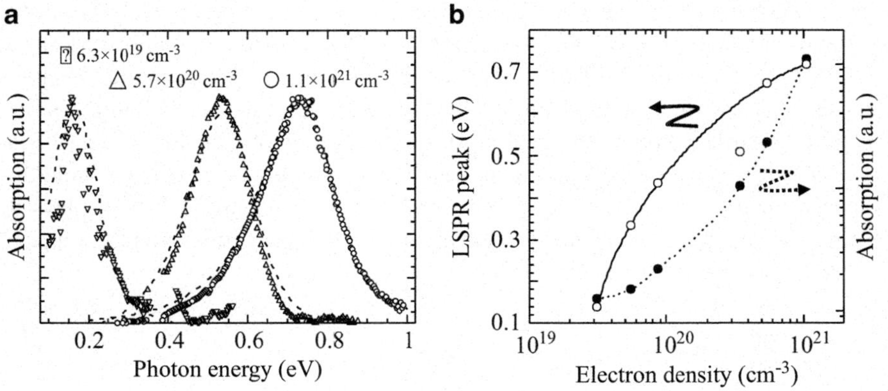

Fig. 3 (**a**) Absorption spectra of ITO NPs with different electron density. Doping with Sn contents of 0.02, 1 and 5% into the NPs induced electron density of 6.3×10^{19} cm^{-3}, 5.7×10^{20} cm^{-3}. and 1.1×10^{21} cm^{-3}, respectively. Dot lines indicate theoretical calculations based on the modified Mie theory. (**b**) LSPR peak and absorption intensity of ITO NPs as a function of electron density

$$\Gamma(\omega) = f(\omega)\Gamma_L + [1 - f(\omega)]\Gamma_H \left(\frac{\omega}{\Gamma_H}\right)^{-3/2} \quad (3)$$

where $f(\omega)$ can be described by $f(\omega) = [1+\exp\{(\omega-\Gamma_x)/\sigma\}]^{-1}$. Γ_H and Γ_L represent the high- ($\omega = \infty$) and low-frequency ($\omega = 0$) damping, respectively. Γ_X and σ represent the change-over frequency and width of the function, respectively.

Calculated absorption spectra were very close to the experimental data. ITO NPs doped with Sn content of 0.02%, 1% or 5% provided electron density of 6.3×10^{19}, 5.7×10^{20} and 1.1×20^{21} cm^{-3}, respectively [Fig. 3a]. We summarized the LSPR resonant peak and absorption intensity as a function of n_e [Fig. 3b]. The LSPR resonant peak gradually showed a redshift from the near-IR to mid-IR range with decreasing n_e. Additionally, the absorption intensity decreased markedly with decreasing n_e. No plasmon excitation was observed in the low n_e region below 10^{19} cm^{-3}. The Mott critical density (N_c) of ITO is estimated as $N_c = 6\times10^{18}$ cm^{-3}. Below the Mott critical density, the impurity band does not overlap with the Fermi energy (E_F) level. ITO NPs show semiconducting behavior. However, the E_F level combined with the impurity band in the middle n_e region from 10^{19} to 10^{20} cm^{-3}. At the high n_e region above 10^{20} cm^{-3}, the E_F level is placed in a highest occupied state in the conduction band. As a consequence, ITO NPs show metallic behavior. These results indicated that a large amount of free carriers were required to excite highly efficient plasmon excitations.

2.3 Field Distributions

Figure 4 shows theoretical evaluations of *E*-field distributions monitored at different wavelengths. We used the three-dimensional Finite-difference time-domain (3D-FDTD) method to calculate *E*-field distributions. The experimental absorbance was close to the theoretical absorbance [Fig. 4a]. The *E*-field distribution of a single ITO NP was examined at different wavelengths. A typical field distribution is shown in Fig. 4e. The *E*-field strength around the NP surface calculated at 1.2 μm was very small [Fig. 4b]. However, the peak position at 1.8 μm corresponding to an LSPR resonant peak position showed the highest *E*-field strength [Fig. 4d]. The field distance from the NP surface was as small as 10 nm. This strong *E*-field plays an important role in a field enhancement at an interparticle gap, resulting in the appearance of resonant reflectance in the IR range.

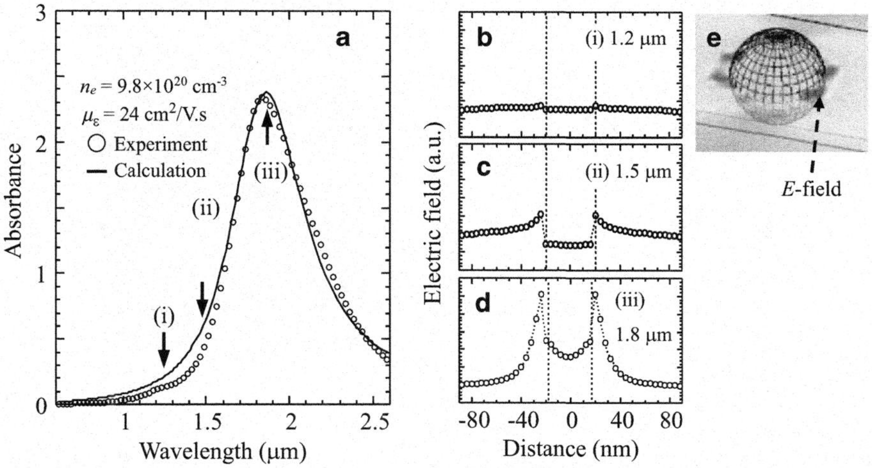

Fig. 4 (**a**) Experimental and simulated absorbance spectra of ITO NPs. Cross-sectional *E*-field distributions of ITO NPs monitored at 1.2 μm (**b**), 1.5 μm (**c**) and 1.8 μm (**d**). (**e**) Inset picture indicates an *E*-field distribution excited on the NP surface

3 Structural and Optical Properties of ITO NP Films

3.1 Structural Evaluations

Assembled NP sheets were deposited on IR transparent CaF_2 substrates by a spin-coating technique. The ITO NP films were investigated by small-angle X-ray scattering (SAXS), which revealed an interesting insight into the scattering vector (q) of the SAXS intensity. The maximum SAXS peak (q_{max}) provides structural information concerning spatial ordering of NPs calculated using $l = 2\pi/q_{max}$ with a spatial period (l). In Fig. 5a, the SAXS pattern had a maximum peak at $q = 0.33$ nm^{-1} followed by clear interferences, which corresponded to an l value of 20 nm being close to the edge-to-edge between NPs. The surface scanning microscopy (SEM) image showed a close-packed structure since spin-coating produces self-organizations of colloidal NPs arranged in a hexagonally close-packed (HCP) structure based on shear and capillary forces on substrates [37, 38] [Fig. 5b]. Additionally, a cross-sectional SEM image also exhibited a close-packed structure [Fig. 5c]. However, the particle alignment showed disordered packing feature between NPs, as indicated from the fast-Fourier transform (FFT) pattern of the SEM image [inset in Fig. 5c]. Disorder of the local structure in the NP film produced the broad interferences of the SAXS pattern, which influences the optical properties of the NP films.

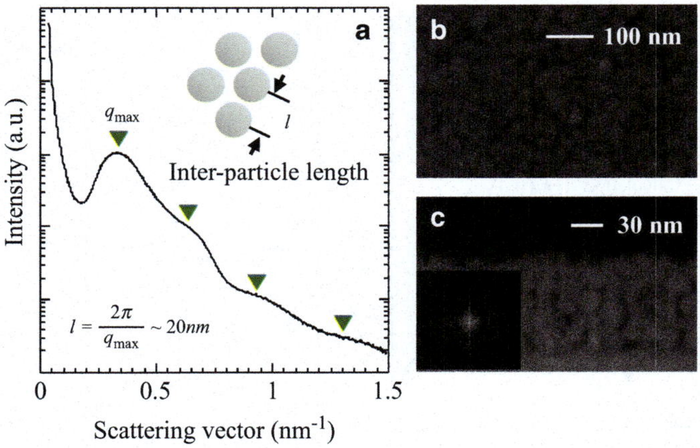

Fig. 5 (**a**) SAXS pattern of a 96-nm-thick NP sheet. Inset shows a schematic picture to indicate the interparticle length (l). (**b**) In-plane surface (**b**) and cross-sectional SEM image (**c**) of 96-nm-thick NP sheets. Inset represents an FFT pattern of a cross-sectional SEM image. (Figs. 5 and 7 from [39]). (Copyright ©2016 American Chemical society)

3.2 Optical Characteristics in the IR Range

The optical properties of a mono-layered ITO NP film are shown in Fig. 6. Transmittance spectra showed a resonant dip at 2.64 μm, indicating a redshift of the resonant wavelength due to a collective plasmon resonance (CPR) effect compared to those NPs dispersed in toluene [Fig. 6a] [40]. In contrast, reflectance at the resonant wavelength was very small. The optical responses were mainly dominated by light absorptions. Furthermore, we evaluated the optical properties of ITO NP films from the viewpoint of theoretical simulations based on the 3D-*FDTD* method. The simulated NP layer was assumed to have a hexagonally close-packed (HCP) structure with an interparticle distance (r) of 2 nm along the in-plane (x–y) and out-of-plane (y–z) directions [Figs. 6c and d]. The sample was illuminated with light directed in the z-direction from the air side. The E-field was parallel to the x-direction. Periodic boundary conditions were applied to the X- and Y-directions, and the bottom and top of the simulated domain in the Z-direction were analyzed using perfectly matched layer (PML) boundary conditions.

The refractive index (real part: 1.437) of capric acid was used for the medium between the NPs. An ellipsometric measurement of an ITO film was conducted to obtain the complex dielectric constants within the visible–IR range. The simulated layer thickness was systematically controlled by the number of NP layers ($N = 1$ to 20). Figure 6b shows simulated optical properties of the mono-layered NP layer, corresponding to $N = 1$, revealing that the simulated transmittance and reflectance spectra were consistent with the experimental results. It was found that the obtained resonant peaks based on the CPR effect were excited due to long-range coherences

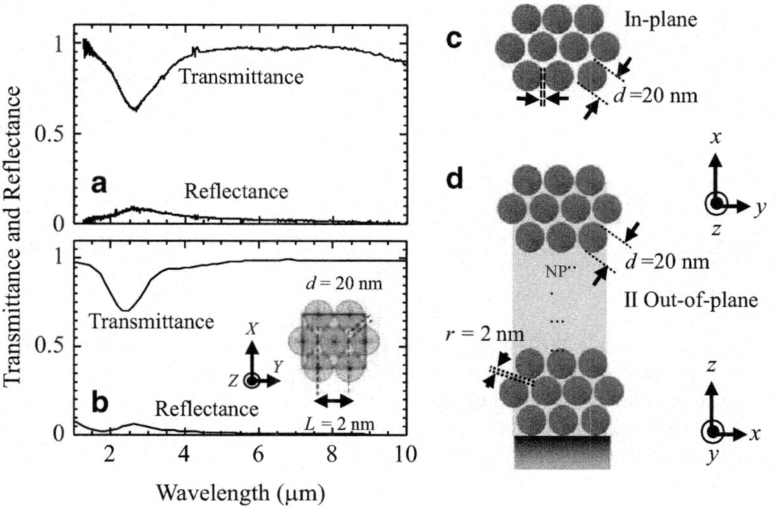

Fig. 6 (**a**) Experimental transmittance and reflectance of a mono-layered NP sheet with a HCP structure. Inset shows a model of a mono-layered NP film. Structural representation of simulated NP layers along the in-plane (x–y) (**c**) and out-of-plane (x–z) directions (**d**). The modeled structure was assumed to have a HCP structure with an interparticle distance (r) of 2 nm and was illuminated with light directed in the z-direction from the air side. The E-field was parallel to the x-direction. (Fig. 4 of [39]). (Copyright 2016 by the American Chemical Society)

of E-field interactions between NPs, which was supported from the two-dimensional (2D) image of the E-field distribution [inset of Fig. 6b].

3.3 *Three-Dimensionally Stacked ITO NP Films*

Three-dimensionally (3D) stacked NP films showed a remarkable change in optical properties (transmittance and reflectance spectra) [Fig. 7a and b]. Transmittance with a resonant wavelength at 2.20 μm decreased to a level close to zero with increasing film thickness. In contrast, reflectance was enhanced at a close proximity of 0.6 in terms of the film thickness. The single peak observed at a 22-nm-thick NP film gradually separated into lower and higher wavelengths with increasing film thickness. We found two types of resonant peaks (I and II) at 2.13 and 4.02 μm in the near- and mid-IR region on the 216-nm-thick NP film, respectively. The ratio R/A of reflectance (R) to absorbance (A) increased rapidly to a large value with increasing film thickness, indicating that the assembled films of ITO NPs showed reflectance-dominated optical properties. 3D-*FDTD* calculations were conducted in order to understand the experimental results. The systematic change in the number of NP layers from $N = 2$ to 20 was capable of reproducing the experimental spectra [Fig. 7c and d], which could clearly reproduce the optical properties of the NP films.

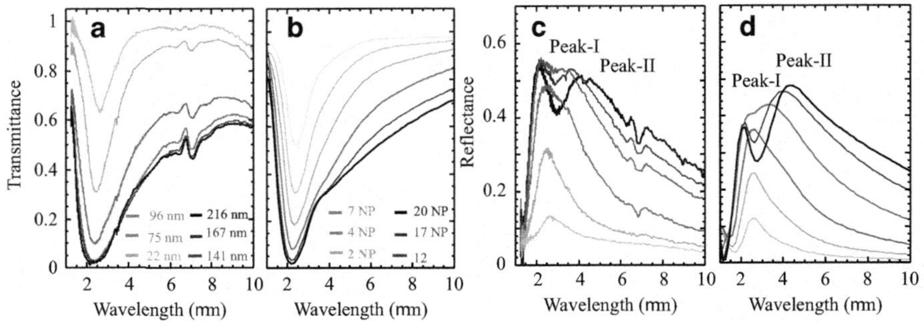

Fig. 7 (**a**) Experimental and (**b**) simulated transmittance spectra of NP sheets with different thicknesses. (**c**) Experimental and (**d**) reflectance spectra of NP films with various NP layers. The modeled NP layer was illuminated with light directed in the *Z*-direction from the air side. The direction of the *E*-field was perpendicular to the light and parallel to the *X*-direction. (Fig. 5 of [39]). (Copyright ©2016 American Chemical Society)

The increase in number of NP layers provided the resonant peaks in transmittance and the peak separations in reflectance. The 3D-*FDTD* simulation was a powerful tool for investigation of the mechanism of the optical properties of the NP films.

3.4 Nanoparticle Gap and Resonant IR Reflectance

The thermal behaviors of the NP films were investigated by TG-DTA in a N_2 atmosphere with a heating rate of 10 °C/min. The weight loss up to 250 °C might be related to the loss of physically or chemically absorbed water. There was an obvious weight loss in the temperature range from 270 to 320 °C because of the generation of organic species confirmed by *m*/*z* peaks at 18 (H_2O) and 44 (CO_2, C_3H_8, C_2H_4O, *etc.*) [Fig. 8a and b]. The decomposed species resulted from thermal removal of the surface ligands composed of fatty acids on the NP surfaces. These phenomena can be identified by FTIR measurements. The relation between surface ligand molecules and optical properties in the NP films was clarified by the spectral changes after annealing at different temperatures (T_A = 150–550 °C). Figure 8c shows temperature-dependent reflectance spectra taken in an inert atmosphere for a 216-nm-thick NP film, revealing remarkable spectral changes with increasing temperature. The two resonant peaks at 150 °C were weakened gradually following the change in spectral shape with the temperature. Above all, the reflection at peak I gradually shifted to longer wavelengths at temperatures above 300 °C corresponding to removal of the surface ligands. Additionally, the annealing effects of NP films were examined by the electrical resistivity in the films. Electrical resistivity was in the order of 10^4 Ω.cm at annealing temperatures below 250 °C because the presence of the surfactant layers on the NPs impedes carrier transport in the NP films. The surface ligands act as interparticle insulating layers in NP networks. However, this was markedly reduced at temperatures above 350 °C due to the removal of surface

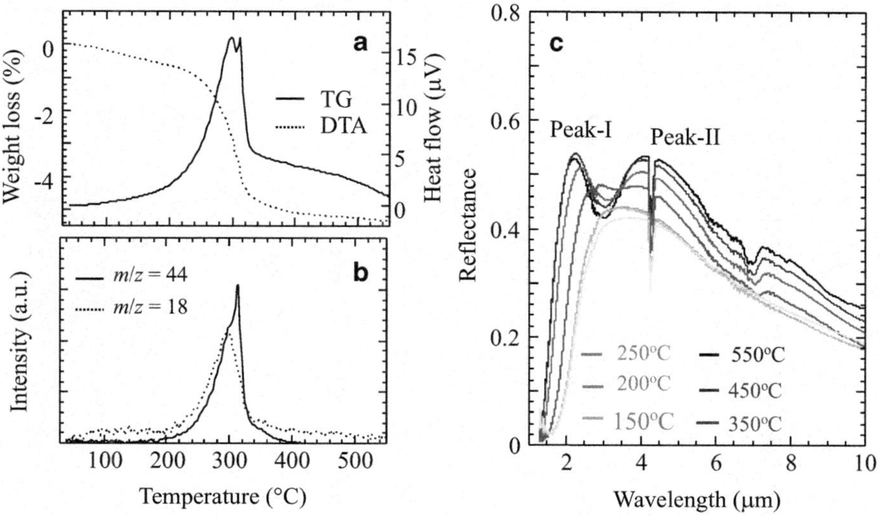

Fig. 8 (**a**) TG-DTA curves of ITO NP samples in an N_2 atmosphere. (**b**) TOF Mass spectroscopy combined with TG-DTA. *m/z* signals at 18 and 44 were detected in the range 27–550 °C. (**c**) Dependence of reflectance spectra on annealing temperature for a 216-nm-thick NP film

ligands. Therefore, the removal of surface ligand molecules from the NPs strongly affected the whole reflective phenomena, which also clarified the importance of nanoparticle gap in obtaining high reflectance performance.

4 Plasmon Coupling and Reflectance Mechanism

4.1 Polarized Spectroscopy

The character of *E*-field coupling in the NP films was investigated to clear mechanism of the resonant reflectance by polarized reflectance measurements. Two types of light polarized perpendicular (*s*-polarized) and parallel (*p*-polarized) to the plane of incidence were introduced the sample at an incident angle of 75°. Figure 9a shows polarized reflectance spectra for *s*- and *p*-polarization configurations for the 216-nm-thick NP film. In the case of *s*-polarization, peak I and peak II were simultaneously obtained since the electric vector of the radiation produced electron oscillations in NPs parallel to the in-plane of the sample. On the other hand, only peak I was found for the *p*-polarization case. The electric vector excites electron oscillations in NPs normal to the plane of the sample and suppresses *E*-field coupling along the in-plane direction. These results suggested that peak II was excited by the field coupling along the in-plane directions. In contrast, the reflectance at peak I was essential for *E*-field coupling along the out-of-plane direction. That is, it is considered that the near- and mid-IR reflectance of the NP sheets is attributed to the field

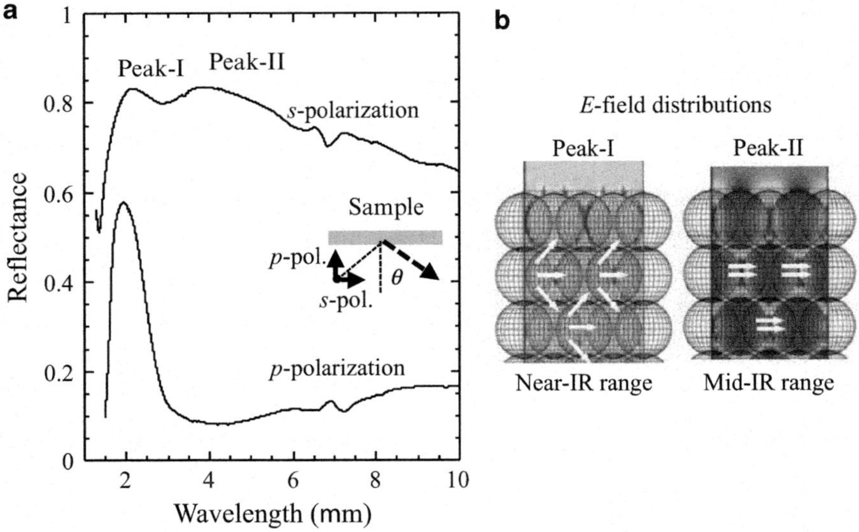

Fig. 9 (**a**) Reflectance spectra of a 216-nm-thick NP sheet for s- and p-polarized lights. Inset indicates the direction of the electric vector of incident light in addition to *E*-field distribution under *p*-polarization. (**b**) Images of *E*-field distributions and charge vectors at peak I and peak II

coupling along the out-of-plane and in-plane directions. The film thickness-dependent plasmon splitting was attributed to the formation of field coupling along the out-of-plane direction, thereby leading to enhanced reflectance in the near-IR range.

Plasmon coupling between NPs produces large enhancements of *E*-fields at interparticle gaps. We typically investigated the *E*-field distributions at peak I and peak II for a 20 NP layer with $d = 20$ nm. Figure 9b shows the *E*-field distributions along the $x-z$ (out-of-plane) direction at peak I and peak II. For peak II, the *E*-fields between the NPs were strongly localized along the *x*-direction when an electric field of light was applied along this direction. In contrast, peak I showed *E*-fields along the out-of-plane direction in addition to those along the in-plane direction. The difference in field distribution between peak I and peak II resulted in change in reflectance spectra of the ITO NP film under *s*- and *p*-polarizations. It was found that the reflectance mechanism at peak I (near-IR range) was different from that at peal-II (mid-IR range).

4.2 *Stretch-Induced Spectral Changes*

The *E*-fields at internanoparticle gaps in the NP films were important in determining the resonant reflectance in the IR range. It is known that plasmon coupling between NPs is sensitive to interparticle distance. Therefore, changes in the optical response

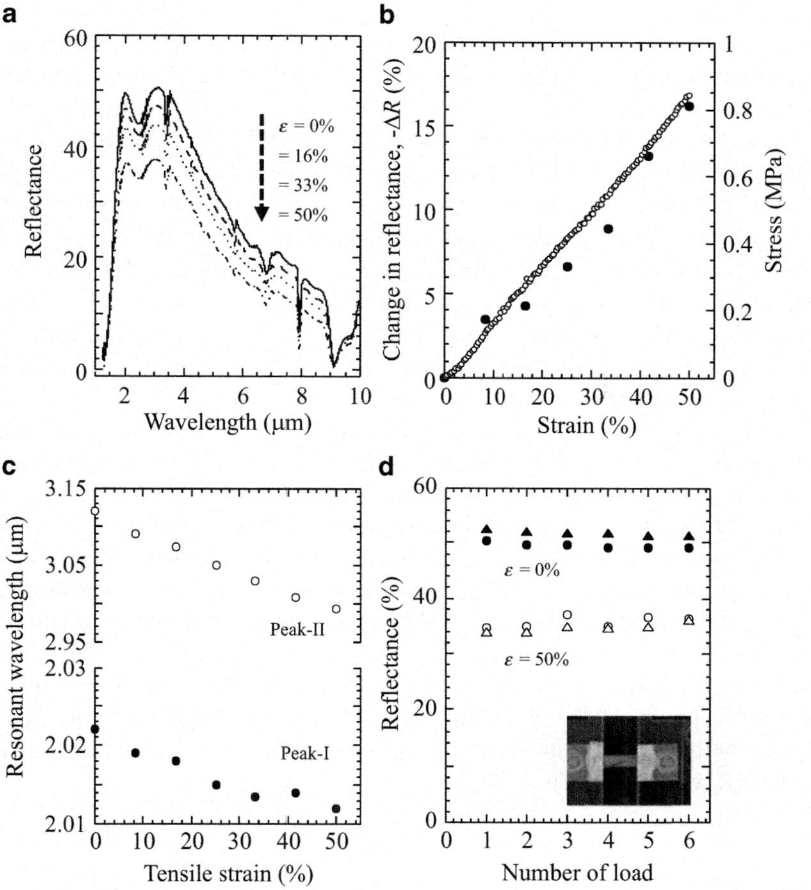

Fig. 10 (**a**) Reflectance spectra of 218-nm-thick ITO NP films ($d = 20$ nm) under different tensile strains ($\varepsilon = 0$, 16, 33, and 50%). (**b**) Reflectance change ($-\Delta R$) and strain as a function of tensile strain. (**c**) Dependence of resonant wavelengths of peaks I (red dots) and II (black dots) on tensile strain. (**d**) Repeated mechanical tests between two tensile stresses of $\varepsilon = 0$ and 50%. The inset shows a photograph of a mechanical test. An E-field was applied along the tensile direction. Light was incident along the direction orthogonal to the NP film. (Fig. 8 of [41]). (Copyright 2018 by the American Chemical Society)

at various strain levels were further investigated, which directly clarified the importance of plasmon coupling between NPs in the NP film. Figure 10a shows resonant reflectance spectra under uniaxial tensile strains from 0 to 50%. The incident light was polarized along the tensile direction. The NP film ($d = 20$ nm) with a thickness of 218 nm was deposited on an elastomeric PDMS substrate that allowed for active tuning of the interparticle gap. Stretching decreased in the resonant peaks when comparing the peaks of the reflectance spectra. A linear reflectance change ($-\Delta R$) occurred by varying the tensile strains, which was to the correlation of the strain and stress curve of the PDMS substrate [Fig. 10b]. A stretch-induced change in the reflectance of the substrate did not affect the plasmonic coupling. Clear blueshift trends followed by a decrease in reflectance were observed in the resonance wavelength with increasing strains, suggesting that peaks I and II were governed by dipole-type plasmon coupling [Fig. 10c]. These peak shifts were consistent with theoretical simulations. We confirmed that the FDTD calculations also showed blue-shifted resonant wavelengths (peaks I and II) with increasing interparticle distance. These results may be attributed to the decay of field strength between NPs due to increased interparticle separation under tensile strain. Spectral changes were further reproduced when the elastomeric substrate was relaxed back within the wide tensile strain range (0–50%). Long-term stability analysis on reversible stretching cycles of the NP sheet clearly demonstrated the effective mechanical tuning of the plasmonic resonance [Fig. 10d]. Mechanical strains of the NP sheets experimentally indicated the importance of plasmon coupling.

The resonant origins of the reflectance of peak I and peak II were theoretically examined as a function of interparticle distance between NPs, which provided further evidence of the importance of plasmon coupling from a theoretical viewpoint. Figure 11a shows the calculated reflectance of NP films with different l values on the basis of a 20 NP layer model, revealing that reflectance monotonically enhanced with decreasing l. Peak II showed a redshift to longer wavelengths with decreasing l from 10 to 1 nm. In contrast, peak I shifted only slightly as compared with peak II [Fig. 11b]. That is, the interparticle distance between NPs directly affected reflectance strength. These theoretical results could describe the experimental results of mechanical tuning of the NP films under the tensile strains. Therefore, the field interactions based on plasmon coupling between NPs played an important i enhancement of IR reflectance of the NP films.

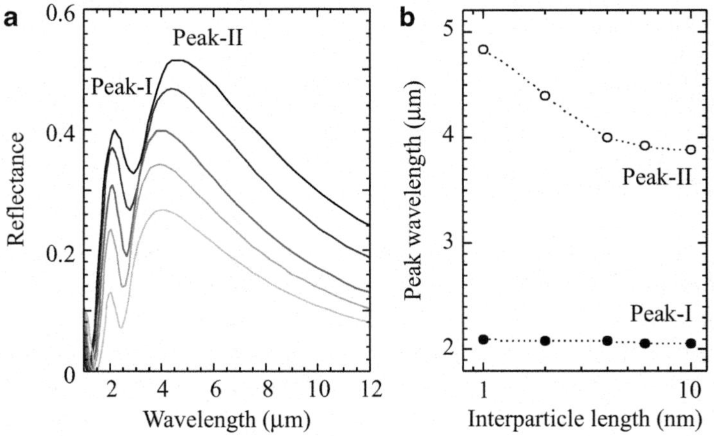

Fig. 11 (**a**) Calculated reflectance of NP sheets with different interparticle lengths. (**b**) Resonant peaks of peak I and peak II as a function of interparticle length. (Fig. 9 of [39]). (Copyright 2014 by the American Chemical Society)

5 Electromagnetic Properties of ITO NP Films

Electromagnetic (EM) properties are briefly discussed for a NP sheet in the microwave range from 0.5 to 40 GHz. This range is an important frequency range for telecommunications. Transparent solar thermal shielding is an effective technique to prevent room heat in order to realize comfortable environment in vehicles. However, there is a requirement for vehicles to be able to transmit EM waves in the microwave range through windows to effect radio communications, such as those necessary for Electronic Toll Collection (ETC) and general Traffic Information Systems (ITS). Therefore, it is important to measure the EM properties of NP sheets in addition to evaluating their optical properties in the IR range.

A 250-nm-thick NP sheet with A4 size was fabricated on a PET substrate (thickness: 0.2 mm) using a roll-coating method [inset of Fig. 12a]. High reflectance with a close proximity of 0.6 was also obtained on a flexible PET substrate [Fig. 12a]. The shielding effectiveness (SE) of the flexible NP sheet was almost zero, which differed from that of an RF-sputtered ITO film [Fig. 12b]. The difference between the two materials is related to electrical conductance (σ) in the sheets, which was in the order of 10^{-5} and 10^{-3} S/cm for the NP film and the sputtered film, respectively. If the shielding material is thin, SE is mainly dominated by EM reflection as follows:

$$SE = 20\log\left(\sqrt{2}\beta_0 / 2\rho\right) \tag{4}$$

where β_0^2 is μ_0/ε_0, and ρ is the sheet resistivity ($=1/\sigma$). The significant obstruction of electron carrier transport between NPs produced low electrical conductance because

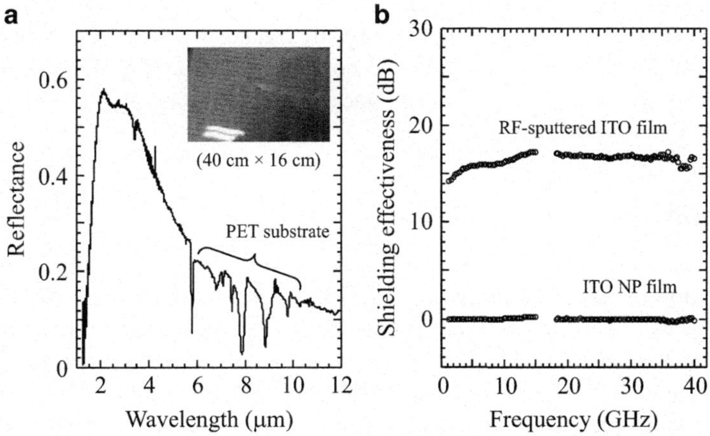

Fig. 12 (**a**) Reflectance of a NP sheet on a PET film. Inset image represents a photograph of the fabricated sheet sample. (**b**) Shielding effectiveness (SB) in the microwave range 0.5–40 GHz for a NP sheet and RF-sputtered ITO film. (Fig. 11 of [39]). (Copyright 2014 by the American Chemical Society)

of the presence of surface ligands on the NPs and allowed for high microwave transmissions.

ITO NPs were used to obtain assembled NP sheets with small interparticle lengths by the presence of ligand molecules on the particle surfaces. This situation produced effective *E*-field coupling along the in-plane and out-of-plane directions. This resulted in the plasmon hybridization for the quadrupole and dipole modes, which played an important role in producing the high reflectance in the near- and mid-IR range. In addition, the *E*-field enhancements between NPs simultaneously caused a remarkable reduction in electrical contacts between the NPs, contributing to the high microwave transmissions. Plasmonic control of assemblies of NPs represents promising potential in the area of structural and optical designs used to fabricate a flexible thermal shielding film with reflection-type based on transparent oxide semiconductors.

The EM properties in the microwave range are closely related to electron transport of ITO NP films. Electron transport of ITO NP films is attributed to tunneling and hopping conductance between NPs due to the presence of organic ligands on the NP surfaces. Figure 13a shows the temperature dependence of electrical resistivity for a 260-nm-thick ITO NP film. The electrical resistivity remarkably increased at low temperatures below $T = 100$ K, indicating a semiconducting behavior. The electrical properties of ITO NP films can be accounted for by Efros–Shklovskii variable hopping (VRH) transport, which can be expressed by $\rho(T) \propto \exp(T_0/T)^{1/2}$, where T_0 indicates the hopping temperature ($T_0 = Be^2/\varepsilon_r k_B \xi$, $B = 9.6$). k_B and ε_r are the Boltzmann constant and effective permittivity, respectively. Electron transport of an ITO NP film at 300 K was derived from Efros–Shklovskii VRH conductance with a hopping temperature of $T_0 = 320$ K [Fig. 13b]. Additionally, electron localization estimated from the hopping temperature was $\xi = 72$ nm. The electron carrier was confined into the nanoscale, which originated from the fact that the electronic state $g(E)$ of the conduction band related to electron transport exponentially decreased [Fig. 13c]. The presence of organic ligands on the NP surfaces affected the spatial separation between NPs and contributed to the high EM transmittance due to the decrease in electron conductance.

6 Summary

Investigation of the crystallinity and local structures of oxide semiconductor NPs (ITO NPs) were performed by XRD and TEM measurements as outlined in Sect. 2. Plasmonic resonances of ITO NPs were clearly obtained in the near-IR range with respect to optical and EELS signals. In particular, electron–impurity scattering contributed toward plasmon damping as one factor that is absent in metal NPs on the basis of theoretical and experimental approaches, which was discussed in Sect. 3. In Sects. 4 and 5, we described IR plasmonic applications in ITO NP sheets for solar thermal shielding technologies. Above all, *E*-field coupling between NPs produced interesting plasmonic coupling because of the creation of narrow crevices in the

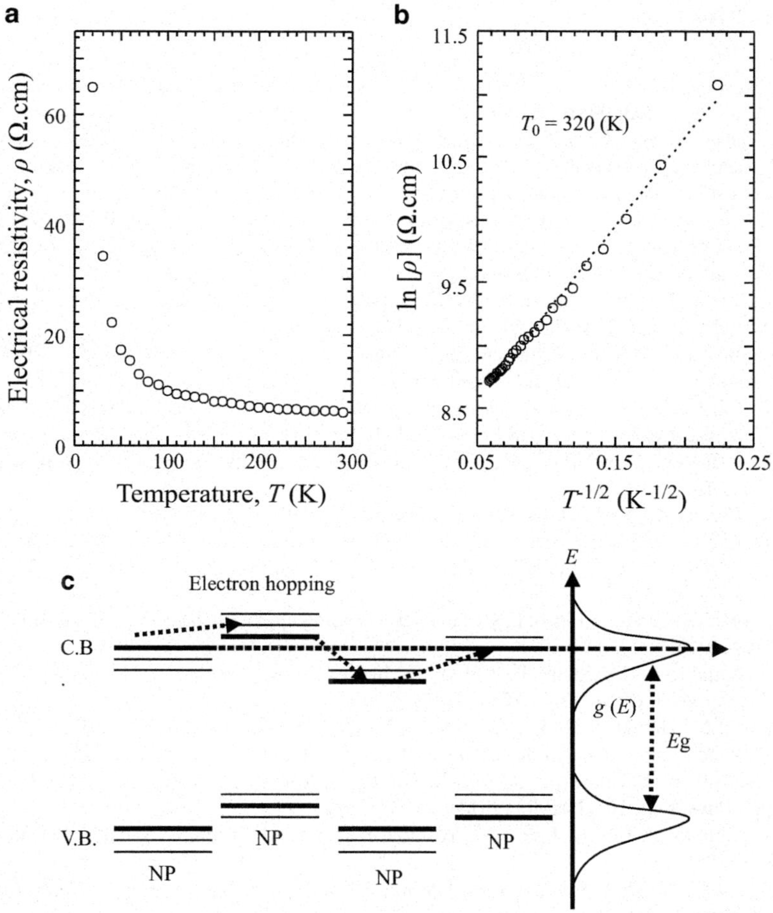

Fig. 13 (**a**) Temperature dependence of electrical resistivity of a 260-nm-thick ITO NP film. (**b**) Linear relationship between ln[ρ] and $T^{-1/2}$. T_0 indicates a hopping temperature. (**c**) Schematic picture of hopping conductance of an electron carrier between ITO NPs

interparticles. 3*D* field interactions along the in-plane and out-of-plane directions caused high reflectance in the near- and mid-IR regions. Finally, ITO NP films could be extended to generate large-sized flexible films on PET substrates, which simultaneously showed microwave transmittance essential for telecommunications.

The abovementioned results provide important insights for basic science and practical applications based on plasmonic investigations of oxide semiconductor NPs. However, further advancements are required in the area of plasmonic properties and applications. Future investigations will focus on plasmonic phenomena of oxide semiconductor NPs with a view to arriving at new concepts concerning optical manipulations in the IR range.

References

1. C. Rhodes, M. Cerruti, A. Efremenko, M. Losego, D.E. Aspnes, J.-P. Maria, S. Franzen, J. Appl. Phys. **103**, 093108 (2008)
2. H. Matsui, W. Badalawa, A. Ikehata, H. Tabata, Adv. Opt. Mater. **1**, 397 (2003)
3. W. Badalawa, H. Matsui, A. Ikehata, H. Tabata, Appl. Phys. Lett. **99**, 011913 (2011)
4. H. Matsui, A. Ikehata, H. Tabata, Appl. Phys. Lett. **106**, 019905 (2015)
5. E. Sachet, M.D. Losego, J. Guske, S. Franzen, J.P. Maria, Appl. Phys. Lett. **102**, 051111 (2013)
6. Y. Zhao, H. Pan, Y. Lou, X. Qiu, J.J. Zhu, C. Burda, J. Am. Chem. Soc. **131**, 4253 (2009)
7. G.V. Naik, V.M. Shalaev, A. Boltasseva, Adv. Mater. **25**, 3264 (2013)
8. G. Garcia, R. Buonsanti, E.L. Runnerstrom, R.J. Mendelsberg, A. Lordes, A. Anders, T.J. Richardson, D.J. Milliron, Nano Lett. **11**, 4415 (2011)
9. T. Yamada, H. Makino, N. Yamamoto, T. Yamamoto, J. Appl. Phys. **107**, 123534 (2010)
10. C. Sonnichesen, T. Franzl, T. Wilk, G. von Plessen, J. Feldmann, New. J. Phys. **4**, 93 (2002)
11. S.Q. Li, P. Guo, B. Buchholz, W. Zhou, Y. Hua, T.W. Odom, J.B. Ketterson, L.E. Ocola, K. Sakoda, R.P.H. Chang, ACS Photonics **3**, 163 (2014)
12. D.B. Tice, S.Q. Li, M. Taglazucchi, D.B. Buccholz, E.A. Weiss, R.P.H. Chang, Nano Lett. **14**, 1120 (2014)
13. J.M. Hamm, A. Pusch, O. Hess, Laser Photonics Rev. **9**, 256 (2015)
14. S.D. Lounis, E.L. Runnerstrom, A. Llordes, D.J. Milliron, J. Phys. Chem. Lett. **5**, 1564 (2014)
15. A.M. Schimpf, S.D. Lounis, E.L. Runnerstrom, D.J. Milliron, D.R. Gamelin, J. Am. Chem. Soc. **137**, 518 (2014)
16. D. Solis Jr., A. Paul, J. Plson, L.S. Slaughter, P. Swanglap, W.S. Chang, S. Link, Nano Lett. **13**, 4779 (2013)
17. N. Tognalli, A. Fainstein, E. Calvo, C. Bonazzola, L. Pietrasanta, M. Campoy-Quiles, P. Etchegoin, J. Chem. Phys. **123**, 044707 (2005)
18. M. Fan, M. Thompson, M.L. Andreade, G. Brolo, Anal. Chem. **82**, 6350 (2010)
19. A. Furube, T. Yoshinaga, M. Kanehara, M. Eguchi, T. Teranishi, Angew. Chem. **51**, 2640 (2012)
20. H. Matsui, W. Badalawa, T. Hasebe, S. Furuta, W. Nomura, T. Yatsui, M. Ohtsu, H. Tabata, Appl. Phys. Lett. **105**, 041903 (2014)
21. K. Okamoto, B. Lin, K. Imazu, A. Yoshida, K. Toma, M. Toma, K. Tamada, Plasmonics **8**, 581 (2012)
22. F. Le, D.W. Brandl, Y.A. Urzhumov, H. Wang, J. Kundu, N.J. Halas, J. Aizpurua, P. Nordlander, ACS Nano **2**, 707 (2008)
23. C.A. Gueymard, D. Myers, K. Emery, Sol. Energy **73**, 553 (2002)
24. A.A. Günay, H. Kim, N. Nagarajan, M. Lopez, R. Kantharaj, A. Alsaati, A. Marconnet, A. Lenert, N. Miljkovic, ACS Appl. Mater. Interfaces **10**, 12603 (2018)
25. M. Wang, Y. Gao, C. Cao, K. Chen, Y. Wen, D. Fang, L. Li, X. Guo, Ind. Eng. Chem. Res. **53**, 18462 (2014)
26. C. Nakamura, K. Manabe, M. Tenjmbayashi, Y. Tokura, K.H. Kyung, S. Shiratori, ACS Appl. Mater. Interfaces **10**, 22731 (2018)
27. J. Kim, G.K. Ong, Y. Wang, G. LeBlanc, T.E. Williams, T.M. Mattox, B.A. Helms, D.J. Milliron, Nano Lett. **15**, 5574 (2015)
28. C. Chen, Z. Wang, K. Wu, H. Chong, Z. Xu, H. Ye, ACS Appl. Mater. Interfaces **10**, 14886 (2018)
29. G. K. Dalapati, S. Masudy-Panah, S. T. Chua, M. Sharma, T. I. Wong, H. R. Tan, D. Chi, Sci. Rep. **6**, 20182 (2016)
30. X. Fang, C.L. Mak, J. Dai, K. Li, H. Ye, C.W. Leung, ACS Appl. Mater. Interfaces **6**, 15743 (2014)
31. I. Hamberg, C.G. Granqvist, J. Appl. Phys. **60**, R123 (1986)
32. S.Y. Li, G.A. Niklasson, C.G. Granqvist, Appl. Phys. Lett. **101**, 071903 (2012)
33. K. Katagiri, R. Takabatake, K. Inumaru, ACS Appl. Mater. Interfaces **5**, 10240 (2013)

34. P. Tao, A. Viswanath, L.S. Schadler, B.C. Benicewicz, R.W. Siegel, ACS Appl. Mater. Interfaces **3**, 3638 (2011)
35. H. Matsui, S. Furuta, H. Tabata, Appl. Phys. Lett. **104**, 211903 (2014)
36. Z. Sun, J. He, A. Kumbhar, J. Fang, Langmuir **26**, 4246 (2010)
37. G.V. Hartland, Chem. Rev. **111**, 3858 (2011)
38. A. Pflug, V. Sittinger, F. Ruske, E. Szyszka, G. Dittmar, Thin Solid Films **455/456**, 201 (2004)
39. H. Matsui, S. Furuta, T. Hasebe, H. Tabata, ACS Appl. Mater. Interfaces **8**, 11749 (2016)
40. S.M. Yang, S.G. Jang, D.G. Chio, S. Kim, H.K. Yu, Small **2**, 458 (2006)
41. H. Matsui, T. Hasebe, N. Hasuike, H. Tabata, ACS Appl. Nano Mater. **1**, 1853 (2018)

Part II
Nanostructured Perovskite Films

Flexible Organic Field-Effect Transistors Using Barium Titanate as Temperature-Sensitive Dielectric Layer

Suman Mandal and Dipak K. Goswami

1 Introduction

Over the last few decades, organic field-effect transistors (OFETs) have drawn significant interests among the researchers due to their potential applications in radio frequency identification tag (RFID) [1, 2], nonvolatile memory device [3–5], display technology [6, 7], integrated circuit [8–10], electronic skin [11, 12], and in various sensors and detectors [13–15]. Flexibility and solution processability make the devices unique for many cost-effective large-scale innovative wearable applications. However, these devices suffer from poor field-effect carrier mobility due to weak Van der Waals interaction between organic semiconducting molecules and the interface roughness. Moreover, the device performance depends on several other parameters, like Schottky barrier height, the thickness of the dielectric and semiconducting layers, annealing temperature, and growth temperature [16–19]. Many researchers have made efforts to improve device performance and the stability by proper selection of materials by optimizing the growth of various films. Besides, when OFETs are used as sensors, either the semiconducting or gate dielectric layer is used as a sensing layer [20–22]. In this work, we have used a temperature sensing material in the gate dielectric layer. OFETs-based sensors are efficient due to their multiparameter accessibility and amplification capability of the sensing signals, unlike chemisorption-based sensors. Here, we have focused on the fabrication of ultrafast, high-precession flexible temperature sensor through engineering the semiconductor/dielectric interface by controlling different growth parameters.

S. Mandal · D. K. Goswami (✉)
Organic Electronics Laboratory, Department of Physics, Indian Institute of Technology Kharagpur, Kharagpur, West Bengal, India
e-mail: dipak@phy.iitkgp.ac.in

A. G. Roca et al. (eds.), *Surfaces and Interfaces of Metal Oxide Thin Films, Multilayers, Nanoparticles and Nano-composites*,
https://doi.org/10.1007/978-3-030-74073-3_5

Temperature is one of the essential physiological parameters in the human body. Continuous monitoring of the body temperature is a preliminary step to track different health-related critical problems, such as febrile seizure and epilepsy, requiring more accurate measurement of body temperature at a fast rate [18–20, 23]. Even the precise monitoring of the body temperature change in women during the complete menstruation cycle can predict the ovulation time more accurately [24, 25]. Besides, the highly ambient-stable flexible temperature sensor can find applications in food and medicine preservation technology [26, 27]. Different types of flexible temperature sensors suitable for wearable applications are also demonstrated in the literature. Such sensors are mainly capacitive or resistive types, where a temperature-sensitive material is incorporating into the devices. The changes in capacitance or resistance due to temperature variation are measured to sense temperature [28–33]. However, to improve the sensitivity and precision, OFETs are introduced as another platform, mainly in different wearable applications. OFET integrated with thermistor also has been widely used for wearable temperature sensing applications [34–36]. However, the integration of thermistor can increase the complexity of the fabrication process. For the fabrication of the OFET-based temperature sensor, a temperature-sensitive material is used as a dielectric layer in the device fabrication. The semiconducting channel layer can also be used as sensing materials in OFETs [37, 38]. However, such sensors suffer from higher response and recovery time that limits their applications. Nevertheless, these devices are operated at high voltage which prevents its usage from wearable applications [34–38].

We have introduced hexagonal barium titanate nanocrystals (h-BTNCs) as temperature sensing materials for the fabrication of OFET-based temperature sensors. Barium titanate ($BaTiO_3$, BTO) is a ferroelectric material that exhibits high dielectric constant and low dielectric loss due to its spontaneous electric polarization. BTO is known to have different crystallographic structures, like rhombohedral, orthorhombic, tetragonal, and cubic at different temperature ranges. The dielectric constant of this material depends on the crystallographic structures. The difficulty to obtain these crystallographic phases is that high-temperature annealing process. It is not easy to grow smooth thin films with these materials as high-temperature annealing makes larger crystallite sizes and film grown with such materials make the film rough. Therefore, BTO in the crystalline phase has not been used for the fabrication of high-performance OFETs. In this work, we have synthesized the hexagonal nanocrystals of BTO within the amorphous phase. This material requires a low processing temperature (~80 °C). Besides, we can grow highly smooth interfaces and can fabricate OFETs with low operation voltage.

2 Experimental Section

Materials Synthesis The solgel method was adopted to synthesis the h-BTNCs solutions, where barium acetate and titanium butoxide were used as precursor materials. Glacial acetic acid and 2-methoxyethanol material were used as the solvent.

Initially, barium acetate was vigorously stirred to dissolved in the solvent for 3 h using a magnetic stirrer at room temperature. Titanium butoxide was added to the reaction mixture drop-by-drop by maintaining the stoichiometric ratio. The solution was stirred again for another 3 h at room temperature. Finally, the ambient-stable h-BTNCs solution was prepared and used as a dielectric material in OFETs.

OFETs Fabrications In conventional OFETs, two metal electrodes such as source and drain are used to connect the organic semiconductor (channel). The other electrode (gate) capacitively coupled through a dielectric layer is used to modulate the current that flows in between source to drain electrode through the channel. The OFETs were fabricated in bottom-gate-top-contact (BGTC) configurations on the glass as well as in the PET substrate. A thin film of aluminum with thickness 125 (±10) nm was deposited on a clean substrate using a thermal evaporation chamber ($\sim 4\times10^{-6}$ mbar). The aluminum surface was anodized at room temperature by using an anodic oxidation method to get a 15 nm thin alumina film used together with h-BTNCs as a bilayer dielectric system of the devices. In this anodization process, citric acid was used as electrolyte material and platinum wire as a cathode material. Source meter (Keithley 2450) was connected to the anode and cathode terminal to provide a constant current density for this anodic oxidation process. After anodization, the anodic surface was flushed with DI water and subsequently dried using hot air. The h-BTNCs sol was spin-coated on top of alumina surface at 3000 rpm for 60 s and subsequently annealed at various substrate temperatures (from 40 °C to 120 °C) for 90 min to get a thin film of around 50 nm thickness. After that, a pentacene film of thickness around 20 nm was deposited under a high vacuum condition ($\sim 2\times10^{-6}$ mbar) using organic molecular beam depositing (OMBD) system. Finally, Copper (Cu) film with thickness 60 nm was deposited on top of pentacene film as a source–drain electrode through a shadow mask to get the channel Length (L) and channel width (W) of 30 μm and 2 mm, respectively.

2.1 *Device Characterizations*

Atomic force microscopy (AFM, Bruker) was used to characterize the surface morphology of each layer of the devices. The thickness of each layer was measured using a 3D surface profilometry. The devices' capacitance and leakage current were characterized in a simple metal–insulator–metal (MIM) structure using a LCR meter (Keysight) and source meters (Keithley 2450). Finally, the variation of drain current with temperature was carried out inside a custom design probe station using two source meters (Keithley 2450). The typical current resolution of the systems is around 1 pA in 1 mA scale and 10 pA in 10 mA current scale.

The field-effect carrier mobility of the devices was extracted from the saturation region by using the following equation.

$$I_D = \frac{W}{2L}.C.\mu.(V_G - V_T)^2 \quad (1)$$

where I_D, C, μ, V_G, and V_T are the drain current, the capacitance of the oxide layer, mobility, gate voltage, and threshold voltage, respectively.

3 Results and Discussions

3.1 Characterization of Temperature-Sensitive h-BTNC Dielectric Material

The dielectric material plays a significant role in deciding the device's performance. The gate capacitance is dependent on the thickness of the dielectric layer and the dielectric constant of the material. The operating voltage is directly related to this parameter. For OFETs, the charge conduction occurs mostly through the semiconductor dielectric interface. Therefore, the roughness of the dielectric layer defines the conducting channel. Besides, the leakage current of the device is decided by the quality of the dielectric film. We have used the solution-processable h-BTNCs materials as one of the dielectric layers of the bilayer dielectric system. The h-BTNCs sol was spin-coated on alumina film and subsequently annealed at 80 °C to prepare a thin film. The elemental composition of the film was analyzed from x-ray photoelectron spectroscopy (XPS). The emission spectra of Ba 3d, Ti 2p, and O 1s state from the h-BTNC thin film are shown in Fig. 1a–d. The corresponding binding energy as observed from Ba3d5/2, Ba3d3/2, Ti2p3/2, Ti2p1/2, and O1s are 778.99 eV, 794.22 eV, 457.53 eV, 463.25 eV, and 529.02 eV, respectively. The values are summarized in Table 1. The presence of residual carbon contamination also has been observed in the O1s spectra at 530.26 eV [19, 39].

High-resolution transmission electron microscopy (HRTEM) measurements were carried out to study the structural information of h-BTNC films. We have observed hexagonal nanocrystalline flakes within the amorphous phase. The low magnification bright-field TEM image of the BTO film is shown in Fig. 2a [19]. The inset Fig. 2a shows the crystallites present in the h-BTNC film taken from Fig. 2a. The typical dimensions of the crystallites are about ~ 50 nm in length and 5–10 nm in diameter. The Fig. 2b and 2c present the selected area diffraction pattern (SAD) taken from the circular regions, marked by I and J, which indicate the amorphous phase of the film. The indexing of the SAD pattern has been done by using the lattice parameters of hexagonal close-packed structure (lattice parameters, a = 5.73 Å, and c = 14.01 Å). The corresponding interplanar spacings (d-spacing) observed from the SAD pattern are 3.50 Å and 2.05 Å, and these are corresponding to (004) and (106) plane of hexagonal barium titanate (JCPDS # 89-8119). The HRTEM image of h-BTNC film is shown in Fig. 2d, and the inset Fig. 2d shows the Fourier-filtered image taken from a dotted region in Fig. 2a. The measured d-spacing

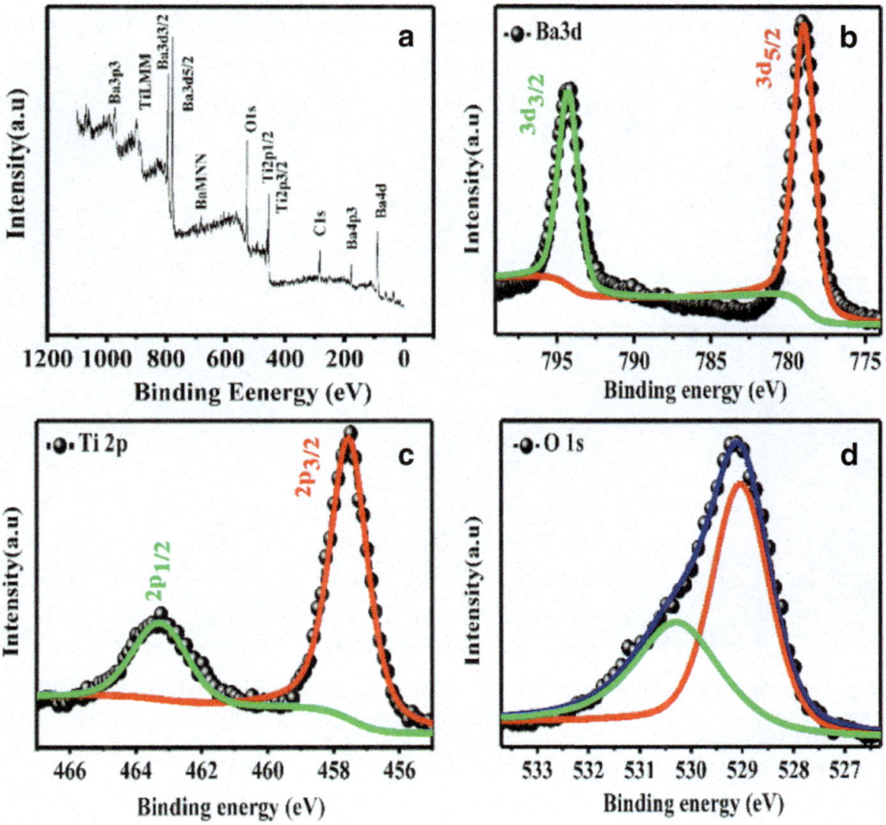

Fig. 1 XPS spectra taken from the h-BTNCs film. (**a**) presents survey scan, and (**b**), (**c**), and (**d**) depict the high-resolution scan of Ba3d, Ti2p, and O1s, respectively. (Reprinted from [19] © 2019 American Chemical Society)

Table 1 Binding energy of various elements presents in h-BTNC film

	States	Binding energy (eV)
Ba	$3d_{5/2}$	778.99
	$3d_{3/2}$	794.22
Ti	$2p_{3/2}$	457.53
	$2p_{1/2}$	463.25
O	1s	529.02

calculated from the HRTEM image is around 3.50 Å that corresponds to the (004) plane.

Characterization of Pentacene Layer Pentacene film grown on BTO film has been used as a semiconducting layer. The molecular ordering of pentacene molecules significantly depends on the roughness of the surfaces on which the film is

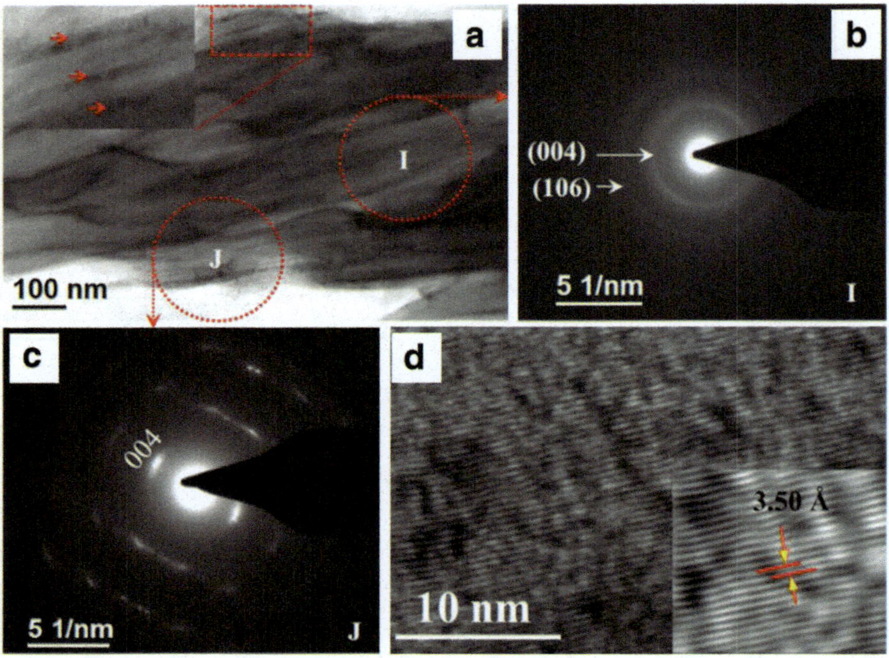

Fig. 2 (**a**) Presents the TEM image of h-BTNC films. (**b**) and (**c**) shows the selected area diffraction pattern (SAD) of it from the circular regions I and J, respectively. (**d**) presents the corresponding HRTEM image with a Fourier-filtered image in the inset taken from a dotted box marked in (**a**). (Reprinted from [19] © 2019 American Chemical Society)

growing. The formation of crystalline dendrite structure for the pentacene films is typical. However, it has been shown that pentacene films become amorphous if the surface roughness of the substrate is very high [40, 41]. X-ray diffraction (XRD) measurements were carried out on pentacene/h-BTNC/Al_2O_3/Al/glass and pentacene/h-BTNC/Al_2O_3/Al/PET films. The corresponding diffraction patterns are shown in Fig. 3a–b, respectively. The inset of the figure shows a narrow scan, where 2θ was varying from 5 to 20 degrees. The dendrite structures have two different crystallographic phases, such as thin film and bulk phases. The crystallite size of the thin-film phase and bulk phase has been calculated from Fig. 3c and 3d,, which are the first peak of Fig. 3a and 3b, respectively. The peak (001') and (001) correspond to the thin-film and bulk phase of the pentacene film's polymorphous nature [42]. We have calculated the mean crystallite size (τ) using the Scherrer equation for both films. The crystallites size for the thin-film phase was 9.22 Å and 11.88 Å for the films grown on h-BTNC/Al_2O_3/Al/ glass and h-BTNC/Al_2O_3/Al/ PET, respectively, as given in Table 2.

It has been found that the pentacene films grown on h-BTNC/Al_2O_3/Al/ PET have higher τ in comparison to the samples grown on the glass substrates. It implies that the PET substrate-based film has higher grain boundaries. Therefore, the grain

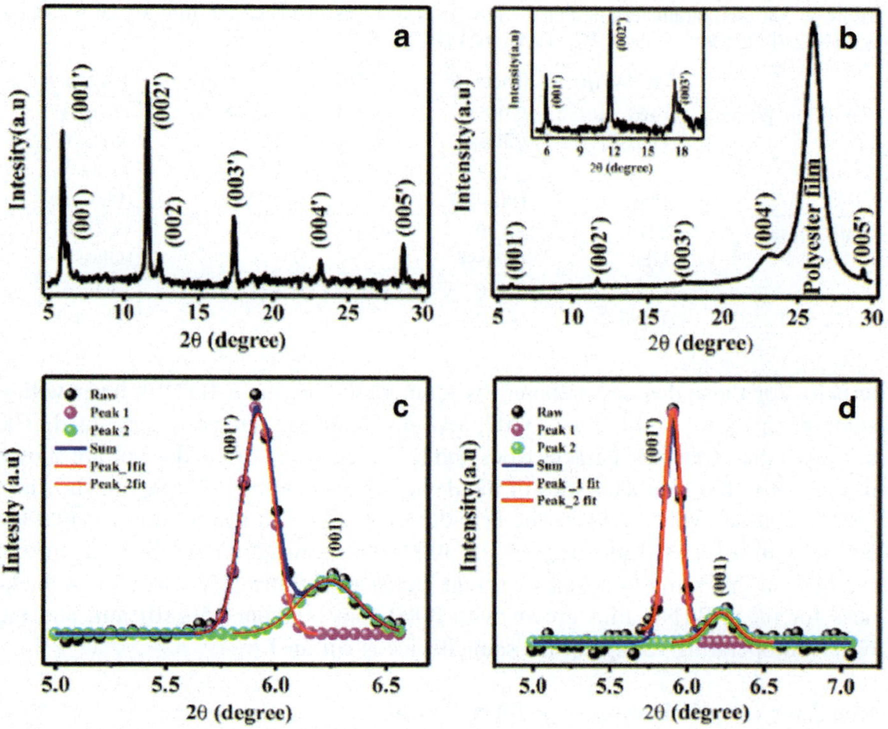

Fig. 3 (**a**) and (**b**) show x-ray diffraction pattern from pentacene/$BaTiO_3$/Al_2O_3/Al/glass and pentacene/$BaTiO_3$/Al_2O_3/Al/PET, respectively. The inset Figure shows a narrow 2θ scan from 5 to 20 degrees; (**c**) and (**d**) depict the first peak corresponding to the (001') and (001) plane of Figure (**a**) and (**b**), respectively. (Reprinted from [19] © 2019 American Chemical Society)

boundary-induced density of defect states is expected to be less in PET substrates. Besides, the higher d-spacing (~14.95 Å) for the thin-film phase indicates that the tilt angle of pentacene molecules is more elevated than in the bulk phase [42].

4 Optimization of Growth of h-BTNC Films

Thickness Optimization of h-BTNC Film The charge accumulation at the conducting channel depends on the gate dielectric capacitance, which crucially depends on the layer's thickness. The device's output current decreases for a particular applied gate bias by increasing the thickness of the gate dielectric. However, a thin dielectric layer increases the leakage current of the device. We have optimized the thickness of the dielectric layer to achieve higher carrier mobility. The film's thickness has been varied by varying the spinning speed of spin coater from 1000 rpm to 4000 rpm. Following the growth, we annealed the films at 80 °C for 90 min. These films were used for the fabrication of the OFETs. The output charac-

Table 2 Various parameters obtained from XRD analysis of pentacene film grown on top of h-BTNC/Al_2O_3/Al/glass and h-BTNC/Al_2O_3/Al/PET

	FWHM (β)	Theta (θ)	τ (glass)	τ (plastic)	d_{001} spacing
Thin-Film phase	0.167 (Glass)	2.955 (Glass)	9.22 Å	11.88 Å	14.95 Å (Glass)
	0.129 (Plastic)	2.953 (Plastic)			14.97 Å (Plastic)
Bulk phase	0.313 (Glass)	3.130 (Glass)	4.93 Å	7.37 Å	14.12 Å (Glass)
	0.209 (Plastic)	3.110 (Plastic)			14.12 Å (Plastic)

teristics for those devices prepared by spin coating of the h-BTNCs layer with a spinning speed of 1000, 2000, 3000, and 4000 rpm are shown in Fig. 4a–d. The corresponding transfer characteristics and leakage current of the devices are shown in Fig. 4e–f. The characteristics of the devices grown with 3000 rpm are shown in Fig. 4c and 4e, which delivers the best device property. It has been found that the device with h-BTNC films prepared at 3000 rpm shows minimum leakage current but beyond that rpm, the leakage current increases significantly. The typical thickness for the h-BTNC film grown with 3000 rpm is about 55(±10) nm, and the observed operating voltage of the same device is within 1 V.

Roughness Optimization of h-BTNC Films The surface roughness (σ) of the dielectric layer can introduce traps in the conducting channel. Therefore, it is necessary to grow a dielectric film with lower σ. We have optimized the σ of the film to achieve better charge conduction through the channel. Nevertheless, pentacene film's growth on the h-BTNC layer crucially depends on the σ [43, 44]. To control the σ, we have varied the annealing temperature of the film. The surface morphology of this film at different annealing temperatures is shown in Fig. 5a–d. The measured surface roughness at 40 °C, 80 °C, 120 °C, and 160 °C is 0.5 nm, 0.7 nm, 3.0 nm, and 5.3 nm, respectively. We observed that the σ of the film significantly increased at 120 °C and above annealing temperature. In this work, we have considered the annealing temperature as 80 °C, which is also the pentacene film's growth temperature.

Frequency and Temperature-Dependent Capacitance Measurements In this section, we have discussed the effect of frequency and temperature on the gate capacitance. A simple metal–insulator–metal (MIM) structure was fabricated along with the OFET to study the capacitance of the gate dielectric layer. The variation of capacitance with frequency in the bilayer dielectric system is shown in Fig. 6a. We have covered the frequency range from 10 kHz to 500 kHz. The capacitance of the device is almost constant up to about 200 kHz. However, it decreased at a relatively faster rate in the frequency range above 200 kHz. Therefore, the devices can be operated within 200 kHz frequency range with constant capacitance value. The dielectric property of h-BTNC material is sensitive to temperature. Consequently,

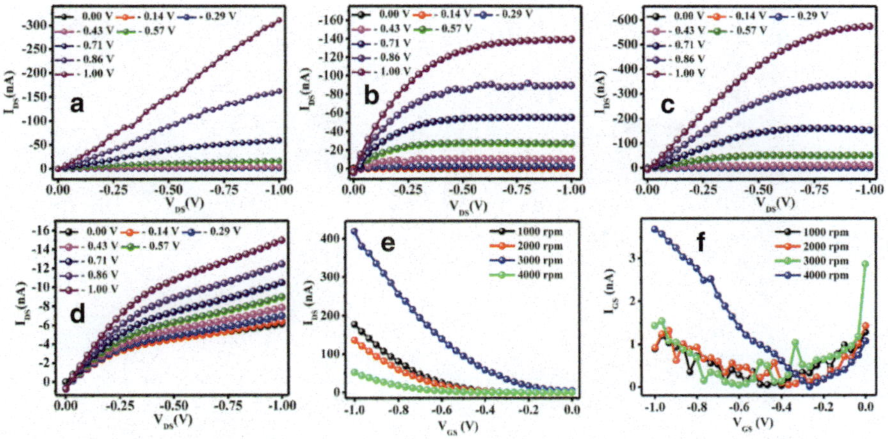

Fig. 4 Thickness optimization of the h-BTNC layer. The output characteristics of the h-BTNC layer deposited at the various spinning speed of (**a**) 1000, (**b**) 2000, (**c**) 3000, and (**d**) 4000 rpm. (**e**) and (**f**) depict transfer characteristics and leakage current of the device, respectively

we measured the temperature-dependent capacitance of the bilayer dielectric system within the temperature range from 20 °C to 50 °C at 100 kHz, as shown in Fig. 6b. The linear temperature dependency has been observed over a range of temperatures from 20 °C to 50 °C with a slope of 2.67 (±0.2) $\times 10^{-2}$ (nF/cm^2)/°C. This temperature dependency of the h-BTNC film can be exploited to the OFET-based temperature sensors.

5 Optimization of Pentacene Film for OFETs Fabrication

In this section, we have discussed the optimization of pentacene film growth in terms of substrate temperature and film thickness to achieve a better device performance.

Effect of Substrate Temperature on the Growth We have carried out detailed studies of the growth of pentacene film at different substrate temperatures. We have used an organic molecular beam deposition (OMBD) system to grow pentacene films with a fixed growth rate of 3 Å/s at various substrate temperatures, such as 60 °C, 80 °C, 100 °C, and 120 °C. The output and transfer characteristics of typical OFETs grown on glass substrates are shown in Fig. 7a–d and 7e–h. It has been found that at 80 °C, the device offers better performance in terms of device current. We observed maximum carrier mobility for the device prepared at 80 °C substrate temperature. The variation of the device carrier mobility with growth temperature is shown in Fig. 7i.

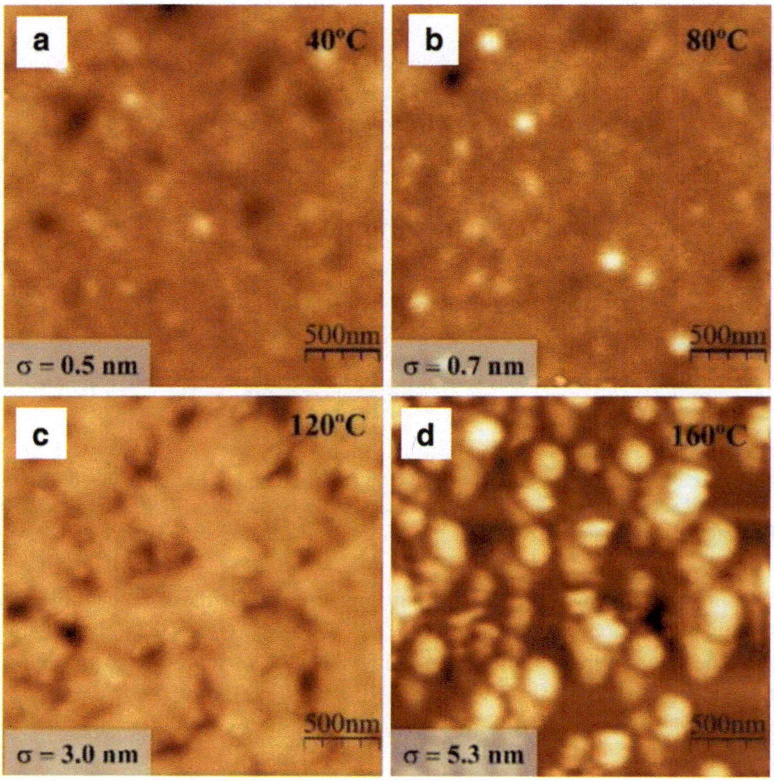

Fig. 5 Presents the effect of annealing on the surface roughness (σ) of h-BTNC film at different temperatures (**a**) 40 °C, (**b**) 80 °C, (**c**) 120 °C, and (**d**) 160 °C on Al_2O_3/Al/glass. A sudden increase in σ is observed at 120 °C annealing temperature

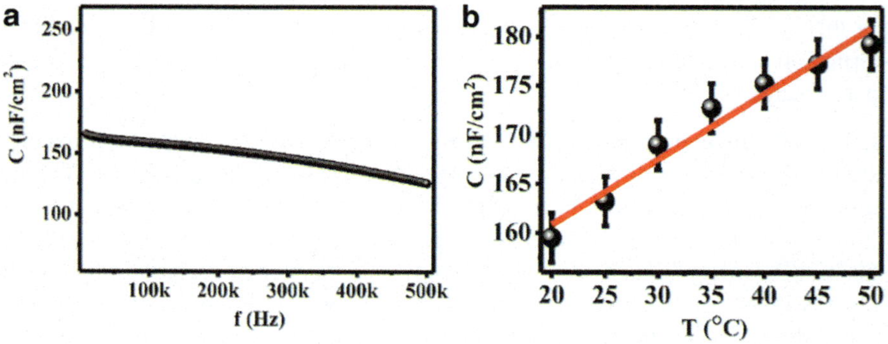

Fig. 6 Presents the capacitance variation of the bilayer dielectric system with (**a**) frequency from 10 kHz to 500 kHz at room temperature and (**b**) temperature from 20 °C to 50 °C at 100 kHz. (Reprinted from [19] © 2019 American Chemical Society)

Effect of Pentacene Film Thickness The OFETs were fabricated using bottom-gate-top-contact configuration. The active region for charge conduction comprises about a few nm of the dielectric/semiconductor interface. The channel resistance of the device increases with the thickness of the pentacene film. Figure 8a shows the schematic design of a specific device. The thickness of the pentacene layer has been varied from 10 nm to 60 nm during the growth. The devices' transfer characteristics with various thicknesses of pentacene film are presented in Fig. 8b. We have found that the device mobility increases with decreasing the thickness of the pentacene layer up to a thickness of 20 nm. As the thickness increases, the resistances R_1 and R_2 increase as electrons have to travel longer path. Besides, the trap density across the film increases with the thickness. Therefore, charge injection from the contact to the conducting channel is reduced by increasing the thickness of the pentacene film. We have extracted the device mobility from these transfer characteristics. The thickness-dependent mobility variation is shown in Fig. 8c. We have observed that the device's mobility drastically reduced with a pentacene film of thickness above 20 nm. This could be due to poor charge injection to channel from the metal contacts. Therefore, we have considered the thickness of the pentacene layer as 20 nm.

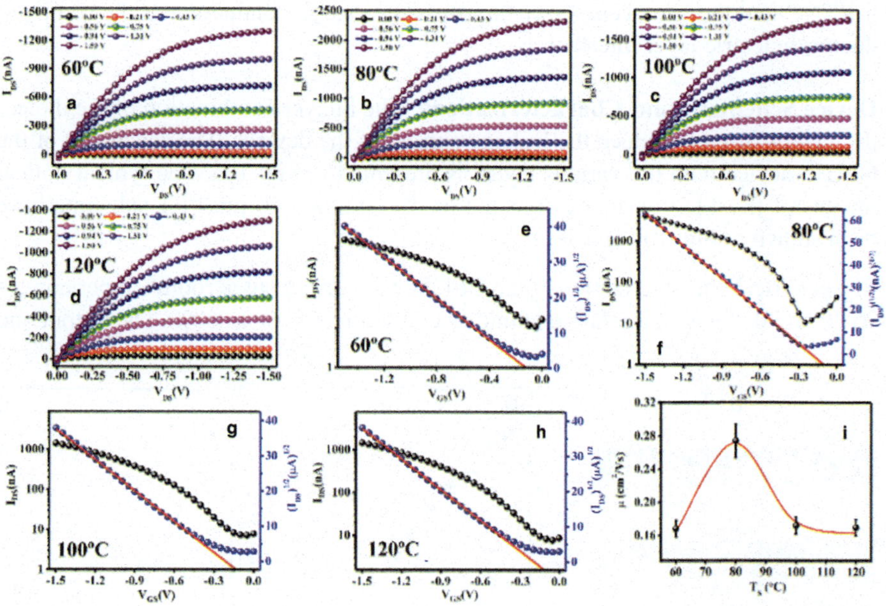

Fig. 7 Substrate temperature optimization during the growth of pentacene films. (**a–d**) and (**e–h**) present output and transfer characteristics for substrate temperature variation from 60 °C to 120 °C with an interval of 20 °C, respectively. (**i**) depicts substrate temperature-dependent mobility variation of those devices

6 Flexible Organic Field-Effect Transistors and Its Characterization

Surface Morphology Characterization of Each Layer The surface morphologies of Al, Al_2O_3/Al, h-BTNC/Al_2O_3/Al, and pentacene/h-BTNC/Al_2O_3/Al grown on the glass and PET substrates have been characterized by using AFM images. The typical surface morphologies of these surfaces are shown in Fig. 9a–d for glass substrates and Fig. 9e–h for PET substrates. The standard root-mean-square (RMS) roughness (σ) of thermally evaporated Al film on glass and PET surfaces is ~ 9.7 and ~ 2.6 nm, respectively, whereas the σ of the Al_2O_3 layer, grown by anodization of Al films, was increased to 11.5 and 3.9 nm for glass and PET substrates, respectively. The h-BTNC film has been spin-coated and subsequently annealed for 90 min at 80 °C for glass and 60 °C for PET substrates. Therefore, the spin-coated h-BTNC films on the rough Al_2O_3 surfaces showed a significant reduction in σ to 0.7 nm and 0.4 nm for glass and PET substrates, respectively. Table 3 summarizes the σ of each of the layers present in glass and PET substrate-based OFETs. The morphology of pentacene films follows the dendrite structures, as observed on any other inert substrate surfaces [40, 41]. The σ of pentacene film was found to be ~7 nm and 6.3 nm for glass and PET substrate, respectively. However, the smooth interface between h-BTNC and the pentacene layer improves the charge conduction at the semiconductor/dielectric layer interface.

Device Schematic and Characterizations The bilayer combination of the dielectric layer is used to reduce the leakage current of the devices. The schematic of the OFETs device with the various layers is presented in Fig. 10a. The typical optical images of flexible transistors prepared on 100 μm and 10 μm PET substrate have been shown in Fig. 10b and 10c.

The electrical characterization of the OFETs was carried out at room temperature (~ 25 °C to 30 °C) with a relative humidity condition of 40% to 60%. The output and

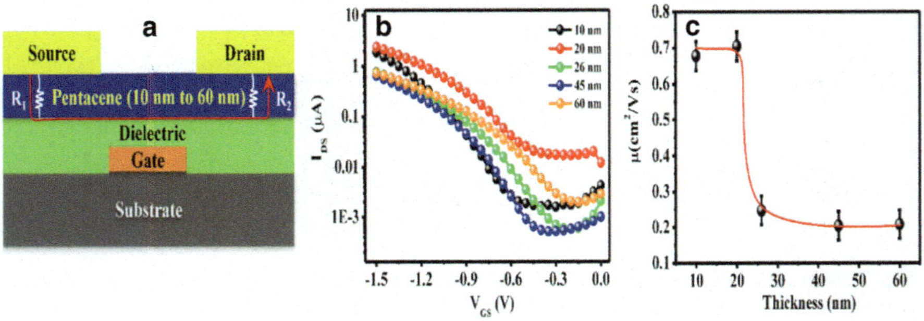

Fig. 8 Thickness dependence of the pentacene layer in OFETs. (**a**) presents the schematic of the device where the thickness of the pentacene layer varied from 10 nm to 60 nm. (**b**) shows the pentacene film thickness-dependent transfer characteristics. (**c**) depicts mobility variation with the thickness of the pentacene layer

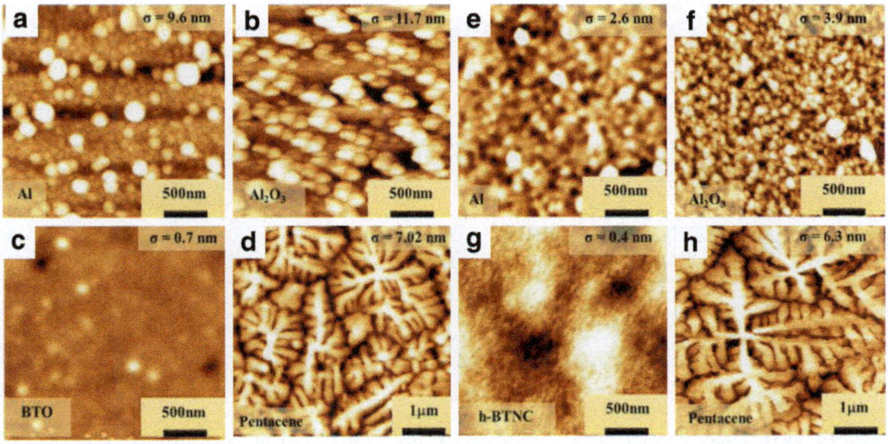

Fig. 9 Surface topology of different layer in OFETs (**a**) Al, (**b**) Al_2O_3/Al, (**c**) h-BTNC/Al_2O_3/Al and (**d**) pentacene/h-BTNC/Al_2O_3/Al film on glass substrate (**e**) Al, (**f**) Al_2O_3/Al, (**g**) h-BTNC/Al_2O_3/Al, and (**h**) pentacene/h-BTNC/Al_2O_3/Al film on PET substrate. (Figure (**e**–**g**) reprinted from [19] © 2019 American Chemical Society)

Table 3 Surface roughness (RMS) of Al, Al_2O_3/Al, h-BTNC/Al_2O_3/Al, and pentacene /h-BTNC/ Al_2O_3/Al film deposited on glass and PET substrates

Substrate	Al (nm)	Al_2O_3(nm)	h-BTNC (nm)	Pentacene (nm)
Glass	9.7	11.5	0.7	7.0
PET	2.6	3.9	0.4	6.3

transfer characteristics of the best devices fabricated on glass and PET (100 μm) substrates are shown in Fig. 11a,e and 11b, e, respectively. Device parameters, such as field-effect mobility (μ), the threshold voltage (V_{th}), and on–off ratio, were calculated from the transfer characteristics. The typical values of these parameters are found to be 0.89(±04) cm^2/Vs, –0.99(±09) V and 10^3, and 1.46(±09) cm^2/Vs, –1.05(±06) V and 10^3 for glass and PET substrate-based OFETs, respectively. The observed operating voltage is about 1.5 V, sufficiently lower and suitable for wearable applications for both set devices. The OFETs fabricated on PET substrates have shown higher carrier mobility. Due to the lower σ of the dielectric layer, the pentacene films exhibit higher crystallinity at initial stage of film growth. The synthesis of h-BTNC sol has been optimized so that we can use it as a dielectric ink for screen printing technology. We have fabricated OFETs on PET substrate using screen printing h-BTNC film on the alumina surfaces. The output and transfer characteristics of screen-printed dielectric-based OFETs are shown in Fig. 11c, f. The mobility, threshold voltage, and on–off ratio calculated from transfer characteristics are 0.70 (±08) cm^2/Vs, –1.26 (±06) V, and 10^3, respectively. For these devices, the carrier mobility decreases with a slight increase in threshold voltage due to the dielectric layer's thickness variation during screen printing. It implies that the

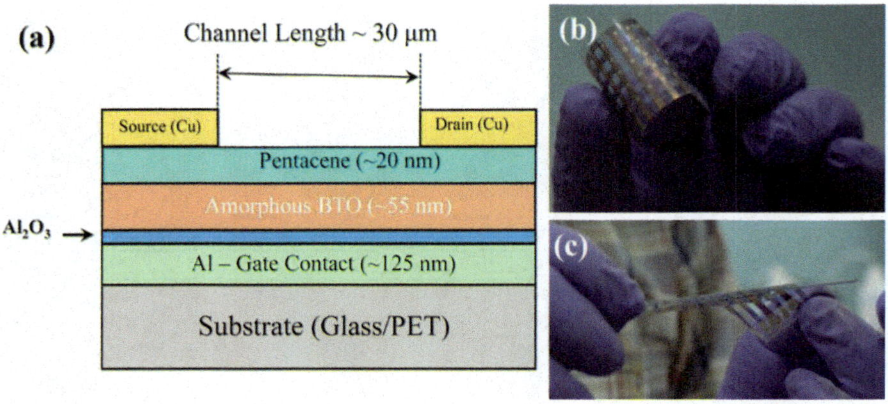

Fig. 10 (**a**) presents the schematic of OFETs; (**b**) and (**c**) depict pictures of the flexible devices on 100 μm and 10 μm thick PET substrates, respectively

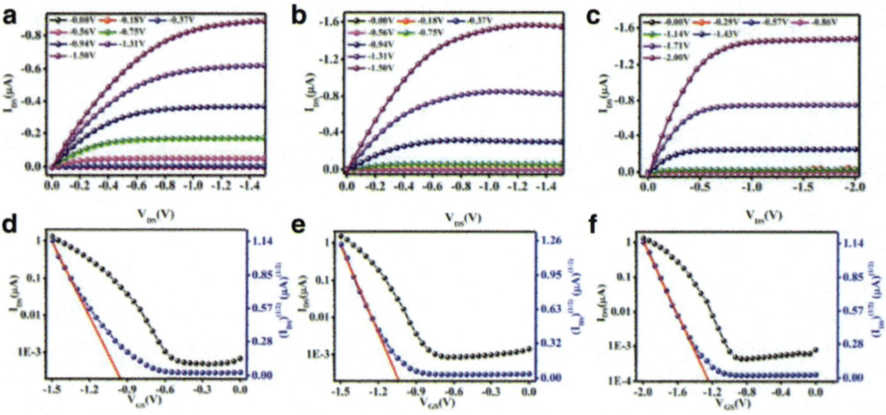

Fig. 11 (**a**) and (**b**) present output, and (**d**), (**e**) depict transfer characteristics of OFETs on glass and PET substrate, respectively. (**c**) and (**f**) show the screen-printed $BaTiO_3$ film-based OFET on PET substrate measured in ambient condition

h-BTNC sol is also a promising material as a dielectric ink for future printed electronics.

We have fabricated and characterized about over 200 devices on each of the glass and PET substrates. The mobility distribution of these devices is shown in Fig. 12a, b for these devices. We have found that most of the devices offer mobility around ~0.2 cm^2/Vs and ~0.25 cm^2/Vs for glass and PET substrate-based devices, respectively.

Device Stability in Ambient Condition The devices showed exceptional ambient stability. The carrier mobility of a single OFETs, kept under ambient conditions at a temperature between 25–30 °C and relative humidity of 50–70%, was monitored over 8 months. We selected a better device with 0.5 cm^2/Vs carrier mobility for this

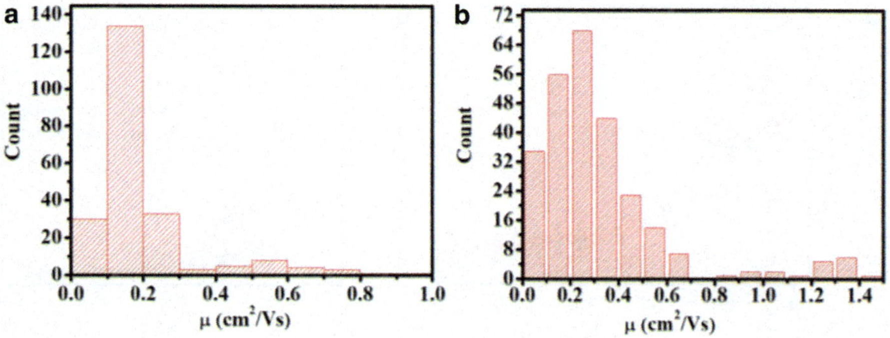

Fig. 12 Histogram of the mobility distribution of devices on top of (**a**) glass substrates and (**b**) PET substrates

study. The variation of carrier mobility with time in days is shown in Fig. 13a. We have shown the corresponding transfer characteristics for this device in Fig. 13b. We observed about 36% reduction in carrier mobility to 0.2 cm^2/Vs at the end of 120 days with a degradation rate of 26.90(±2.48) $\times 10^{-4}$ cm^2/Vs-day. However, the observed degradation rate significantly reduced from 120 days to 240 days. Such device degradation could be due to the continuous exposure of different gaseous molecules in ambient conditions and the oxidation of metal contacts.

Device Flexibility Test The flexibility of devices is a crucial requirement for many wearable applications [45]. The PET substrate-based OFETs are mechanically flexible. The device characteristics of the encapsulated OFETs fabricated on 10 μm PET substrates have been measured under tensile and compressive bending conditions in ambient conditions. The respective transfer characteristics of devices are shown in Fig. 14a and 14b. The variation of device mobility with different bending radius is presented in Fig. 14c. The device field-effect mobility is the same up to 4 mm bending condition. However, the devices work until the bending radius of 2 mm at the cost of about 60% reduction in carrier mobility. The device's performance during bending is degraded due to the formation of defect states or cracks generated on bending conditions [46].

Effect of Temperature on Device Performances The OFETs have been developed for sensing temperature. To check the temperature dependency, we have taken transfer characteristics at various temperatures from 20 °C to 50 °C with an interval of 5 °C. The corresponding transfer curves for glass and PET substrate-based devices have been shown in Fig. 15a and b, respectively. The I_{DS} increases linearly with temperature. Such variation of device current for maximum V_{GS} and V_{DS} (for glass $V_{GS} = -1.0$ V and $V_{DS} = -1.0$ V and PET substrate $V_{GS} = -1.5$ V and $V_{DS} = -1.5$) with temperature is shown in Fig. 15c and15d. The measured temperature range covers the human body temperature. Therefore, the OFET-based temperature sensors can be used for monitoring body temperature. The thermal sensitivity of the

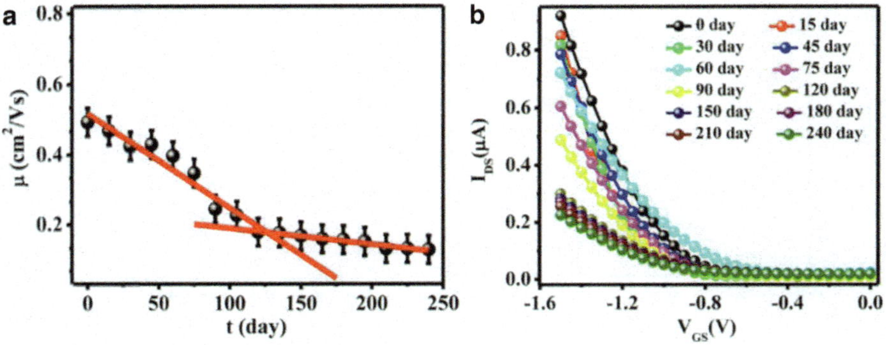

Fig. 13 presents the time-dependent (**a**) mobility and (**b**) transfer characteristics variation in days. (Reprinted from [19] © 2019 American Chemical Society)

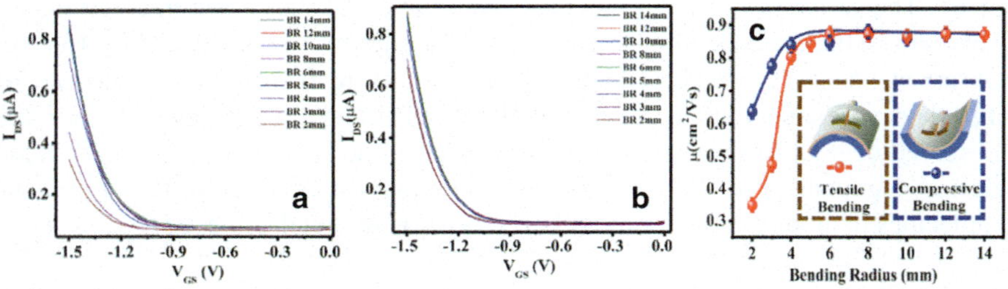

Fig. 14 shows transfer characteristics of the device (**a**) under tensile bending condition and (**b**) under compressive bending condition. (**c**) depicts the variation of device mobility during tensile as well as compressive bending conditions. (Reprinted from [19] © 2019 American Chemical Society)

sensors calculated from the slope of Fig. 15c and 15d for glass and PET substrates-based device is 36.8 (±2.9) nA/°C and 30.6 (±1.1) nA/°C, respectively, within the temperature range of 20 °C to 50 °C.

We have measured the temperature-dependent device current responses for both sets of devices fabricated on glass and PET substrates. Figure 16a showed a typical response at different temperatures at a constant bias of V_{DS} = –1.5 V and V_{GS} = –1.5 V for glass substrates. Similar responses for the devices fabricated on PET substrates-based device with V_{DS} = –1.2 V and V_{GS} = –1.2 V are shown in Fig. 16c. To confirm the sensitivity of the thermal sensor around body temperature (~37 °C), we have measured the responses at a temperature range covering body temperature. The typical responses are shown in Fig. 16b and 16dfor glass and PET substrate-based devices, respectively. These results confirm the suitability of the OFETs as temperature sensors for wearable healthcare applications.

We have encapsulated the devices with a thick Polydimethylsiloxane (PDMS) film to prevent degradation in ambient conditions. The extra PDMS layer introduces heat loss while measuring temperature using this sensor. We have measured the

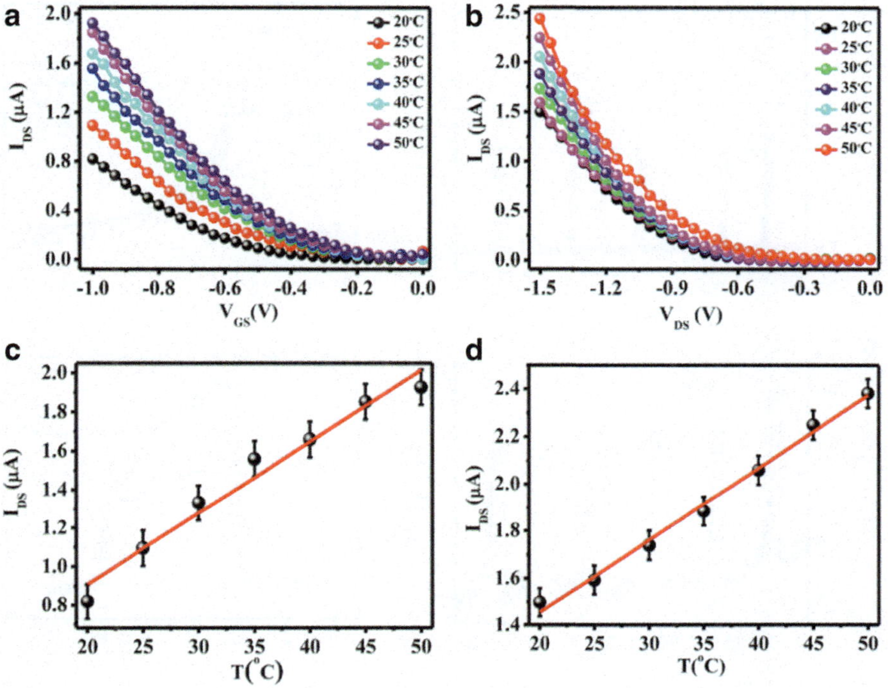

Fig. 15 Temperature-dependent transfer characteristics (**a**) for glass substrate working within 1.0 V and (**b**) for plastic substrate working within 1.5 V. The extracted device current from transfer characteristics as a function of temperature for (**c**) glass substrate at $V_{DS} = -1.0$ V and $V_{GS} = -1.0$ V and (**d**) plastic substrate at $V_{DS} = -1.5$ V and $V_{GS} = -1.5$ V over a range from 20 °C to 50 °C

temperature response from the back and front sides of 100 μm PET substrate-based OFETs. Figure 17a and 17b shows the typical temperature responses taken from the devices for touching it on the PDMS layer and the PET substrate, respectively. The temperature response seems to be the same from both backsides of PET substrates and the PDMS coated side. This effect is mainly due to the similar thermal conductivity of 0.15 W/mK of PDMS and PET materials [47, 48]. So, the heat loss for both the materials would be similar, as observed in this case. The corresponding response and recovery process of such 70 repeated cycles are shown in Fig. 17b. The inset of the figure shows one individual response. The optical image of that device is shown in Fig. 17c, during the performance of the above measurements. Figure 17d presents the array of flexible OFET suitable for measuring special temperature variation in wearable applications.

The response time and recovery time are essential parameters in this OFET-based temperature sensors. We used the sensor to monitor the temperature difference of inhaling and exhaling air during breathing to measure response and recovery time. The temperature of the inhaled air is mainly room temperature (26.5 °C). However, the exhaled air is primarily due to CO_2 coming out of the body. We have observed an almost 2 °C temperature difference between inhaled and exhaled air.

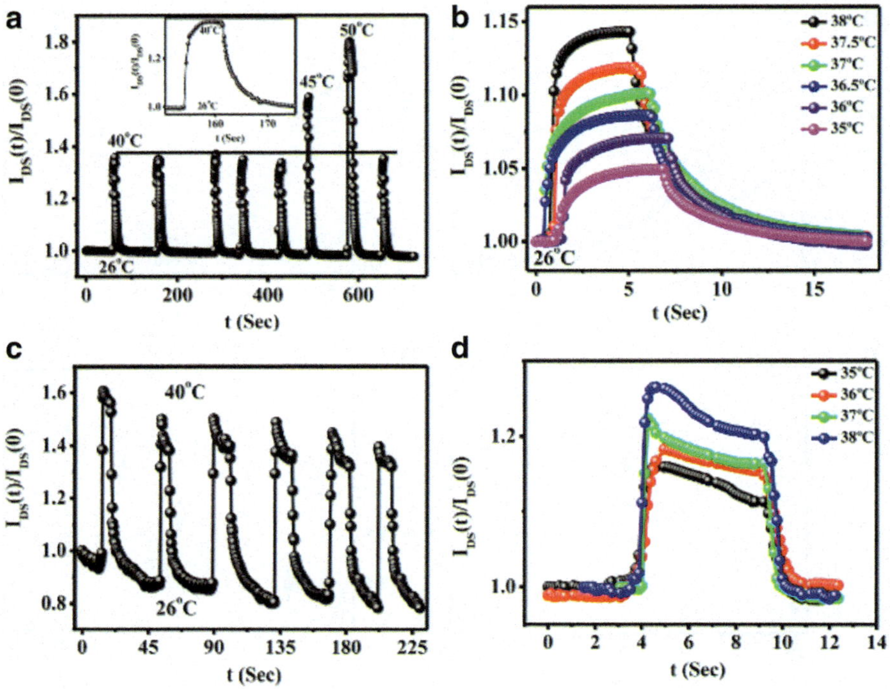

Fig. 16 The variation of device current with time measured during the change in temperature from (**a**) 26 °C to 40 °C, 45 °C, and 50 °C where Inset shows individual temperature response from 26 °C to 40 °C and (**b**) 26 °C to 35 °C, 36 °C, 36.5 °C, 37 °C, 37.5 °C, and 38 °C for glass substrate at constant $V_{DS} = -1.5$ V and $V_{GS} = -1.5$ V. Device current response for the temperature change from (**c**) 26 °C to 45 °C (d) 26 °C to 35 °C, 36 °C, 37 °C, and 38 °C for PET substrate-based device at constant $V_{DS} = -1.2$ V and $V_{GS} = -1.2$ V

We have measured the response of the devices from the speedy breathing exercises, as shown in Fig. 18a. One such individual temperature response profile is shown in Fig. 18b. The measured response time and recovery time from this device are about 24 ms and 51 ms, respectively.

To demonstrate the functionality of the temperature sensors for continuous monitoring of the body temperature, we have measured the variation of body temperature of mice during mild anesthesia. The body temperature of mice also was continuously monitored by this sensor under a mild dose of anesthesia. It is well known that mammals' body temperature falls by a few degrees when anesthesia is given. We mounted the sensor on the cleaned chest of mice. Two mild doses of anesthesia were given within a short time duration. We observed how the mice recover the body temperature with time. The corresponding variation of body temperature of the mice is shown in Fig. 18c. We have observed that the mice's body temperature decreased by 3–4 °C due to the anesthesia dose. The recovery process also has been monitored continuously, as shown in Fig. 18c. It has been found that the sensor can measure ultralow temperature variation of the mice while recovering

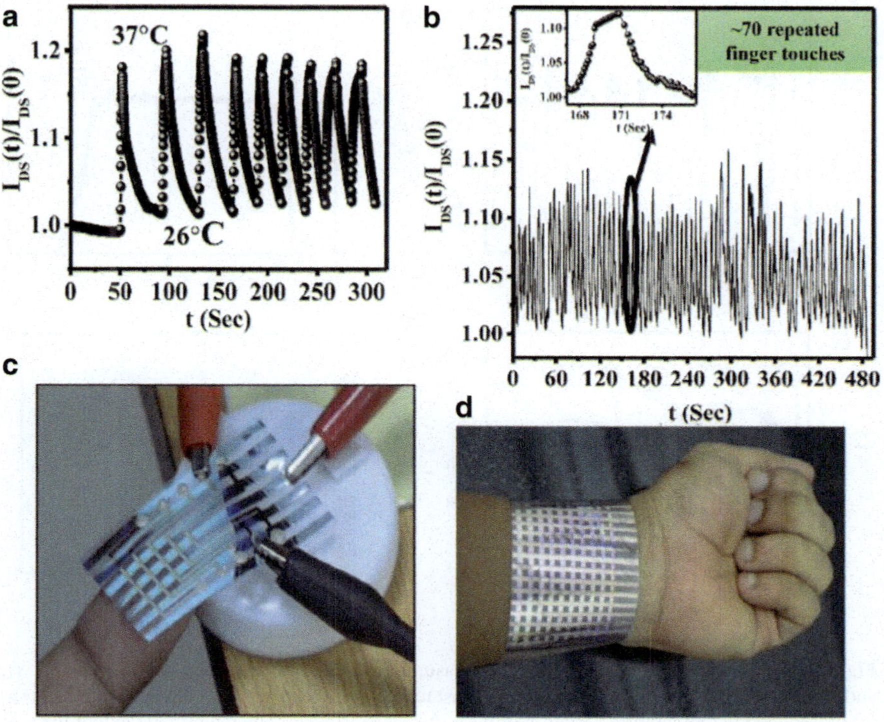

Fig. 17 Body temperature response from the encapsulating device upon touching the sensor from (**a**) front (PDMS) and (**b**) back (PET) side. One such cycle is shown in the inset Figure (**b**). (**c**) presents the optical image of the device during the measurement time. (**d**) shows the image of a flexible OFET-based array. (Reprinted from [19] © 2019 American Chemical Society)

from anesthesia. The sensor's precession was calculated from the linear part of A and B regions, as marked in the inset of Fig. 18c. The measured precession in temperature for this temperature sensor is 4.3 mK. The typical optical image of the temperature sensor mounted on the mice's chest during the experiment is shown in Fig. 18d.

7 Conclusions

This article discussed the synthesis of temperature-sensitive h-BTNCs sols using a low-cost solution process and has been used it for the fabrication of OFET-based flexible temperature sensors. We have covered from material characterization to fabrication of OFET through the various steps of optimizations of device parameters. The devices were found to be highly sensitive temperature sensors with 24 ms response time and 4.3 mK precession. Besides, the devices' power consumption is very low (~1 μW), working at 1.5 V operating voltage. We have demonstrated the

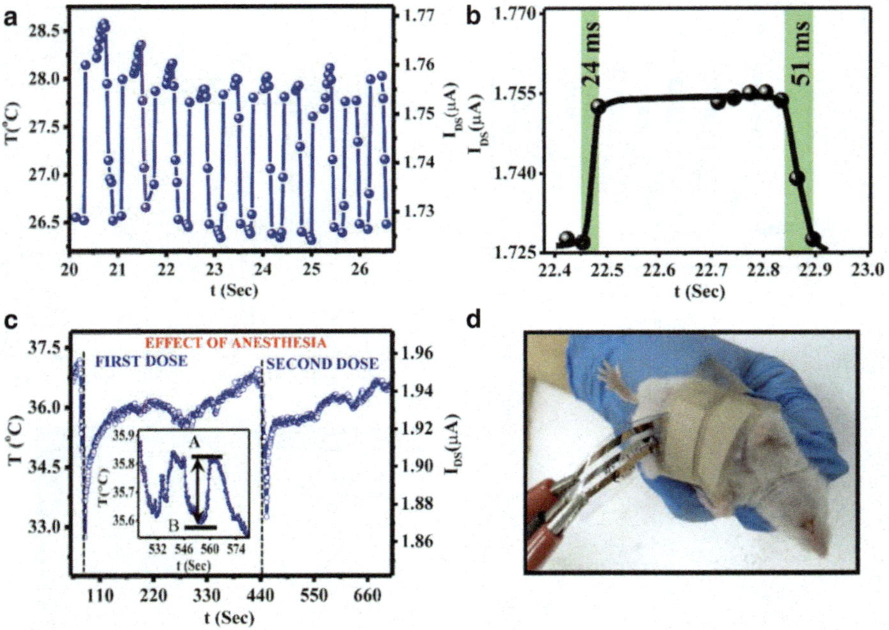

Fig. 18 (**a**) depicts fast breathing response measurements of the temperature sensor biased at V_{GS} and V_{DS} of −1.20 V. (**b**) shows one individual fast temperature response was taken from Figure (**a**). (**c**) presents body temperature variation of mice during anesthesia. The magnified view of the temperature variation taken in a different scan with higher time resolution shown in the inset of Figure (**c**). (**d**) illustrates a typical picture of a temperature sensor mounted in the mice during the experiment. (Reprinted from [19] © 2019 American Chemical Society)

use of the sensors for measuring body temperature. The flexibility of the device has been tested over 100 bending cycles with the smallest bending radius of 4 mm within which device property remains unchanged. We have also demonstrated that devices can monitor human beings' respiratory rate and can be suitable for detecting diseases like sleep apnea, asthma, and COPD.

Acknowledgments This work has been partially supported by various funding agencies of Government of India with sanction no: (MeitY) 5(1)/2017-NANO, (DST) DST/NM/NNetRA/2018(G) – IIT-KGP.

References

1. B.K.C. Kjellander, W.T.T. Smaal, K. Myny, J. Genoe, W. Dehaene, P. Heremans, G.H. Gelinck, Optimized circuit design for flexible 8-bit RFID transponders with active layer of ink-jet printed small molecule semiconductors. *Org. Electron.* ***14***, 768–774 (**2013**)
2. V. Subramanian, P.C. Chang, J.B. Lee, S.E. Molesa, S.K. Volkman, Printed organic transistors for ultra-low-cost RFID applications. *IEEE Trans. Adv. Packag.* ***28***, 742–747 (2005)

3. S.J. Kim, J.S. Lee, Flexible organic transistor memory devices. *Nano Lett.* ***10***, 2884–2890 (2010)
4. S. Lee, H. Seong, S.G. Im, H. Moon, S. Yoo, Organic flash memory on various flexible substrates for foldable and disposable electronics. *Nat. Commun.* ***8***, 725 (2017)
5. T. Sekitani, T. Yokota, U. Zschieschang, H. Klauk, S. Bauer, K. Takeuchi, M. Takamiya, T. Sakurai, T. Someya, Organic nonvolatile memory transistors for flexible sensor arrays. *Science* ***326***, 1516–1519 (2009)
6. H. Fukagawa, T. Sasaki, T. Tsuzuki, Y. Nakajima, T. Takei, G. Motomura, M. Hasegawa, K. Morii, T. Shimizu, Long-lived flexible displays employing efficient and stable inverted organic light-emitting diodes. *Adv. Mater.* ***30***, 1706768 (2018)
7. C.D. Sheraw, L. Zhou, J.R. Huang, D.J. Gundlach, T.N. Jackson, M.G. Kane, I.G. Hill, M.S. Hammond, J. Campi, B.K. Greening, J. Francl, J. West, Organic thin-film transistor-driven polymer-dispersed liquid crystal displays on flexible polymeric substrates. *Appl. Phys. Lett.* ***80***, 1088–1090 (2002)
8. J. Smith, R. Hamilton, M. Heeney, D.M. de Leeuw, E. Cantatore, J.E. Anthony, I. McCulloch, D.D.C. Bradley, T.D. Anthopoulos, High-performance organic integrated circuits based on solution processable polymer-small molecule blends. *Appl. Phys. Lett.* ***93***, 253301 (2008)
9. B. Crone, A. Dodabalapur, Y.Y. Lin, R.W. Filas, Z. Bao, A. LaDuca, R. Sarpeshkar, H.E. Katz, W. Li, Large-scale complementary integrated circuits based on organic transistors. *Nature* ***403***, 521–523 (2000)
10. Y. Takeda, Y. Yoshimura, Y. Kobayashi, D. Kumaki, K. Fukuda, S. Tokito, Integrated circuits using fully solution-processed organic TFT devices with printed silver electrodes. *Org. Electron.* ***14***, 3362–3370 (2013)
11. A. Chortos, J. Liu, Z.A. Bao, Pursuing prosthetic electronic skin. *Nat. Mater.* ***15***, 937–950 (2016)
12. X.D. Wang, L. Dong, H.L. Zhang, R.M. Yu, C.F. Pan, Z.L. Wang, Recent progress in electronic skin. *Adv. Sci.* ***2***, 1500169 (2015)
13. C. Xie, F. Yan, Flexible photodetectors based on novel functional materials. *Small* ***13***, 1701822 (2017)
14. X.D. Liu, Y.W. Lin, Y.J. Liao, J.Z. Wu, Y.H. Zheng, Recent advances in organic near-infrared photodiodes. *J. Mater. Chem. C* ***6***, 3499–3513 (2018)
15. S. Khan, S. Ali, A. Bermak, Recent developments in printing flexible and wearable sensing electronics for healthcare applications. *Sensors (Basel)* ***19***, 1230 (2019)
16. G. Nan, Q. Shi, Z. Shuai, Z. Li, Influences of molecular packing on the charge mobility of organic semiconductors: From quantum charge transfer rate theory beyond the first-order perturbation. *Phys. Chem. Chem. Phys.* ***13***, 9736–9746 (2011)
17. H. Zhang, X. Guo, J. Hui, S. Hu, W. Xu, D. Zhu, Interface engineering of semiconductor/dielectric heterojunctions toward functional organic thin-film transistors. *Nano Lett.* ***11***, 4939–4946 (2011)
18. J. Liu, H.T. Zhang, H.L. Dong, L.Q. Meng, L.F. Jiang, L. Jiang, Y. Wang, J.S. Yu, Y.M. Sun, W.P. Hu, A.J. Heeger, High mobility emissive organic semiconductor. *Nat. Commun.* ***6***, 10032 (2015)
19. S. Mandal, M. Banerjee, S. Roy, A. Mandal, A. Ghosh, B. Satpati, D.K. Goswami, Organic field-effect transistor-based ultrafast, flexible, physiological-temperature sensors with hexagonal barium titanate nanocrystals in amorphous matrix as sensing material. *ACS Appl. Mater. Interfaces* ***11***, 4193–4202 (2019)
20. T.Q. Trung, N.E. Lee, Flexible and stretchable physical sensor integrated platforms for wearable human-activity monitoring and personal healthcare. *Adv. Mater.* ***28***, 4338–4372 (2016)
21. X. Wu, Y. Ma, G. Zhang, Y. Chu, J. Du, Y. Zhang, Z. Li, Y. Duan, Z. Fan, J. Huang, Thermally stable, biocompatible, and flexible organic field-effect transistors and their application in temperature sensing arrays for artificial skin. *Adv. Funct. Mater.* ***25***, 2138–2146 (2015)
22. H. Li, W. Shi, J. Song, H.J. Jang, J. Dailey, J.S. Yu, H.E. Katz, Chemical and biomolecule sensing with organic field-effect transistors. *Chem. Rev.* ***119***, 3–35 (2019)
23. J.N. Curti, K.L. Adriance, E.C. Land, in *Adaptive Temperature Sensor for Breath Monitoring Device*. US 8740808 (2017)

24. A.L. Dunlop, R. Schultz, E. Frank, Interpretation of the BBT chart: Using the "Gap" technique compared to the coverline technique. *Contraception* ***71***, 188–192 (**2005**)
25. H.W. Su, Y.C. Yi, T.Y. Wei, T.C. Chang, C.M. Cheng, Detection of ovulation, a review of currently available methods. *Bioeng. Transl. Med.* **2**, 238–246 (2017)
26. B. Kudlak, N. Jatkowska, P. Kubica, G. Yotova, S. Tsakovski, Influence of storage time and temperature on the toxicity, endocrine potential, and migration of epoxy resin precursors in extracts of food packaging materials. *Molecules* ***24***, 4396 (2019)
27. A.M. Karow, W.B. Gilbert, J.B. Black, Effects of temperature, potassium concentration, and sugar on human spermatozoa motility: a cell preservation model from reproductive medicine. *Cryobiology* ***29***, 250 (1992)
28. S.M. Abdullah, Z. Ahmad, K. Sulaiman, A solution-based temperature sensor using the organic compound CuTsPc. *Sensors* ***14***, 9878–9888 (2014)
29. F.T. Silva, B. Sorli, V. Calado, C. Guillaume, N. Gontard, Feasibility of a gelatin temperature sensor based on electrical capacitance. *Sensors* ***16***, 2197 (2016)
30. H.Y. Ma, Q.A. Huang, M. Qin, T. Lu, A micromachined silicon capacitive temperature sensor for wide temperature range applications. *J. Micromech. Microeng.* ***20***, 055036 (2010)
31. J. Jeon, H.B.R. Lee, Z. Bao, Flexible wireless temperature sensors based on Ni microparticle-filled binary polymer composites. *Adv. Mater.* ***25***, 850–855 (2013)
32. S. Harada, K. Kanao, Y. Yamamoto, T. Arie, S. Akita, K. Takei, Fully printed flexible fingerprint-like three-axis tactile and slip force and temperature sensors for artificial skin. *ACS Nano* ***8***, 12851–12857 (2014)
33. W. Honda, S. Harada, S. Ishida, T. Arie, S. Akita, K. Takei, High-performance, mechanically flexible, and vertically integrated 3D carbon nanotube and InGaZnO complementary circuits with a temperature sensor. *Adv. Mater.* ***27***, 4674–4680 (2015)
34. T. Yokota, Y. Inoue, Y. Terakawa, J. Reeder, M. Kaltenbrunner, T. Ware, K.J. Yang, K. Mabuchi, T. Murakawa, M. Sekino, W. Voit, T. Sekitani, T. Someya, Ultraflexible, large-area, physiological temperature sensors for multipoint measurements. *Proc. Natl. Acad. Sci. U. S. A.* ***112***, 14533–14538 (2015)
35. S.Y. Hong, Y.H. Lee, H. Park, S.W. Jin, Y.R. Jeong, J. Yun, I. You, G. Zi, J.S. Ha, Stretchable active matrix temperature sensor array of polyaniline nanofibers for electronic skin. *Adv. Mater.* ***28***, 930–935 (2016)
36. X. Ren, K. Pei, B. Peng, Z. Zhang, Z. Wang, X. Wang, P.K.L. Chan, A low-operating-power and flexible active-matrix organic-transistor temperature-sensor array. *Adv. Mater.* ***28***, 4832–4838 (2016)
37. T.Q. Trung, S. Ramasundaram, S.W. Hong, N.E. Lee, Flexible and transparent nanocomposite of reduced graphene oxide and P(VDF-TrFE) copolymer for high thermal responsivity in a field-effect transistor. *Adv. Funct. Mater.* ***24***, 3438–3445 (2014)
38. T.Q. Trung, N.T. Tien, D. Kim, J.H. Jung, O.J. Yoon, N.E. Lee, High thermal responsiveness of a reduced graphene oxide field-effect transistor. *Adv. Mater.* ***24***, 5254–5260 (2012)
39. C. Miot, E. Husson, C. Proust, R. Erre, J.P. Coutures, Residual carbon evolution in BaTio3 ceramics studied by XPS after ion etching. *J. Eur. Ceram. Soc.* ***18***, 339–343 (1998)
40. S. Steudel, D.S. Vussera, D.S. Jonge, D. Janssen, S. Verlaak, J. Genoe, P. Heremans, Influence of the dielectric roughness on the performance of pentacene transistors. *Appl. Phys. Lett.* ***85***, 4400 (2004)
41. S.E. Fritz, T.W. Kelley, C.D. Frisbie, Effect of dielectric roughness on performance of pentacene TFTs and restoration of performance with a polymeric smoothing layer. *J. Phys. Chem. B* ***109***, 10574–10577 (2005)
42. M. Girtan, S. Dabos-Seignon, A.L. Stanculescu, On morphological, structural and electrical properties of vacuum deposited pentacene thin films. *Vacuum* ***83***, 1159–1163 (2009)
43. Y. Jung, R.J. Kline, D.A. Fischer, E.K. Lin, M. Heeney, I. McCulloch, D.M. DeLongchamp, The effect of interfacial roughness on the thin film morphology and charge transport of high-performance polythiophenes. *Adv. Funct. Mater.* ***18***, 742–750 (2008)

44. X.N. Sun, C.A. Di, Y.Q. Liu, Engineering of the dielectric-semiconductor interface in organic field-effect transistors. *J. Mater. Chem.* ***20***, 2599–2611 (2010)
45. S. Patel, H. Park, P. Bonato, L. Chan, M. Rodgers, A review of wearable sensors and systems with application in rehabilitation. *J. Neuroeng. Rehabil.* ***9***, 21 (2012)
46. V. Raghuwanshi, D. Bharti, S.P. Tiwari, Flexible organic field-effect transistors with TIPS-pentacene crystals exhibiting high electrical stability upon bending. *Org. Electron.* ***31***, 177–182 (2016)
47. K. Uetani, T. Okada, H.T. Oyama, Crystallite size effect on thermal conductive properties of nonwoven nanocellulose sheets. *Biomacromolecules* ***16***, 2220–2227 (2015)
48. P. Yi, R.A. Awang, W.S.T. Rowe, K. Kalantar-zadeh, K. Khoshmanesh, PDMS nanocomposites for heat transfer enhancement in microfluidic platforms. *Lab Chip* ***14***, 3419–3426 (2014)

Tailoring Strategies to Enhance the Photoelectrocatalytic Activity of Perovskite Oxide Surfaces ABO_3 for Efficient Renewable Energy Generation

J. Manuel Mora-Hernandez and Leticia M. Torres-Martínez

1 Introduction

The requirement to develop novel materials drives several worldwide research groups to focus their efforts on the synthesis and enhancement of advanced materials with application in the power generation field. Metals, ceramics, polymers, organics, and composites are involved in all the producing energy processes [1]. In the last decades, those novel materials related to renewable and sustainable methods to generate energy have attracted the attention of technological and power generation industries due to their feasibility to produce energy without causing damage as serious as that generated by the exploitation of non-renewable sources [2, 3]. Among the common renewable sources such as solar energy, wind energy, sea energy, bioenergy, geothermic, and hydrogen energy [4], those related with the exploitation of solar energy present a special interest, since the total amount of energy reaching our planet surface each year is estimated to be 3,400,000 EJ, this is between 7000 and 8000 times the annual global primary energy consumption [5]. Photovoltaic and photocatalysis are two important and arising technologies classified as branches of

J. M. Mora-Hernandez (✉)
CONACYT – Universidad Autónoma de Nuevo León, UANL, Facultad de Ingeniería Civil, Departamento de Ecomateriales y Energía, Av. Universidad S/N Ciudad Universitaria, San Nicolás de los Garza, Nuevo León, México
e-mail: jmorah@conacyt.mx

L. M. Torres-Martínez
Universidad Autónoma de Nuevo León, UANL, Facultad de Ingeniería Civil, Departamento de Ecomateriales y Energía, Av. Universidad S/N Ciudad Universitaria, San Nicolás de los Garza, Nuevo León, México

Centro de Investigación en Materiales Avanzados, S.C. (CIMAV), Chihuahua, Chih, México
e-mail: leticia.torresgr@uanl.edu.mx

A. G. Roca et al. (eds.), *Surfaces and Interfaces of Metal Oxide Thin Films, Multilayers, Nanoparticles and Nano-composites*,
https://doi.org/10.1007/978-3-030-74073-3_6

frontier sciences [6] since they involve sunlight harvesting and the constant development of advanced materials for energy generation purposes [7, 8]. Regarding the photocatalytic technology, this one is widely used in decontamination processes [9–12] and sustainable power generation, through the generation of H_2 as fuel, [13–15] and photoreduction processes to convert the CO_2 in solar-based fuels [16–18]. In this way, advanced photocatalytic materials such perovskite semiconductors with formula ABO_3 containing metal transition ions in the structural sites A or B have generated great attention to develop numerous investigations, since such materials present a bandgap between 1.5 and 3.0 eV, which makes them suitable for use in the visible region of the electromagnetic spectrum for photocatalytic applications. Titanates such as $ZnTiO_3$, $CdTiO_3$, and $PbTiO_3$ show favorable characteristics for the use of visible light with application potential for the production of hydrogen. Despite this, the study of such titanates perovskites has been reported in few publications [19–22]. Most research works focus their studies in the final catalytic activity of materials, and the synthesis process used to synthesize the ABO_3 perovskites (solid-state, hydrothermal, microwave-assisted hydrothermal, and sol-gel); such reports do not establish well a relationship between the synthesis method, the crystalline structure, and catalytic activity caused by the substitution of the A or B perovskite cations and the photocatalytic activity. Lanthanum-based perovskites such as $LaMnO_3$ and $LaCoO_3$ also have been reported as suitable compounds to perform the degradation of polluting substances such as methyl orange [23], direct Green [24], and bisphenol A [9] using visible light as the excitation source. $LaFeO_3$ is another perovskite which has shown excellent activity for the removal of organic pollutants from water [25] and NO_x from the air [26]. However, there are few reports on the photocatalytic activity of $LaFeO_3$, $LaMnO_3$, and $LaCoO_3$ as photocatalysts to carry out the photoelectrocatalytic hydrogen evolution reaction. In particular, it has been reported that $LaCoO_3$ exhibits a deactivation process during the photocatalytic reaction. Such behavior was also reported for $Cu/LaFeO_3$ and $LaMnO_3$ powders, in all cases this deactivation process leads to low photocatalytic efficiency. In the same way that $ATiO_3$ and $LaBO_3$ present a variation in the photocatalytic activity as the cation A or B is substituted, the utilization of a secondary material to generate heterojunctions is another strategy to improve the photocatalytic activity. The addition of a second semiconductor (SB) to the bare main semiconductor structure (SA) modifies the levels of the valence and conduction bands of each material, overlapping in some cases the energetic levels and enhancing the separation of charge carrier's and avoiding the recombination processes, thus increasing the photocatalytic activity [27]. $NaTaO_3$ is a perovskite material widely used for photocatalytic degradation and energy generation [28–31]; its structure modification can lead to better photocatalytic performance. In addition to the bands overlapping, the doping or couple of $NaTaO_3$ with other compounds can also produce the formation of intermediate energy levels; this feature can reduce the bandgap of the SA, inducing a shift to the red spectrum, and enhancing the visible light harvesting [17, 32–34]. Therefore, the design of highly efficient photoactive semiconductors requires an extensive study of the structural and superficial parameters. The aim of this chapter is focused on analyzing how the cation substitution in the tetrahedral and octahedral sites, the heterojunction formation, and the doping of the SA as structural and morphological methodology affects the psychochemical, optical, and photoelectrocatalytic properties of inorganic semiconductor perovskites ABO_3 toward photocatalytic reduction processes. It is necessary a critical analysis of the physicochemical characteristics of bare perovskites and the existing relationship between the substitution of different cations, doping and heterojunction formation, in order to clearly identify the processes that enhance or limiting the photocatalytic activity of these advanced materials.

2 Scientific and Technologic Interest in Perovskites ABO_3

The constant improvement of new energy conversion systems has allowed the development of new technologies to produce energy by the use of non-expensive and abundant materials, thus reducing the materials manufacturing and energy production costs [35]. An emergent technology in the branch of novel processes for the power generation field is photocatalysis, nowadays classified as a branch of frontier science [6], which means it is a promising technology still in development. The photocatalytic technology began its technological breakthrough in the 1970s when Fujishima and Honda reported for the first time the water molecule splitting by a photocatalytic process using titanium dioxide (TiO_2) as a photoactive material in 1972 [36]. From this first approach, many research groups have focused their efforts on the development of new and more efficient semiconductor materials to perform photocatalytic processes. Among an extensive variety of semiconductors, perovskite materials have played a remarkable role in different conversion energy applications. Perovskite materials with general formula ABO_3 induce a high charge carrier's production (electron and holes), capable to carry out electrochemical redox processes, decreasing the charge carrier's recombination to improve photocatalytic reactions to produce new renewable energy sources. It is worth highlighting that in the last decade, perovskite materials have played a fundamental role in the generation of energy utilizing solar radiation sources with applications in photovoltaic panels and photocatalytic processes. The development of silicon-based solar panels (Crystalline Si cells) and thin films (thin-film technologies) started their production in the 1970s. In the first years, such materials presented photovoltaic efficiencies less than 15%. In general, the silicon-based panels have presented an efficiency average increase of 12% in 45 years, showing energy efficiencies around 28% in 2020. Despite this efficiency increase represents a significative technological development, the current energy demand requires high impact and well-developed strategies to supply the energetical requirements. The photovoltaic hybrid-perovskite materials have become the materials with the highest energy efficiency performance, increasing from 14% to 26% in the last 5 years. Similarly, the cell efficiency of perovskite-silicon-based tandems has shown an increase from 25% to 28% in the last 3 years. In this way, there is enough evidence that has demonstrated that perovskites are emerging as an exemplary candidate for the manufacture of photovoltaic devices.

In this context, it is very important to continue with the exhaustive study and development of perovskite materials, to modify their physicochemical and optical properties to increase their efficiency in photovoltaic processes. In this context, perovskite semiconductors have also shown exceptional improvement in the development of photocatalytic systems with application in the branch of solar-based fuels production, such as the generation of hydrogen via photocatalytic water splitting process, or the production of low carbon-containing compounds and fuels from the photocatalytic carbon dioxide reduction reaction (CO_2RR). Regarding the photocatalytic process and its applications, complex oxides with general formula ABO_3, such as the $SrTiO_3$ [37, 38], $NaTaO_3$ [32, 34] and $KTaO_3$ [39], have shown outstanding efficiency for the hydrogen generation; Huerta et al. reported a hydrogen production of 5672 μmol g^{-1} h^{-1} onto a $NaTaO_3$ laser assisted chemical vapor deposition (LCVD) film [33]. Tedsuda Kida et al. [40] reported a photocatalytic hydrogen production of 29 μmol h^{-1} for the heterostructured material $LaMnO_3$/CdS, employing Na_2SO_3/Na_2S as a sacrificial agent. Meizini et al. [41] obtained a H_2 production of 60 μmol h^{-1} with the compound $LaCoO_3$/SnO_2, employing also Na_2SO_3 as a sacrificial agent. Another research publication reported the use of Cu/$LaFeO_3$ to produce 187 μmol g^{-1} h^{-1} of H_2 using a solution containing water + 10 wt.% of triethanolamine [42]. Some studies report that $LaFeO_3$ coupled with g-C_3N_4 formed a heterostructure capable to perform properly the photocatalytic hydrogen evolution reaction in presence of sacrificial agents [43, 44]. While these works highlight the use of different sacrificial agents to improve the photocatalytic reactions, they do not emphasize the use of rare earth elements to substitute cations, which promotes better characteristics for the sunlight harvesting compared to transition metals, due to a greater number of 4f energy levels which maximize the photon absorption during photocatalytic processes [45]. Similarly, Jin Luo et al. recently reported the coupling of g-C_3N_4/$LaCoO_3$, and this time was applied to carry out the degradation of methyl orange under visible light irradiation [46]. Even though most of this research works attribute the enhanced photocatalytic activity to the heterostructure formation, it is not fully understood how the main physicochemical characteristics of such heterostructures affect the photocatalytic mechanisms in order decrease the charge carrier's recombination, which improves the catalytic properties of the based g-C_3N_4 heterostructures.

Focusing on the structural aspects of perovskite-based materials, and its relationship with the photocatalytic activity, these materials present a propitious electronic band structure to perform photoelectrocatalytic reduction processes; besides, the substitution of the A and B cations, which conform the crystalline structure in the ABO_3 formula, provides a broad panorama for designing and modifying the crystal and its electronic structure, which affects directly the physicochemical properties [47]. Many of these perovskite oxides contain metal transition ions d^0 o d^{10} (Ti^{4+}, Zr^{4+}, V^{5+}, Nb^{5+}, Ta^{5+}, W^{6+}) in their electronic structure [48], which induced a higher charge carrier's generation, responsible for carrying out the water splitting process. Most of these transition metals such as titanium, iron, and manganese, among others are earth abundant and non-toxic elements. Also, because they can easily lose electrons in their *d* orbitals, they can be found in nature with different oxidation (mostly a valence state +2), which bring them the characteristic to be easily substituted, and more importantly, the capability to be photocatalytic active in the region of the visible electromagnetic spectrum [49]. Among different perovskites reported as good photocatalytic materials to perform the hydrogen evolution reaction, tantalates ($NaTaO_3$ [32, 34], $KTaO_3$ [39], $K_3Ta_3Si_2O_{13}$ [50], $BaTa_2O_6$ [51], $SrTa_2O_6$ [52]), and titanates ($SrTiO_3$ [37, 38], $BaTiO_3$ [53], $CaTiO_3$ [54]) have attracted great attention due to their high photocatalytic performance, mainly those composed by alkali and alkaline earth metals.

3 Strategies for the Development of New Functional and Efficient Photocatalytic Materials

Despite the great effort that research groups worldwide have dedicated to the development of efficient materials for obtaining hydrogen by photocatalysis, the current efficiencies are still below the desired level. For this reason, different strategies have been implemented to increase the efficiency of the photocatalysts used in hydrogen production and other photoreduction processes. In this way, the photocatalytic activity depends on the cumulative effect of the following processes: light absorption, charge separation, and charge carrier's migration and transport [55, 56]. The current bottleneck in the photocatalysis technology lies in its low quantum yield, which is affected by the rapid recombination of the hole-electron pairs on the surface, but also in the bulk. The stability of the photocatalysts is also a very important parameter to improve the photocatalytic activity. The overall efficiency of photocatalytic hydrogen production depends directly on the thermodynamic and kinetic equilibrium, which depend on the electronic and crystalline structure [56]. The most important factors and characteristics that influence the efficiency of the photocatalytic processes, as well as their corresponding improvement strategies, are an effective separation and transfer of the electron-hole charge carriers, the light harvesting, an adequate bandgap, low cost and minimum toxicity of the photocatalyst, as well as an optimized kinetic reaction on the surface to carry out the redox processes. There are important strategies reported in the literature to improve the photocatalytic processes, such as the band structure modification, the design of micro/nanometric structures, the modification of the surface, and the modification of the semiconductors interfaces [57]. However, it is also important to consider the physicochemical properties, particularly the crystalline and electronic structure, and its relationship with the photocatalytic activity. These two characteristics are notably influenced in perovskite-type structure ABO_3, due to the effects of the substitution of cations A and B in the cationic sites [47, 48]. These properties are attributed to the position of the semiconductor conduction bands, a band with higher energy (more negative potential), thermodynamically favors the evolution of hydrogen [58]; in the same way, distortions in the crystal structure due to bond angles close to 180° (for example Ta-O-Ta in the $KTaO_3$) promote better mobility and transference of the photogenerated charges [59]. Considering all the aforementioned factors, a deep understanding of these materials is necessary for the adequate development of photocatalytic systems capable to perform efficiently the hydrogen production and other photoreduction processes.

3.1 Crystalline Structure Modification: Cation A Substitution

To understand how the physicochemical, optical, and catalytic properties of some perovskites are related to the structural modifications depending on the substitution of the cation A in the ABO_3 formula, Fig. 1a shows the XRD patterns for the perovskite $ATiO_3$, where A = Zn, Cd, Pb. The three perovskites were synthesized by a solvo-combustion methodology [32]. The $ZnTiO_3$ presents a rhombohedral structure with the spatial group *R-3* [60]. Besides, a TiO_2 rutile phase (3 wt.%) is present in this compound. Similarly, the cadmium titanate ($CdTiO_3$) presents a rhombohedral structure with the same spatial group *R-3*. However, for the $PbTiO_3$ it presents a tetragonal structure with the spatial group *p4/mmm* (01-074-2495) (Fig. 1b). Since we are substituting the cation A, the difference in the crystalline structures is directly related to the difference between the atomic radii of the A = Zn, Cd, and Pb cations. According to the atomic radii reported in the literature: Zn^{2+} (0.74 Å) < Cd^{2+} (0.97 Å) < Pb^{2+} (1.19 Å) [61], the materials containing cations with shorter atomic radii present crystalline structures with higher distortion rates, meanwhile larger cationic

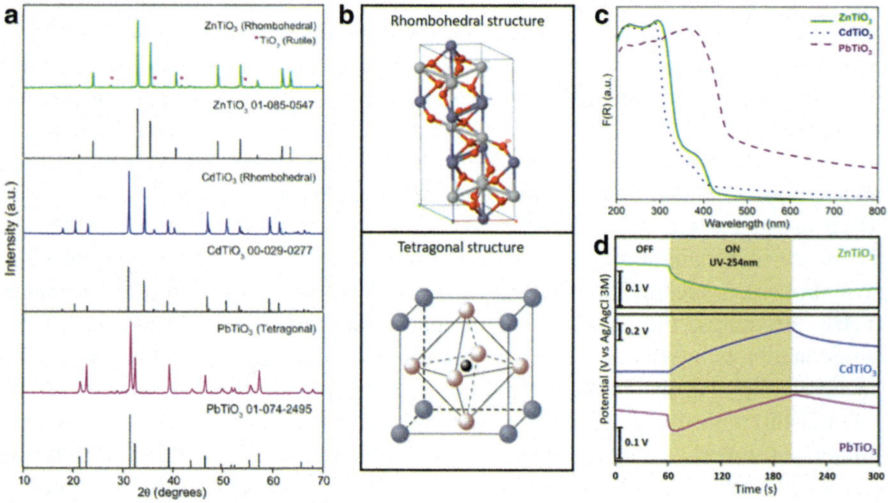

Fig. 1 (**a**) X-ray diffraction patterns for $ZnTiO_3$, $CdTiO_3$ and $PbTiO_3$. (**b**) Crystalline structure for $ZnTiO_3$ and $CdTiO_3$ (rhombohedral) and $PbTiO_3$ (tetragonal). (**c**) Diffuse reflectance spectroscopy measurements for $ATiO_3$, A = Zn, Cd, Pb. (**d**) Electrochemical potentiometry measurements. ((**a**) Reprinted with permission from O. Carrasco-Jaim et al. J Photoch Photobio A, 371, 98–108. Copyright (2019) Elsevier)

radii promotes the formation of a more-symmetric structural phase. Research works have reported that the cubic structure of different perovskite materials, particularly metallic tantalates present an enhanced photocatalytic activity compared with distorted structures such as the orthorhombic. This difference in photocatalytic activity relies on the Ta-O-Ta angle of the crystalline structure, which is closer to 180° in the cubic structure which induces an easy transport of the photogenerated charge carriers through the bulk crystalline structure, thus improving the photocatalytic activity [59, 62].

Figure 1c shows the diffuse reflectance UV-Vis spectra for the metal transition titanates perovskites. From this figure, it is important to notice that $ZnTiO_3$ presented an absorbance signal at 410 nm, which can be associated with the rhombohedral phase [60]. Additionally, it is possible to observe a signal around 300 nm, which corresponds to the rutile phase TiO_2 [63] previously identified. The $CdTiO_3$ spectrum depicts a signal transition around 300 nm, which is attributed to the rhombohedral phase [64], and an additional band at 380 which corresponds to an energetic value around 3.26 eV, this link energy is associated to intrinsic defects caused by oxygen vacancies in the crystalline structure of $CdTiO_3$ [65]. Regarding the $PbTiO_3$, this perovskite shows an extended absorption signal in the visible range (200–450 nm) which indicates a better visible light harvesting. It is worth to mention that $ZnTiO_3$ and $CdTiO_3$ show energy transitions below 300 nm ($\geq$ 4.13 eV). According to the DOS calculation reported for these materials, such transitions correspond to the *d* orbitals from the level Ti-*3d* from 3 to 9 eV. Such energetic levels are responsible for the major contribution in the conduction band for these perovskites [66]. On the other hand, it is possible to observe a similar absorption band for $ZnTiO_3$ and $CdTiO_3$, and this effect is attributed to the *d*-states in their respective conduction bands (Zn-*3d* and Cd-*5d*), in comparison to the Pb-*6p* state for $PbTiO_3$ [66, 67].

A piece of clear evidence that how the substitution of the cation A in the $ATiO_3$ structure modifies the photocatalytic activity is described as follows; when a material presents a lower conduction band energy level, as in the case of $PbTiO_3$, it is possible to observe an absorption rate extended to the visible light range [68]. In this way, the substitution of the cation of the site A from an electronic configuration with lower energy values induces the obtention of photocatalytic materials active to harvest light in the visible region. Another significant effect observed when the cation of the site A is substituted in materials $ATiO_3$ (A = Zn, Cd, Pb) is related to the nature of the generation and transference of the photogenerated charges. This effect can be explained by the variation of the electrochemical potential under dark and light conditions (UV light, 254 nm). Since the electrochemical potential of $ZnTiO_3$ presents a shift toward negative potentials (Fig. 1d), which means an electron accumulation in the conduction band [69], we can classify this material as an n-type semiconductor. On the other hand, $CdTiO_3$ and $PbTiO_3$ present a contrary behavior under light irradiation, the displacement to positive potentials indicates a p-type semiconductor. In particular, $PbTiO_3$ presents a slight negative shift when it is irradiated, this effect is associated with the accumulation of negative charges on the material surface (charge accumulation layer) which compensates the excess of positive charges present in a p-type semiconductor. Once the negative charges are consumed, the plot presents a potential change caused by the movement of the majority charge carriers (holes) [70].

Since we have elucidated how the substitution of the cation A modifies the structural arrangement of the $ATiO_3$ perovskites, these crystallographic modifications affect the photocatalytic activity to perform redox processes. Figure 2 depicts the micrographs and photocatalytic production of molecular hydrogen related to the A cation, A = Zn, Cd, Pb. From this plot, it is clear that $ZnTiO_3$ generates the highest amount of hydrogen, 470 μmol g^{-1} in 3 h of reaction; this value corresponds to a solar-to-hydrogen conversion efficiency (STH) equal to 0.48%. The highest hydrogen production presented by $ZnTiO_3$ is attributed to two main factors: (i) a higher crystallite size which presents a lower number of crystalline defects, thus reducing the charge carrier's recombination process, and (ii) the presence of a secondary phase TiO_2, which acts as an efficient co-catalyst to form the heterostructure $ZnTiO_3/TiO_2$, thus enhancing the separation of the photogenerated charges and minimizing their recombination rate. To compare the hydrogen production rates, as well as the crystallographic characteristics, Table 1 summarizes the most relevant data. Since $PbTiO_3$ generated a higher hydrogen amount than $CdTiO_3$ (140 μmol g^{-1} and 80 μmol g^{-1}, respectively). It is suggested this variation can be attributed to the different crystalline structures presented by each compound. As aforementioned, $ZnTiO_3$ and $CdTiO_3$ present a rhombohedral structure, meanwhile $PbTiO_3$ crystalized with a tetragonal structure. A more ordered structure induces the formation of Ti-O-Ti angles close to 180°, which enhances the light harvesting; this characteristic enhances the photocatalytic activity of $PbTiO_3$ over the $CdTiO_3$ to produce hydrogen.

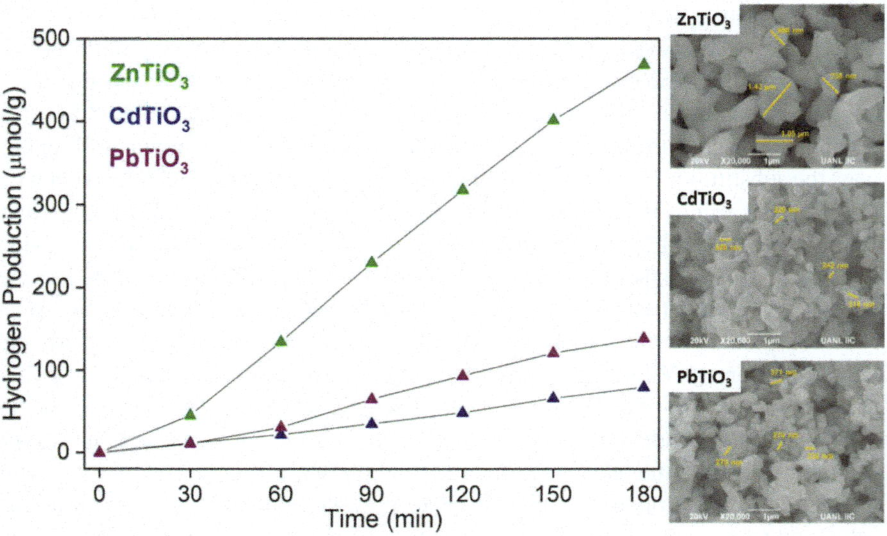

Fig. 2 Photocatalytic hydrogen production and scanning electron micrographs for $ATiO_3$ (A = Zn, Cd, Pb)

Table 1 Crystalline properties and photocatalytic H_2 production for $ATiO_3$ (A = Zn, Cd, Pb)

Material	Phase	Crystallite Size (nm)	H_2 production (μmol g^{-1}) *	H_2 production rate (μmol $m^{-2}h^{-1}$) *	STH (%)
$ZnTiO_3$	Rhombohedral	318	470	130	0.48
$CdTiO_3$	Rhombohedral	243	80	22	0.08
$PbTiO_3$	Tetragonal	65	140	38	0.14

[1]H_2 production quantified after 3 h of reaction, 0.2 g of the photocatalyst in 200 ml of deionized water

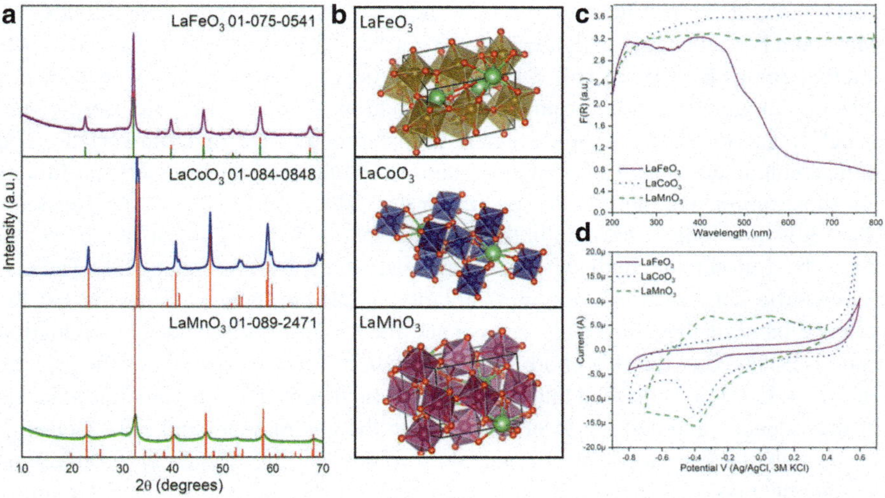

Fig. 3 (**a**) X-ray diffraction patterns for $LaFeO_3$, $LaCoO_3$, and $LaMnO_3$. (**b**) Crystalline structures for $LaBO_3$, B = Fe, Co, Mn. (**c**) Diffuse reflectance spectroscopy measurements for $LaFeO_3$, $LaCoO_3$, and $LaMnO_3$. (**d**) Cyclic voltammetry characterization. ((**a**) Reprinted with permission from L. Ibarra-Rodriguez et al. J Phys Chem Solids, 136, 109189. Copyright (2020) Elsevier)

3.2 *Crystalline Structure Modification: Cation B Substitution*

Section 1.3.1 was dedicated to understanding how the substitution of the cation A in the perovskite ATiO3, A = Zn, Cd, Pb can tailor the photocatalytic activity to perform the hydrogen production since the different atomic radio of each cation promotes the formation of different crystallographic arrangements. This modifies the lattice distortion promoting a redshifting for $PbTiO_3$ and the formation of heterostructures $ZnTiO_3/TiO_2$, which improves in both cases the generation and separation of photogenerated charge carriers and decrease the recombination rate. In this context, it is important to study how the modification of the cation B in the perovskite structure ABO_3 affects the photocatalytic parameters. Figure 3a depicts the X-ray diffraction patterns of three different perovskites $LaBO_3$ (B = Fe, Co, Mn) synthesized by the sol-gel method [34]. The perovskite $LaFeO_3$ presents a cubic phase (JCPDS-01-075-0541); however, it is possible to notice the presence of an orthorhombic phase (JCPDS-01-070-777). Meanwhile, the $LaCoO_3$ semiconductor powders show a double reflection 2θ = 32.8° – 33.3°, feature characteristic of the rhombohedral phase (JCPDS 01-084-0848). Since the perovskite $LaMnO_3$ is easily affected by the oxygen content, this fact can modify the unitary cell structure (Fig. 3b). It is common to find the nomenclature $LaMnO_{3+\delta}$, where δ represents the oxygen excess. Commonly, if δ increases the crystalline structure changes form the orthorhombic arrangement to the rhombohedral structure [71]. The material $LaMnO_3$ described in this section was indexed with the orthorhombic phase JCPDS 01-089-2471, an additional peak is observed at 2θ = 30°, such signal corresponds to La_2O_3 since it is very difficult to obtain the pure phase $LaMnO_3$. Regarding Co^{+3}, Mn^{+3}, and Fe^{+3}, the ionic radii are 2.18 Å < 2.10 Å < 2.01 Å, respectively [71]. Their internal angles B-O-B are 163°< 157°< 155°. As occurs for the family of materials $ATiO_3$, a larger ionic radio promotes the formation of angles B-O-B closer to 180° which makes easier the charge carrier's transference to the material surface. The optical characteristics are also an important parameter to observe the effect of the B ion substitution, and the UV-Vis spectra for the $LaBO_3$ (B: Co, Mn, Fe) perovskites (Fig. 3c) show light harvesting in the visible range. $LaMnO_3$ and $LaCoO_3$ present a wide absorption range in the visible electromagnetic spectrum; meanwhile, $LaFeO_3$ shows different transitions, the most notable at 550 nm. It is well known that the valence band of these materials is assigned the O_2-2*p* orbital, and the conduction band is assigned to the B^{+3} 3*d* orbitals. However, the crystalline field of the orbitals B^{+3}: *d* (B: Co, Mn) is divided into two sets t_{2g}, which contributes to the highest levels of the valence band and the levels e_g added in the conduction band [72]. Hence, it exists a spatial overlapping of the metal *d* orbitals and the oxygen 2*p* orbital, thus leading the presence of $d \rightarrow d$ transitions, which require less energy than $p \rightarrow d$ transitions. Particularly, the $LaBO_3$ (B = Fe, Co, Mn) perovskites present localized states in the levels e_g, which makes them more susceptible to $d \rightarrow d$ transitions, thus inducing a higher visible light absorption [41, 72, 73]. In comparison with the $ATiO_3$ perovskites (A = Zn, Cd, Pb), the substitution of the B cation changes significatively the optical properties, because as above-mentioned, the energy levels in this position conform the semiconductor conduction band.

The active species of cations B^{+3} located onto the material surface allow the performance of the redox reactions due to their multiple oxidation states. Since the electrochemical techniques are very sensible procedures to characterize and observe the oxidation transition states of heterostructured materials, cyclic voltammetry in Fig. 3d allows the cation oxidation transition from B^{+2} to B^{+3}, and also a partial transition in $LaMnO_3$ from carbonated species to MnO. From these results, it is possible to point out that $LaMnO_3$ y $LaCoO_3$ present a higher surface area and a higher concentration of active metal ions B (B = Co, Mn) [74] onto the perovskite surface, these species perform a key role in the capability of such semiconductor to photogenerated molecular hydrogen [75].

Regarding the photocatalytic performance toward the hydrogen evolution reaction, the results show a correlation between the H_2 generation and the B (B = Fe, Co, Mn) cation employed in the $LaBO_3$ structure. $LaMnO_3$ exhibits the highest production rate compared to the perovskites $LaFeO_3$ and $LaCoO_3$ (Fig. 4). A general

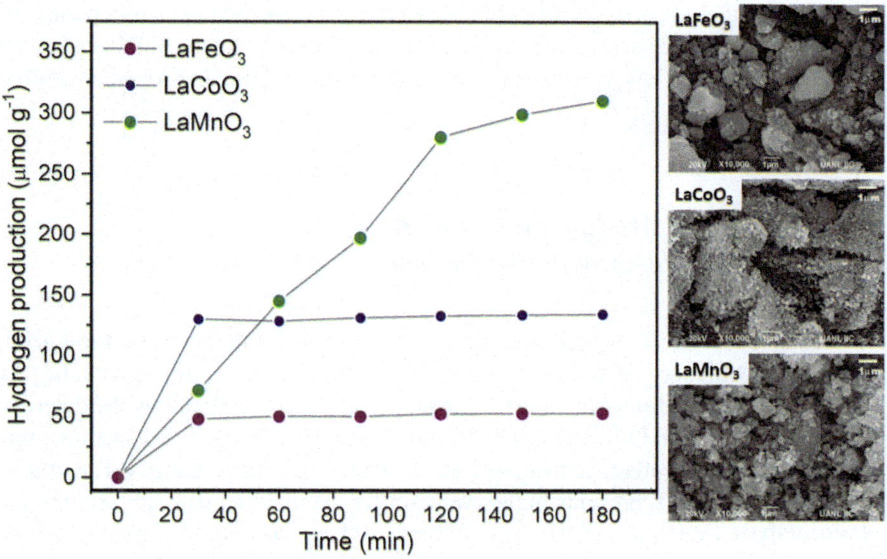

Fig. 4 Photocatalytic hydrogen production and scanning electron micrographs for $LaBO_3$ (A = Fe, Co, Mn). (Reprinted with permission from L. Ibarra-Rodriguez et al. J Phys Chem Solids, 136, 109189. Copyright (2020) Elsevier)

Table 2 Crystalline properties and photocatalytic H_2 production for $LaBO_3$ (B = Fe, Mn, Co)

Material	Phase	Crystallite Size (nm)	Average Grain Size (μm)	S_{BET} (m^2 g^{-1})	Bandgap (eV)	Actividad (μmolg^{-1} h^{-1})
$LaFeO_3$	Orthorhombic / cubic	22	0.429	16	2.39	17
$LaCoO_3$	Rhombohedral	25	0.300	19	1.62	44
$LaMnO_3$	Orthorhombic	10	0.174	26	1.76	103

summary of the photocatalytic activity and structural parameters are displayed in Table 2. Such behavior is attributed to the following suggestions: (i) $LaMnO_3$ presents a higher superficial area; in this way, there exist more actives sites which increase the surface contact and the photocatalytic activity; (i) a higher concentration of metal ions onto the material surface; (iii) the presence of the La_2O_3 phase which improves the charge carrier's transport since its lanthanide 4*f* level plays an important role in the interfacial charge transference and the decrease in the charge carrier's recombination rate [12, 76]. It is worth remembering that titanates family $ATiO_3$ (A = Zn, Cd, Pb), $PbTiO_3$ present the most ordered structure, and a Ti-O-Ti angle closer to 180°, hence the highest photocatalytic activity. In this context, the $LaBO_3$ (B = Fe, Co, Mn) perovskites show a similar result. The $LaMnO_3$ presents an angle B-O-B value of 163°, being this material the one that presents the highest photocatalytic activity. Such results show the influence of the crystalline structure over the photocatalytic performance. A free defect and highly symmetric structure enhance the charge carrier's transport. However, once the photogenerated charges are transported to the surface, the superficial characteristics like the number of metal active ions play a key role to improve the photocatalytic hydrogen evolution reaction.

3.3 *Superficial Modification: Key Role in the Photocatalytic Performance*

To further understand the influence rate of the cation B substituted in the $LaBO_3$ (B = Fe, Co, Mn) perovskites and its relation with the photocatalytic activity, further analyses have to be discussed. Superficial measurements by XPS allow the determination of the relation O-*ads*/O-*lattice* (O-*ads* = adsorbed oxygenated species), this relation is used as a reference to appreciate the material surface changes. The O-*ads*/O-*lattice* for $LaMnO_3$ presents higher values after the material was subjected to the photocatalytic evaluation (1.35 and 1.67 before (Fig. 5a) and after photocatalytic tests (Fig. 5b), respectively), this feature is related with a higher number of

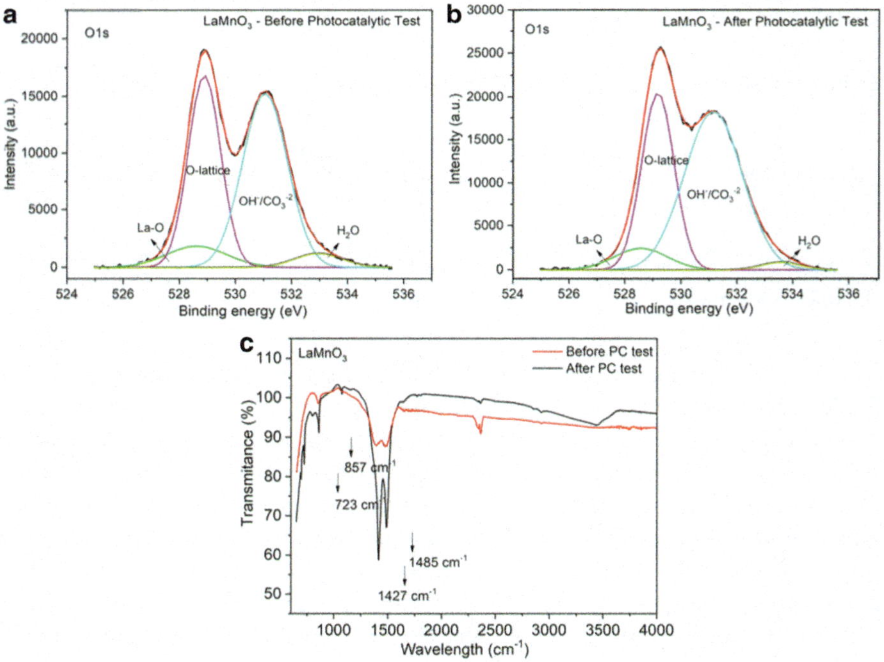

Fig. 5 XPS spectra, O 1s signal for $LaMnO_3$ (**a**) Before the photocatalytic tests and (**b**) After the photocatalytic test. (**c**) FTIR spectra for $LaMnO_3$ before and after the photocatalytic tests. ((**a**), (**b**) and (**c**) Reprinted with permission from L. Ibarra-Rodriguez et al. J Phys Chem Solids, 136, 109189. Copyright (2020) Elsevier)

oxygenated species on the $LaMnO_3$ surface mainly CO_3^{-2} and OH^-. To corroborate the presence of carbonate (CO_3^{-2}) or hydroxyl (OH^-) ions absorbed on the $LaMnO_3$ surface, the infrared FTIR measurements (Fig. 5c) confirm the presence of these species. It is possible to observe two vibrational bands (symmetric curves) at 723 and 857 cm^{-1}, as well as two asymmetric signals at 1427 and 1485 cm^{-1} corresponding to the carbonated anions [77, 78]. These signals become more intense after the photocatalytic reaction, thus indicates the formation of carbonate links. Such behavior can be attributed to the carbon dioxide adsorption over the perovskite surface which is later converted into carbonated species during the photocatalytic reaction. In this way, it is suggested that carbonate ions (CO_3^{-2}) compete for the available active sites over the semiconductor surface, causing in this way the deactivation of catalytic sites to perform the hydrogen generation. Considering the aforementioned explanation, Fig. 6 shows a mechanism to elucidate the role of the oxygenated species in the photocatalytic reaction. Once the light source promotes the photogeneration of the charge carriers (e^- and h^+), different phenomena take place. The photoexcited electrons are displaced to the conduction band, which is commonly composed of the *d* transition levels. From this point, the electrons can take two different transport ways: $d \rightarrow d$ (t2g) or O^{-2}: $2p \rightarrow$ M: d, this suggests that the exposed transition ions over the surface promote the formation or partial charges, increasing

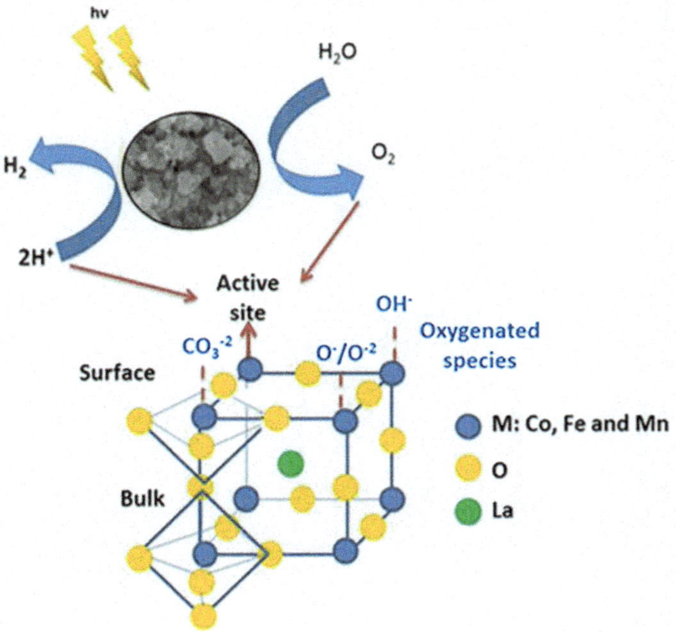

Fig. 6 Photocatalytic hydrogen evolution mechanism for $LaBO_3$ (A = Fe, Co, Mn) semiconductors. (Reprinted with permission from L. Ibarra-Rodriguez et al. J Phys Chem Solids, 136, 109189. Copyright (2020) Elsevier)

the available active sites to perform the redox reactions. However, partial charges are not attracting just water, they also attract another oxygenated species [79], such carbonated and oxygenated species, CO_3^{-2} and OH^- block the active sites decreasing the photoinduced hydrogen generation.

It has been demonstrated that the substitution of cations A or B in the perovskite ABO_3 affects the structural, morphological, and optical characteristics, thus modifies the photocatalytic activity of such materials. However, it is necessary the combination of multiple factors to obtain an adequate synergy. A very important factor is that the compounds $ZnTiO_3$ y $LaMnO_3$, which showed the highest photocatalytic activities, present a secondary phase. It is suggested that these secondary compounds can generate intermediate energetic levels which induce the charge carrier's generation. The theoretical diagrams for the materials $ZnTiO_3$, $CdTiO_3$ y $PbTiO_3$ show that the conduction band for $CdTiO_3$ is more negative than the conduction band for $ZnTiO_3$ and $PbTiO_3$ (Fig. 7a). This condition has been related to a higher photocatalytic activity to perform reduction processes [58, 68]. Although, regarding the $ATiO_3$ (A = Zn, Cd, Pb) materials, $PbTiO_3$ showed the best photocatalytic efficiency producing more hydrogen than $CdTiO_3$. This enhancement is related to the ferroelectric properties of $PbTiO_3$ [80] which induce spontaneous polarizations capable to produce dislocation of electric charges. This charge carrier's separation is favorable to reduce the holes – electrons recombination, enhancing in this way the photocatalytic activity [81]. On the other hand, the band diagram for $LaBO_3$ (B: Co, Mn, Fe) shows that La_2O_3 presents a wide bandgap ~4 eV (Fig. 7b); however, it is suggested a section these energy levels generates transition levels to act as a heterostructure [82].

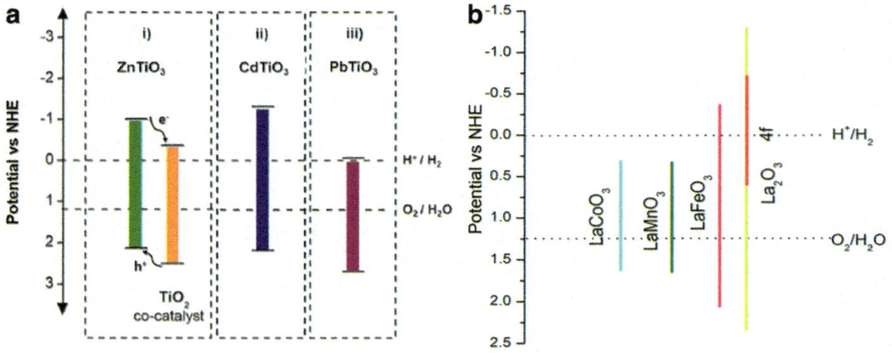

Fig. 7 (**a**) Theoretical band diagram for $ATiO_3$ (A = Zn, Cd, Pb). (**b**) Theoretical band diagram for $LaBO_3$ (A = Fe, Co, Mn) semiconductors. ((**a**) Reprinted with permission from O. Carrasco-Jaim et al. J Photoch Photobio A, 371, 98–108. Copyright (2019) Elsevier)

3.4 *Tailoring the Photocatalytic Activity Through the Heterojunction Formation*

Several strategies have been carried out to increase the photocatalytic activity of semiconductor materials to perform redox processes. The doping with metal and non-metal nanoparticles [54, 83], structural changes to reduce the bandgap values [84], sensitization with visible light materials [85], and the formation of heterojunctions [11, 27, 86] are some of the most used strategies to improve the visible light harvesting and avoid the charge carrier's recombination rate. Regarding these strategies, the heterojunction formation is one of the most efficient methodologies to improve the photocatalytic activity of semiconductors by controlling the electron-hole recombination rate [27]. Semiconductor heterojunction involves the stacking of two photocatalytic materials with different bandgaps. A direct bandgap transition is better than an indirect bandgap. The direct bandgap refers to the highest level of the valence band aligned with the lowest level on the conduction band; in this way, the direct recombination takes place releasing an energy amount equal to the energy difference between the recombining particles [87].

The sodium tantalate $NaTaO_3$ is an extensively studied perovskite reported as a suitable material to perform the photoelectrocatalytic hydrogen production from the water splitting under UV-Vis irradiation [32, 33, 88]. Taking into account the $NaTaO_3$ presents a bandgap ~4 eV [88], this perovskite is capable to generate charge carriers only under UV irradiation, and this fact decreases its effectiveness in technological scaling applications in devices irradiated with sunlight. On the other hand, although barium bismuth ($BaBiO_3$) does not show a high production of H_2 when it is used as a photoanode [89], this semiconductor presents a bandgap ~2 eV [90], which makes it a photocatalytic active material under visible irradiation. In this way, and continuing with the task to obtain better and more efficient photocatalytic materials, the following paragraphs describe the formation and photocatalytic performance of a heterojunction $BaBiO_3/NaTaO_3$ to elucidate the advantages of the heterostructures formation. Fig. 8a–c shows the particle sizes for the heterostructure, and this one decreases its particle size in comparison with bare $NaTaO_3$ and $BaBiO_3$. In turn, the heterojunctions containing 20 and 25% by weight of $BaBiO_3$ present the largest crystal size, which from the photocatalytic point of view reduces the recombination effects of charge carriers, improving the catalytic activity (Fig. 8d).

While it is true a larger crystal size affects and improves the photoelectrocatalytic activity of the heterojunction, this structural modification affects the optical and electrocatalytic parameters. The heterostructured material shows a decrease in the charge carrier's recombination rate as the amount of $BaBiO_3$ increases until 20%-$BaBiO_3/NaTaO_3$, this ratio presents the lowest recombination rate (Fig. 9a, b); therefore, the highest photocatalytic hydrogen production (Fig. 9c) [91]. The photoelectrochemical evaluation reveals the high effectivity in the formation of the heterojunction 20%-$BaBiO_3/NaTaO_3$ to perform the water splitting by the generation of •OH radicals to form oxygen onto the electrode surface and hence the hydrogen

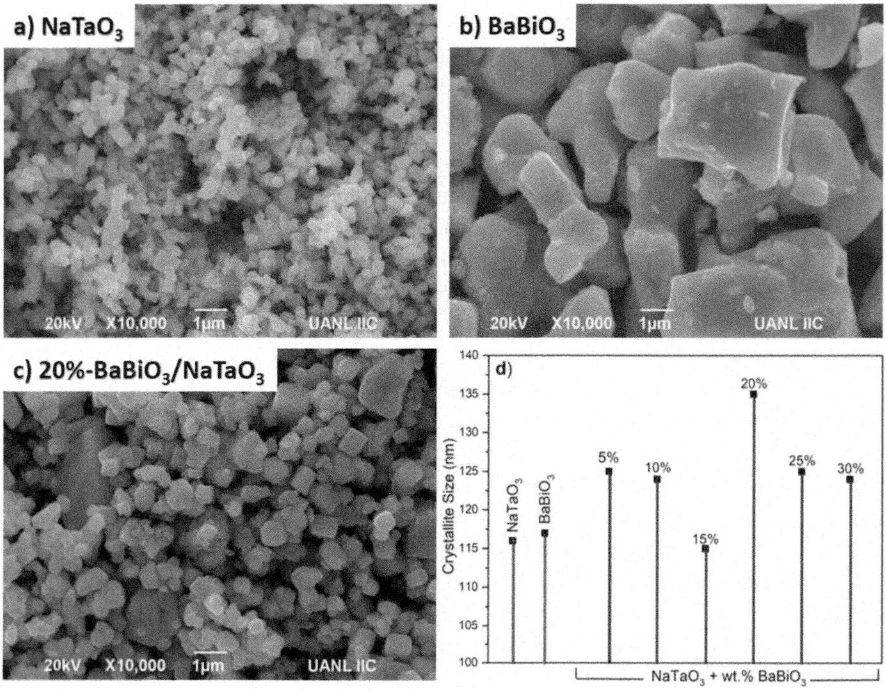

Fig. 8 (**a**) Scanning electron micrographs for (**a**) $NaTaO_3$, (**b**) $BaBiO_3$, and (**c**) 20%-BaBiO3/NaTaO3. Theoretical band diagram for $ATiO_3$ (A = Zn, Cd, Pb). (**d**) Crystallite sizes as wt.% $BaBiO_3$ is added. ((**a**) and (**b**) Reprinted with permission from J.M. Mora-Hernandez et al. J Photoch Photobio A, 391, 112363. Copyright (2018) Elsevier)

generation on the counter electrode (Fig. 9d). These results allow considering this material as an effective compound to perform the photoelectrocatalytic water splitting as an energy vector. It is suggested the photocatalytic enhancement is caused by: (i) an increase in the crystallite size compared to the bare perovskites, (ii) an optimal addition of $BaBiO_3$ (20 wt.%) which enhances the heterostructure synergy and decreases the charge transfer resistance, and iii) La continuous charge carrier's transference induced by energy bands overlapping, thus avoiding the recombination issues and improving the charge transference processes.

As mentioned above, heterojunctions have been developed for several applications, pollutant degradation [92], microbial disinfection [93], energy generation by H_2 production by water splitting [94], and CO_2 reduction [18]. Depending on the application needs, the heterojunction components are selected to generate specific reactive species. Graphitic carbon nitride (g-C_3N_4) is a novel semiconductor material that has been widely used conjointly with several perovskite materials such as $MnTiO_3$ [95], $CoTiO_3$ [96], $SrTiO_3$ [10], and $NaTaO_3$ [97], with application in degradation processes. On the other hand, the family of perovskites $LaMO_3$ (M: Co, Fe, Mn) has been employed in catalysts, sensors, and membranes in syngas applications [98]. There are several reports regarding the use of g-C_3N_4/$LaFeO_3$ in the hydrogen

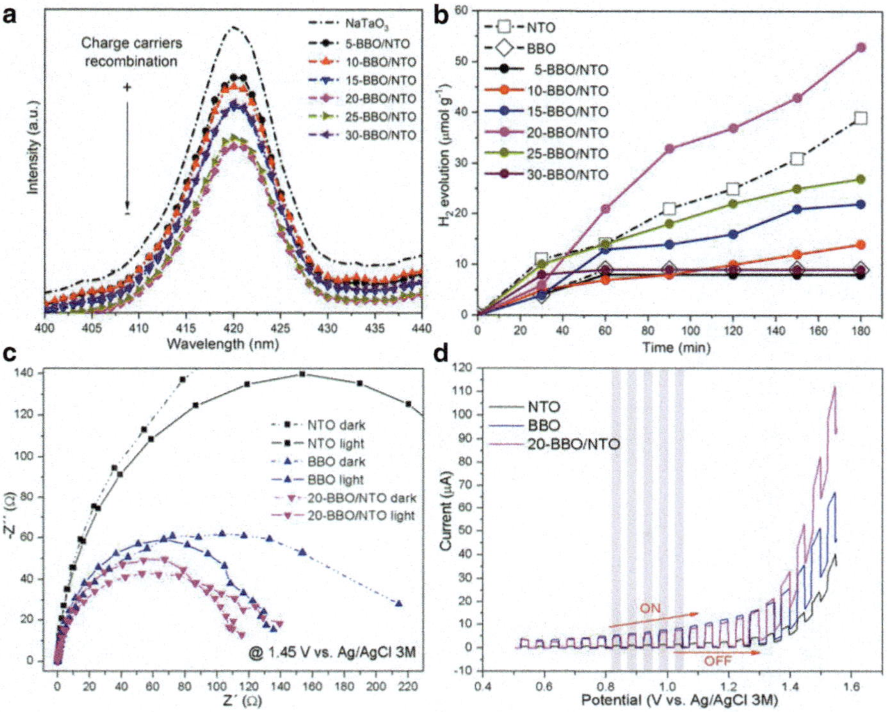

Fig. 9 (**a**) Charge carrier's recombination rate by photoluminescence measurements. (**b**) Photocatalytic hydrogen generation. (**c**) Charge transfer resistance under dark and light conditions. (**d**) Photoelectrochemical activity under oxidation scan potential. ((**a**), (**b**), (**c**) and (**d**) Reprinted with permission from J.M. Mora-Hernandez et al. J Photoch Photobio A, 391, 112363. Copyright (2018) Elsevier)

production in presence of sacrificial agents [43], as well as the employment of g-C_3N_4/$LaCoO_3$ to carry out the methyl orange degradation under visible light [46]. A comparative study of photocatalytic hydrogen evolution employing g-C_3N_4/(2, 5, 10%)$LaMO_3$ (M: Co, Mn, Fe) composites is analyzed to elucidate the mechanism involved in the enhanced charge transfer processes over the heterostructured semiconductor. Table 3 shows a summary of the physicochemical properties and the photocatalytic activity for g-C_3N_4/$LaBO_3$ (B: Co, Mn, Fe) heterostructures. In almost all cases, a $LaBO_3$ percentage above 2 wt.% affects negatively the compound performance. It is worth to mention that samples containing 2 wt.% of $LaBO_3$ (B: Co, Mn, Fe) exhibit the highest hydrogen generation. This feature can be related to a surface saturation where the perovskite particles block the g-C_3N_4 active sites. The addition of the second material enhances the photocatalytic activity, increasing the separation and transference of charge carriers. An energy band diagram (Fig. 10) allows us to observe when the materials are coupled, the electrons are transferred from the g-C_3N_4 conduction band to the perovskite conduction band. The electron transference continues until the Fermi energy level in both compounds is in

Table 3 Physicochemical properties and photocatalytic activity of g-C_3N_4/$LaBO_3$, B = Co, Mn, Fe

Material	wt. % Perovskite	Morphology	S_{BET} (m^2 g^{-1})	Bandgap (eV)	Activity (μmolg^{-1} h^{-1})
g-C_3N_4	***	Irregular particles	19	2.85	10
g-C_3N_4/ $LaMnO_3$	2%		17	2.85	40
	5%		–	2.83	14
	10%		–	2.81	11
g-C_3N_4/ $LaCoO_3$	2%		23	2.83	30
	5%		–	2.77	6
	10%		–	2.75	20
g-C_3N_4/ $LaFeO_3$	2%		16	2.86	140
	5%		–	2.84	2
	10%		–	2.84	11

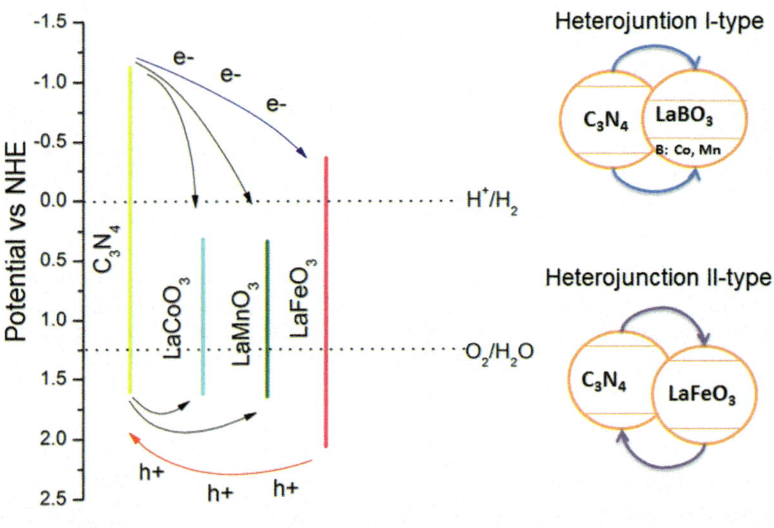

Fig. 10 Band diagrams, charge transference mechanism, and heterojunctions obtained by the coupling of $LaBO_3$ (B = Fe, Co, Mn) and g-C_3N_4

equilibrium, and this improves the charge carrier's separation and increases the photocatalytic activity. As the perovskite co-catalyst percentage increases, the spatial charge region becomes narrower, and the depth of light penetration into carbon nitride exceeds the space charge layer, this promotes the hole – electron recombination (Fig. 11) [99].

At this point, it is worth mentioning the definition and physicochemical and optical characteristics of the most common heterojunctions. Based on the different band and electronic structures, heterojunctions can be typically classified into three groups: type-I (straddling gap), type-II (staggered gap), and type-III (broken gap) (Fig. 12) [27]. For the type-I, the conduction band and the valence band levels for

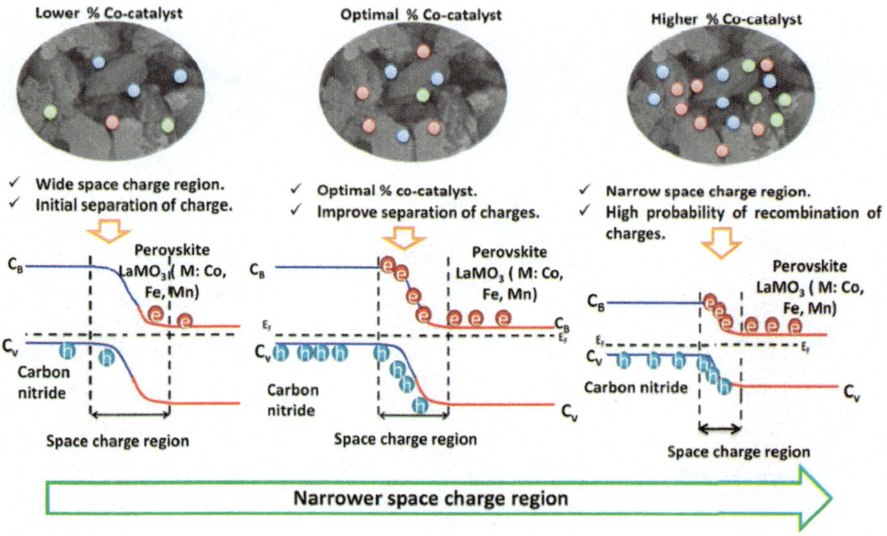

Fig. 11 Surface saturation diagram for the heterojunctions obtained by the coupling of $LaBO_3$ (B = Fe, Co, Mn) and g-C_3N_4. (Reprinted with permission from L. Ibarra-Rodriguez et al. Mater Sci Semicond Process, 103, 104643. Copyright (2019) Elsevier)

the semiconductor A (SA) are higher and lower than the CB and VB of semiconductor B (SB), the charge carriers migrate from the SA to the CB and VB of the SB, the charge carriers cannot separate effectively due to their accumulation on the SB (Fig. 12a). In type-II heterojunctions, the level of both, conduction and valence band of SB, is higher than both band levels in SA (Fig. 12b). In this way, electrons migrate from the to the CB of SA, and holes migrate to the VB of SB which results in effective spatial charge separation and transference. Fig. 12c depicts the band positions of type-III heterojunctions, in this case, the staggering gap becomes so wide that the bandgaps do not overlap. Thus, the synergic charge carrier's migration and separation between SA and SB do not occur. Among these three conventional heterojunctions, the type-II heterojunction is the most effective due to its spatial migration and charge carrier's separation. From the band diagram in Fig. 10, it is possible to observe that only the compounds g-C_3N_4 and $LaFeO_3$ present suitable position bands to form a type-ll heterostructure, which makes it the most efficient material to carry out the photocatalytic hydrogen evolution. On the other hand, the herejunctions generated by g-C_3N_4 and $LaCoO_3$ or $LaMnO_3$ form a type-l heterostructure. These two heterostructures act as light sensitizers; however, this better light harvesting does not improve the photocatalytic activity since all the charge carriers are accumulated on the $LaCoO_3$ or $LaMnO_3$.

The heterojunctions formation is one of the best strategies to enhances the photoelectrocatalytic activity in perovskite semiconductors; however, the coupling of two catalytic materials not always results in a heterojunction formation, sometimes the addition of a second material promotes the formation of intermedium states or

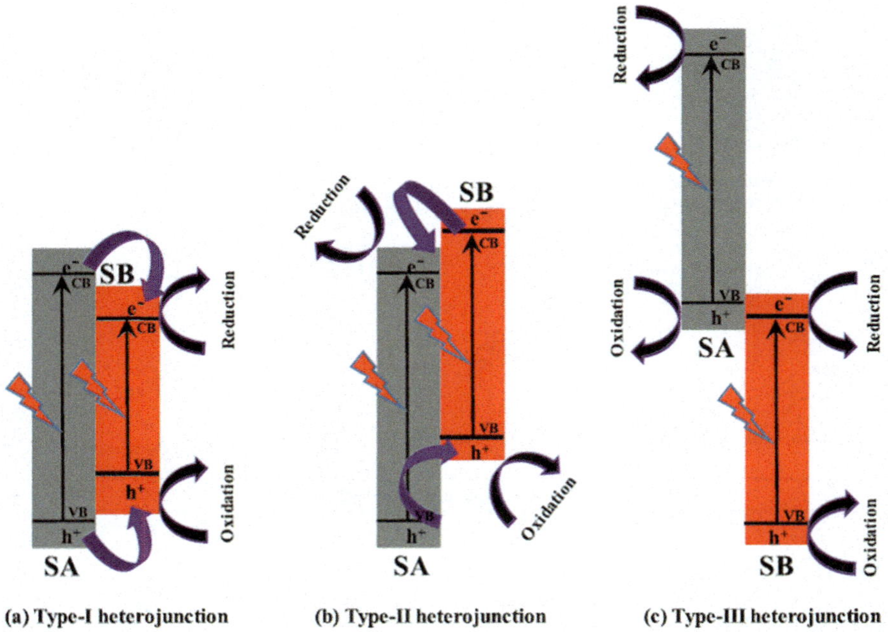

Fig. 12 Semiconductor heterojunctions and charge carriers transfer processes. (**a**) Type-l, (**b**) Type-ll, (**c**) Type-lll. (Reprinted with permission from A. Kumar et al. Water Res, 170, 115356. Copyright (2020) Elsevier)

Table 4 Quantitative composition obtained by EDXS. Carbon contained in the C-$NaTaO_3$ samples

Element (at. %)	As-Prepared	500 °C	550 °C	600 °C	650 °C	700 °C
Carbon K	46.75	37.34	34.93	32.94	29.60	28.89
Oxygen K	38.41	40.35	41.13	41.57	43.82	44.14
Sodium K	5.17	7.22	9.64	8.92	10.86	10.78
Tantalum K	9.67	15.08	14.3	16.39	15.72	16.18

defect levels capable to reduce the semiconductor bandgap [100, 101]. To elucidate this feature, this section presents the tailoring of a $NaTaO_3$ perovskite by the incorporation of amorphous carbon to produce the generation of intermediate states to decrease the bandgap broadening and enhance the photocatalytic activity to perform reduction processes. The carbon-doped $NaTaO_3$ (C-$NaTaO_3$) was obtained by a solvo-combustion method, which involves the use of organic and volatile substances that are responsible for the incorporation of amorphous carbon into the $NaTaO_3$ lattice (Table 4) [17]. The as-prepared C-$NaTaO_3$ contains a 47 at. % of carbon, and this sample was subjected to an annealing process to remove different amounts of carbon. The sample annealed at 650 °C (30 at. % carbon) presented the optimal physicochemical and optical properties to perform photocatalytic reduction processes. Figure 13a shows the DTA/TGA measurements, the mass loss before 200 °C corresponds to the water evaporation, meanwhile the decrease in mass from 250 to

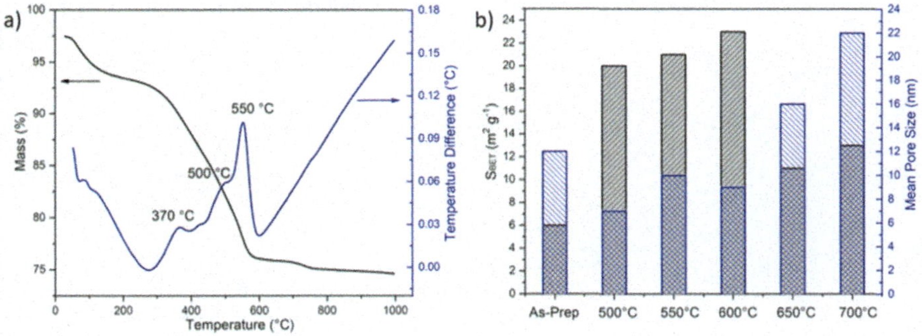

Fig. 13 (**a**) DTA/TGA plots for the as-prepared C-$NaTaO_3$ photocatalyst. (**b**) Specific surface area (SBET) and sample pore size (BHJ) for the as-prepared C-$NaTaO_3$ and samples annealed from 500 to 700 °C. ((**a**) Reprinted with permission from J.M. Mora-Hernandez et al. J CO_2 Util, 27, 179–187. Copyright (2018) Elsevier)

520 °C evidence the carbon removal as the temperature increases. The peak appreciated at 550 °C is attributes to the $NaTaO_3$ crystallization. From Fig. 13b, it is evident that the heat treatment process modifies the specific surface area S_{BET}. Considering that most of the carbon allotropes possess a high specific surface area [102, 103], the temperature increase promotes a significant S_{BET} loss above 600 °C. Additionally, the $NaTaO_3$ recrystallization causes the increase of the grain sizes promoting also the decrease of the S_{BET} [104]. Although, a high crystallinity enhances the photocatalytic activity of semiconductors due to a low number of crystalline defects responsible for the charge carrier's recombination [105]. In this way, an efficient photocatalytic semiconductor must exhibit an optimal relationship between a high crystallinity and a high surface area, and this optimal relationship is found in the sample annealed at 650 °C.

Since the sample C-$NaTaO_3$ annealed at 650 °C show the optimal relationship between surface area and crystallinity. This sample presents the highest photocatalytic activity to carry out the photoconversion of CO_2 to produce formaldehyde (Fig. 14a), overpassing the photocatalytic efficiency of samples annealed at different temperatures, and bare $NaTaO_3$. The high conversion rate achieved by C-$NaTaO_3$ at 650 °C is attributed to three main factors: (i) the reduction of the bandgap caused by the generation of intermediate states in the energy bands, (ii) an enhancement of the electrical conductivity by the addition of carbon in the $NaTaO_3$ structure, and (iii) an optimal ratio between the crystallization degree and the active surface area. Regarding the generation of intermediate states, Fig. 14b depicts the mechanism to carry out the photocatalytic generation and transportation of charge carriers. Typically, $NaTaO_3$ presents a UV light absorption attributed to the band transition for O2*p* (valence band) to the Ta5*d* (conduction band); however, the carbon incorporation induces the apparition of mid-gaps which reduce the original bandgap value [106], and this better light harvesting and a major number of photogenerated charges is indicative of the presence of oxygen vacancies and impurities in the $NaTaO_3$ lattice [107]. Nevertheless, the defects induced by the carbon-doping play

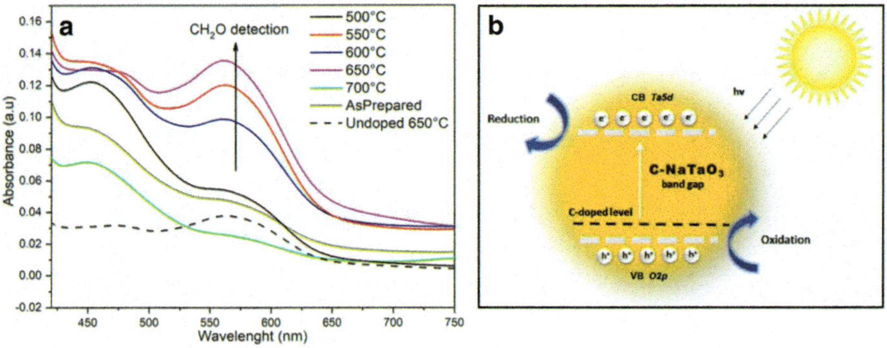

Fig. 14 (**a**) Photocatalytic CO_2 reduction over C-$NaTaO_3$ samples. (**b**) Mechanism diagram for induced visible light activity in C-$NaTaO_3$. (Reprinted with permission from J.M. Mora-Hernandez et al. J CO_2 Util, 27, 179–187. Copyright (2018) Elsevier)

a detrimental role, acting also as recombination centers and decreasing the number of active charge carriers to perform the redox processes. This is the reason why an appropriate combination of surface area, light absorption, and crystalline structure are determinant to improve the photocatalytic activity.

4 Conclusions

In the last decades, inorganic perovskite semiconductors have attracted the attention of the scientific community since these materials have presented remarkable results in the photocatalytic energy generation field. The modification of structural parameters induced by the substitution of cations A or B in the ABO_3 general formula, the heterojunction formation, and doping strategies have been employed as photocatalytic tailoring strategies. Such improvement strategies affect the psychochemical, optical, and photoelectrocatalytic properties of such semiconductors toward photocatalytic reduction processes. This chapter highlights the results obtained for three different perovskite families following the above-mentioned tailoring strategies. Regarding the $ATiO_3$ family, A = Zn, Cd, Pb, the $ZnTiO_3$ showed the best photocatalytic performance toward the hydrogen production (470 μmol g^{-1}), since this compound presented the highest crystallite size, thus a lower number of crystalline defects. The substitution of cation A in the $ATiO_3$ perovskites showed that cations with lower ionic radio generate a significant structure distortion, the compounds with a Ti-O-Ti angle closer to 180 °C presented better light harvesting. Additionally, the cations presenting energy levels lower than Ti^{3+} promote a photocatalysts redshifting. On the other hand, in the perovskites family $LaBO_3$, B = Co, Mn, Fe, the compound $LaMnO_3$ presented the highest photocatalytic hydrogen production (103 μmol g^{-1}), such enhancement is attributed to the presence of the secondary phase La_2O_3 which promotes a 4*f* modification levels improving the charge carrier's transference. Also, this material presented a higher superficial area, thus exists a higher number of photocatalytic active sites, as well as a higher number of metallic ions responsible to generate partial charges where water molecules can be adsorbed and later reduced. Similarly, to the $ATiO_3$ family, the cation B substitution induced promoted a structural distortion. The use of lower cation radii generates less ordered structures, the perovskite structure presents enough internal space to generate higher structural distortions. The transition metals Co, Fe, and Mn present energetic transitions *d*→*d*, which require less energy and allow the activation of the $LaBO_3$, B = Co, Mn, Fe in the visible light region. On the other hand, the heterojunction formation and doping strategies also promote better light harvesting and enhancement in the photocatalytic activity of some perovskites. For the $BaBiO_3/NaTaO_3$ heterojunction, the addition of 20 wt.% $BaBiO_3$ enhanced the visible light absorption and increase the heterostructures crystallite size in comparison with bare semiconductors, this structural modification decreases the grain limit defects number, thus reducing the charge carrier's recombination rate. Such photocatalytic improvement was also presented by the heterojunction $ZnTiO_3/TiO_2$, where the heterostructure formation induced better charge carrier's transference by the modification of the energy levels to generate a type-ll heterojunction. Regarding the carbon-doped $NaTaO_3$ compound (C-$NaTaO_3$), the presence of 30 at. % of amorphous carbon improved the photocatalytic activity of the C-$NaTaO_3$ composite. The carbon amount was a key parameter to improve and reduce the bandgap value and diminish the charge transference resistance by an improvement in the electrical conductivity, but more important, by the generation of intermediate energy levels in the $NaTaO_3$ structure capable to reduce the bandgap. The optimal ratio between the surface area and crystallinity of the sample was achieved by the C-$NaTaO_3$ annealed at 650 °C. These parameters play an important role to explain an improved photocatalytic activity caused by a higher reactive surface and lower recombination of the charges photogenerated onto larger crystals. The photoelectrochemical tests revealed that the material annealed at 650 °C is the only sample with a conduction band value more negative than the potential required to perform the generation of formaldehyde from the photocatalytic reduction of carbon dioxide in aqueous solution. Is it suggested that the appropriate combination and synergy of crystalline structure, light absorption, and surface area properties of the samples are determinant factors in the photocatalytic activity in carbon-doped materials.

References

1. D. Cahen, I. Lubomirsky, Energy, the global challenge, and materials. Mater. Today **11**(12), 16–20 (2008)
2. Mahjabeen et al., Renewable energy, institutional stability, environment and economic growth nexus of D-8 countries. Energ. Strat. Rev. **29**, 100484 (2020)
3. K. Saidi, A. Omri, The impact of renewable energy on carbon emissions and economic growth in 15 major renewable energy-consuming countries. Environ. Res. **186**, 109567 (2020)

4. N.L. Panwar, S.C. Kaushik, S. Kothari, Role of renewable energy sources in environmental protection: A review. Renew. Sust. Energ. Rev. **15**(3), 1513–1524 (2011)
5. P. Breeze, Chapter 13 – Solar power, in *Power Generation Technologies*, 3rd edn, ed. by P. Breeze (Newnes, 2019), pp. 293–321
6. G. Knör, U. Monkowius, Photosensitization and photocatalysis in bioinorganic, bio-organometallic and biomimetic systems, in *Advances in Inorganic Chemistry*, eds. by R.v. Eldik, G. Stochel (Academic, 2011), pp. 235–289
7. O.K. Echendu et al., Electrochemical deposition and characterization of ZnOS thin films for photovoltaic and photocatalysis applications. J. Alloys Compd. **769**, 201–209 (2018)
8. R. Li, Latest progress in hydrogen production from solar water splitting via photocatalysis, photoelectrochemical, and photovoltaic-photoelectrochemical solutions. Chin. J. Catal. **38**(1), 5–12 (2017)
9. J. Guo et al., Synthesis and characterization of $Ag_3PO_4/LaCoO_3$ nanocomposite with superior mineralization potential for bisphenol A degradation under visible light. J. Alloys Compd. **696**, 226–233 (2017)
10. P.-S. Konstas et al., Synthesis, characterization of g-$C_3N_4/SrTiO_3$ heterojunctions and photocatalytic activity for organic pollutants degradation. Catalysts **8**(11), 554 (2018)
11. Parul et al., Photodegradation of organic pollutants using heterojunctions: A review. J. Environ. Chem. Eng. **8**(2), 103666 (2020)
12. V.H.-T. Thi, B.-K. Lee, Effective photocatalytic degradation of paracetamol using La-doped ZnO photocatalyst under visible light irradiation. Mater. Res. Bull. **96**, 171–182 (2017)
13. X. Yu et al., Mesocrystalline Ta_2O_5 nanosheets supported PdPt nanoparticles for efficient photocatalytic hydrogen production. Int. J. Hydrog. Energy **43**(17), 8232–8242 (2018)
14. L.F. Garay-Rodríguez et al., Photocatalytic hydrogen evolution over the isostructural titanates: $Ba_3Li_2Ti_8O_{20}$ and $Na_2Ti_6O_{13}$ modified with metal oxide nanoparticles. Int. J. Hydrog. Energy **43**(4), 2148–2159 (2018)
15. M. Wang et al., Effects of sacrificial reagents on photocatalytic hydrogen evolution over different photocatalysts. J. Mater. Sci. **52**(9), 5155–5164 (2017)
16. M. Akatsuka et al., Preparation of Ga_2O_3 photocatalyst highly active for CO_2 reduction with water without cocatalyst. Appl. Catal. B Environ. **262**, 118247 (2020)
17. J.M. Mora-Hernandez, A.M. Huerta-Flores, L.M. Torres-Martínez, Photoelectrocatalytic characterization of carbon-doped $NaTaO_3$ applied in the photoreduction of CO_2 towards the formaldehyde production. J. CO2 Util. **27**, 179–187 (2018)
18. B. Han et al., The photoelectrocatalytic CO_2 reduction on TiO_2@ZnO heterojunction by tuning the conduction band potential. Electrochim. Acta **285**, 23–29 (2018)
19. R. Li, Y. Zhao, C. Li, Spatial distribution of active sites on a ferroelectric $PbTiO_3$ photocatalyst for photocatalytic hydrogen production. Faraday Discuss. **198**(0), 463–472 (2017)
20. Y. Li et al., Litchi-like CdS/$CdTiO_3$–TiO_2 composite: synthesis and enhanced photocatalytic performance for crystal violet degradation and hydrogen production. RSC Adv. **6**(56), 51374–51386 (2016)
21. S. Liu et al., Photocatalytic activity of $MTiO_3$ (M = Ca, Ni, and Zn) nanocrystals for water decomposition to hydrogen. J. Mater. Res. **29**(11), 1295–1301 (2014)
22. K.H. Reddy, K. Parida, Fabrication, characterization, and photoelectrochemical properties of Cu-doped $PbTiO_3$ and its hydrogen production activity. ChemCatChem **5**(12), 3812–3820 (2013)
23. M. Shaterian et al., Synthesis, characterization and photocatalytic activity of $LaMnO_3$ nanoparticles. Appl. Surf. Sci. **318**, 213–217 (2014)
24. J. Hu et al., Ag modified $LaMnO_3$ nanorods-reduced graphene oxide composite applied in the photocatalytic discoloration of direct green. Solid State Sci. **61**, 239–245 (2016)
25. J. Yang et al., A novel p-$LaFeO_3$/n-Ag_3PO_4 heterojunction photocatalyst for phenol degradation under visible light irradiation. Chem. Commun. **52**(12), 2620–2623 (2016)
26. Q. Zhang et al., Perovskite $LaFeO_3$-$SrTiO_3$ composite for synergistically enhanced NO removal under visible light excitation. Appl. Catal. B Environ. **204**, 346–357 (2017)

27. J. Low et al., Heterojunction photocatalysts. Adv. Mater. **29**(20), 1601694 (2017)
28. H.W. Kang, S.B. Park, Water photolysis by $NaTaO_3$-C composite prepared by spray pyrolysis. Adv. Powder Technol. **21**(2), 106–110 (2010)
29. G.R. Portugal, S.F. Santos, J.T. Arantes, $NaTaO_3$ cubic and orthorhombic surfaces: An intrinsic improvement of photocatalytic properties. Appl. Surf. Sci. **502**, 144206 (2020)
30. H. Sudrajat et al., Stability of La dopants in $NaTaO_3$ photocatalysts. J. Alloys Compd. **775**, 1277–1285 (2019)
31. Y.-X. Zhao et al., Preparation, characterization and photocatalytic activity of N-doped $NaTaO_3$ nanocubes. Powder Technol. **214**(1), 155–160 (2011)
32. C. Gómez-Solís et al., Facile solvo-combustion synthesis of crystalline $NaTaO_3$ and its photocatalytic performance for hydrogen production. Fuel **130**, 221–227 (2014)
33. A.M. Huerta-Flores et al., Laser assisted chemical vapor deposition of nanostructured $NaTaO_3$ and $SrTiO_3$ thin films for efficient photocatalytic hydrogen evolution. Fuel **197**, 174–185 (2017)
34. L.M. Torres-Martínez et al., Enhanced photocatalytic water splitting hydrogen production on RuO_2/La:$NaTaO_3$ prepared by sol–gel method. Catal. Commun. **12**(4), 268–272 (2010)
35. A. Zerrahn, W.-P. Schill, C. Kemfert, On the economics of electrical storage for variable renewable energy sources. Eur. Econ. Rev. **108**, 259–279 (2018)
36. A. Fujishima, K. Honda, Electrochemical photolysis of water at a semiconductor electrode. Nature **238**(5358), 37–38 (1972)
37. R. Konta et al., Photocatalytic activities of noble metal ion doped $SrTiO_3$ under visible light irradiation. J. Phys. Chem. B **108**(26), 8992–8995 (2004)
38. L. Macaraig, S. Chuangchote, T. Sagawa, Electrospun $SrTiO_3$ nanofibers for photocatalytic hydrogen generation. J. Mater. Res. **29**(1), 123–130 (2013)
39. H. Zhou et al., Artificial photosynthesis on tree trunk derived alkaline tantalates with hierarchical anatomy: Towards CO_2 photo-fixation into CO and CH_4. Nanoscale **7**(1), 113–120 (2015)
40. T. Kida et al., Photocatalytic hydrogen production from water over a $LaMnO_3$/CdS nanocomposite prepared by the reverse micelle method. J. Mater. Chem. **13**(5), 1186–1191 (2003)
41. D. Meziani et al., Hydrogen evolution under visible light over $LaCoO_3$ prepared by chemical route. Energy Convers. Manag. **82**, 244–249 (2014)
42. J. Li et al., Synergetic effect of copper species as cocatalyst on $LaFeO_3$ for enhanced visible-light photocatalytic hydrogen evolution. Int. J. Hydrog. Energy **40**(40), 13918–13925 (2015)
43. Y. Wu et al., Quasi-polymeric construction of stable perovskite-type $LaFeO_3$/g-C_3N_4 heterostructured photocatalyst for improved Z-scheme photocatalytic activity via solid p-n heterojunction interfacial effect. J. Hazard. Mater. **347**, 412–422 (2018)
44. K. Xu, J. Feng, Superior photocatalytic performance of $LaFeO_3$/g-C_3N_4 heterojunction nanocomposites under visible light irradiation. RSC Adv. **7**(72), 45369–45376 (2017)
45. Y. Yu et al., Recent advances in rare-earth elements modification of inorganic semiconductor-based photocatalysts for efficient solar energy conversion: A review. J. Rare Earths **33**(5), 453–462 (2015)
46. J. Luo et al., Utilization of $LaCoO_3$ as an efficient co-catalyst to boost the visible light photocatalytic performance of g-C_3N_4. Sep. Purif. Technol. **201**, 309–317 (2018)
47. J. Shi, L. Guo, ABO3-based photocatalysts for water splitting. Prog. Nat. Sci. Mater. Int. **22**(6), 592–615 (2012)
48. A. Kudo, Y. Miseki, Heterogeneous photocatalyst materials for water splitting. Chem. Soc. Rev. **38**(1), 253–278 (2009)
49. L. Theofylaktos et al., Perovskites with d-block metals for solar energy applications. Dalton Trans. **48**(26), 9516–9537 (2019)
50. A. Kudo, H. Kato, Photocatalytic decomposition of water into H_2 and O_2 over novel photocatalyst $K_3Ta_3Si_2O_{13}$ with Pillared structure consisting of three TaO_6 chains. Chem. Lett. **26**(9), 867–868 (1997)
51. T. Xu, X. Zhao, Y. Zhu, Synthesis of hexagonal $BaTa_2O_6$ nanorods and influence of defects on the photocatalytic activity. J. Phys. Chem. B **110**(51), 25825–25832 (2006)

52. H. Kato, A. Kudo, Photocatalytic decomposition of pure water into H_2 and O_2 over $SrTa_2O_6$ prepared by a flux method. Chem. Lett. **28**(11), 1207–1208 (1999)
53. K. Maeda, Rhodium-doped barium titanate Perovskite as a stable p-type semiconductor photocatalyst for hydrogen evolution under visible light. ACS Appl. Mater. Interfaces **6**(3), 2167–2173 (2014)
54. R. Wang et al., Hollow $CaTiO_3$ cubes modified by La/Cr co-doping for efficient photocatalytic hydrogen production. Appl. Catal. B Environ. **225**, 139–147 (2018)
55. K. Maeda, Photocatalytic water splitting using semiconductor particles: History and recent developments. J. Photochem. Photobiol. C Photochem. Rev. **12**(4), 237–268 (2011)
56. K. Maeda, K. Domen, New non-oxide photocatalysts designed for overall water splitting under visible light. J. Phys. Chem. C **111**(22), 7851–7861 (2007)
57. X. Li et al., Engineering heterogeneous semiconductors for solar water splitting. J. Mater. Chem. A **3**(6), 2485–2534 (2015)
58. A. Kudo, H. Kato, S. Nakagawa, Water splitting into H_2 and O_2 on new $Sr_2M_2O_7$ (M = Nb and Ta) photocatalysts with layered perovskite structures: Factors affecting the photocatalytic activity. J. Phys. Chem. B **104**(3), 571–575 (2000)
59. H. Kato, A. Kudo, Water splitting into H_2 and O_2 on alkali tantalate photocatalysts $ATaO_3$ (A = Li, Na, and K). J. Phys. Chem. B **105**(19), 4285–4292 (2001)
60. M. Salavati-Niasari et al., Synthesis, characterization, and morphological control of $ZnTiO_3$ nanoparticles through sol-gel processes and its photocatalyst application. Adv. Powder Technol. **27**(5), 2066–2075 (2016)
61. R.D. Shannon, Revised effective ionic radii and systematic studies of interatomic distances in halides and chalcogenides. Acta Crystallographica Sect. A **32**(5), 751–767 (1976)
62. H. Kato, H. Kobayashi, A. Kudo, Role of Ag^+ in the band structures and photocatalytic properties of $AgMO_3$ (M: Ta and Nb) with the Perovskite structure. J. Phys. Chem. B **106**(48), 12441–12447 (2002)
63. T. Luttrell et al., Why is anatase a better photocatalyst than rutile? Model studies on epitaxial TiO2 films. Sci. Rep. **4**(1), 4043 (2014)
64. B. Pant, M. Park, S.-J. Park, Synthesis, characterization, and photocatalytic performances of electrospun cadmium titanate nanofibers immobilized into the reduced graphene oxide sheets. Mater. Lett. **228**, 365–368 (2018)
65. M. Zheng et al., Identification of substitution mechanism in group VIII metal oxides doped Pb(Zn1/3Nb2/3)O_3–$PbZrO_3$–$PbTiO_3$ ceramics with high energy density and mechanical performance. J. Am. Ceram. Soc. **96**(8), 2486–2492 (2013)
66. Z. Ali et al., Structural and optoelectronic properties of the zinc titanate perovskite and spinel by modified Becke–Johnson potential. Phys. B Condens. Matter **420**, 54–57 (2013)
67. L. Wang et al., First-principles study of tetragonal $PbTiO_3$: Phonon and thermal expansion. Mater. Res. Bull. **49**, 509–513 (2014)
68. W. Zhang, J. Tang, J. Ye, Structural, photocatalytic, and photophysical properties of perovskite $MSnO_3$ (M = Ca, Sr, and Ba) photocatalysts. J. Mater. Res. **22**(7), 1859–1871 (2011)
69. D. Guerrero-Araque et al., SnO_2–TiO_2 structures and the effect of CuO, CoO metal oxide on photocatalytic hydrogen production. J. Chem. Technol. Biotechnol. **92**(7), 1531–1539 (2017)
70. M.L. Rosenbluth, N.S. Lewis, "Ideal" behavior of the open circuit voltage of semiconductor/liquid junctions. J. Phys. Chem. **93**(9), 3735–3740 (1989)
71. Q. Huang et al., Structure and magnetic order in undoped lanthanum manganite. Phys. Rev. B **55**(22), 14987–14999 (1997)
72. L.V. Nomerovannaya et al., The influence of the Co^{3+} spin state on the optical properties of $LaCoO_3$and $HoCoO_3$. J. Phys. Condens. Matter **16**(28), 5129–5136 (2004)
73. S. Gong, B.-G. Liu, Electronic energy gaps and optical properties of $LaMnO_3$. Phys. Lett. A **375**(12), 1477–1480 (2011)
74. V. Celorrio et al., Mean intrinsic activity of single Mn sites at $LaMnO_3$ nanoparticles towards the oxygen reduction reaction. ChemElectroChem **5**(20), 3044–3051 (2018)

75. J.K. Nørskov et al., The nature of the active site in heterogeneous metal catalysis. Chem. Soc. Rev. **37**(10), 2163–2171 (2008)
76. Y. Zhang et al., Surface phase of TiO_2 modified with La_2O_3 and its effect on the photocatalytic H_2 evolution. Mater. Res. Bull. **53**, 107–115 (2014)
77. H. Hu et al., Morphology-controlled hydrothermal synthesis of $MnCO_3$ hierarchical superstructures with Schiff base as stabilizer. Mater. Res. Bull. **46**(11), 1908–1915 (2011)
78. X.-h. Zhang et al., Synthesis, characterization and nonisothermal decomposition kinetics of $La_2(CO_3)_3{\cdot}3.4H_2O$. J. Therm. Anal. Calorim. **119**(3), 1713–1722 (2015)
79. J. Suntivich et al., Design principles for oxygen-reduction activity on perovskite oxide catalysts for fuel cells and metal–air batteries. Nat. Chem. **3**(7), 546–550 (2011)
80. M.-H. Whangbo et al., Tolerance factor and cation–anion orbital interactions differentiating the polar and antiferrodistortive structures of perovskite oxides ABO_3. Z. Anorg. Allg. Chem. **641**(6), 1043–1052 (2015)
81. Y. Cui, J. Briscoe, S. Dunn, Effect of ferroelectricity on solar-light-driven photocatalytic activity of $BaTiO_3$—Influence on the carrier separation and stern layer formation. Chem. Mater. **25**(21), 4215–4223 (2013)
82. S.V. Eremeev et al., Interface induced states at the boundary between a 3D topological insulator Bi_2Se_3 and a ferromagnetic insulator EuS. J. Magn. Magn. Mater. **383**, 30–33 (2015)
83. A.K. Adhikari, K.-S. Lin, Improving CO_2 adsorption capacities and CO_2/N_2 separation efficiencies of MOF-74(Ni, Co) by doping palladium-containing activated carbon. Chem. Eng. J. **284**, 1348–1360 (2016)
84. S. Zhuiykov, Chapter 1 – Structural chemical modification of semiconductor nano-crystals, in *Nanostructured Semiconductors,* 2nd edn, ed. by S. Zhuiykov (Woodhead Publishing, 2018), pp. 1–52
85. A. Soto-Arreola et al., Comparative study of the photocatalytic activity for hydrogen evolution of MFe_2O_4 (M = Cu, Ni) prepared by three different methods. J. Photochem. Photobiol. A Chem. **357**, 20–29 (2018)
86. A. Kumar et al., Recent developments and challenges in practical application of visible–light–driven TiO_2–based heterojunctions for PPCP degradation: A critical review. Water Res. **170**, 115356 (2020)
87. M.A. Mahadadalkar, S.W. Gosavi, B.B. Kale, Interstitial charge transfer pathways in a TiO_2/$CdIn_2S_4$ heterojunction photocatalyst for direct conversion of sunlight into fuel. J. Mater. Chem. A **6**(33), 16064–16073 (2018)
88. B. Modak, K. Srinivasu, S.K. Ghosh, Photocatalytic activity of $NaTaO_3$ doped with N, Mo, and (N,Mo): A hybrid density functional study. J. Phys. Chem. C **118**(20), 10711–10719 (2014)
89. A.M. Huerta-Flores et al., Visible-light-driven $BaBiO_3$ perovskite photocatalysts: Effect of physicochemical properties on the photoactivity towards water splitting and the removal of rhodamine B from aqueous systems. J. Photochem. Photobiol. A Chem. **368**, 70–77 (2019)
90. A.S. Chouhan et al., BaBiO3: A potential absorber for all-oxide photovoltaics. Mater. Lett. **210**, 218–222 (2018)
91. M. Humayun et al., Modification strategies of TiO_2 for potential applications in photocatalysis: A critical review. Green Chem. Lett. Rev. **11**(2), 86–102 (2018)
92. J. Qiu et al., A recyclable nanosheet of Mo/N-doped TiO_2 nanorods decorated on carbon nanofibers for organic pollutants degradation under simulated sunlight irradiation. Chemosphere **215**, 280–293 (2019)
93. P. Ganguly et al., Antimicrobial activity of photocatalysts: Fundamentals, mechanisms, kinetics and recent advances. Appl. Catal. B Environ. **225**, 51–75 (2018)
94. S.A. Rawool et al., pn Heterojunctions in NiO:TiO_2 composites with type-II band alignment assisting sunlight driven photocatalytic H_2 generation. Appl. Catal. B Environ. **221**, 443–458 (2018)
95. X. Li et al., Hydrothermal synthesized novel nanoporous g-C_3N_4/$MnTiO_3$ heterojunction with direct Z-scheme mechanism. Electrochim. Acta **258**, 998–1007 (2017)

96. Q. Wang et al., The flux growth of single-crystalline $CoTiO_3$ polyhedral particles and improved visible-light photocatalytic activity of heterostructured $CoTiO_3$/g-C_3N_4 composites. Dalton Trans. **45**(44), 17748–17758 (2016)
97. T.-H. Kim et al., Synthesis of g-C_3N_4/$NaTaO_3$ hybrid composite photocatalysts and their photocatalytic activity under simulated solar light irradiation. J. Nanosci. Nanotechnol. **15**(9), 7125–7129 (2015)
98. Y. Liu et al., Controlled generation of uniform spherical $LaMnO_3$, $LaCoO_3$, Mn_2O_3, and Co_3O_4 nanoparticles and their high catalytic performance for carbon monoxide and toluene oxidation. Inorg. Chem. **52**(15), 8665–8676 (2013)
99. F. Ling et al., PdO/$LaCoO_3$ heterojunction photocatalysts for highly hydrogen production from formaldehyde aqueous solution under visible light. Int. J. Hydrog. Energy **41**(14), 6115–6122 (2016)
100. W. Körner, D.F. Urban, C. Elsässer, Origin of subgap states in amorphous In-Ga-Zn-O. J. Appl. Phys. **114**(16), 163704 (2013)
101. Q. Tay et al., Defect engineered g-C_3N_4 for efficient visible light photocatalytic hydrogen production. Chem. Mater. **27**(14), 4930–4933 (2015)
102. P. Plerdsranoy et al., Effects of specific surface area and pore volume of activated carbon nanofibers on nanoconfinement and dehydrogenation of $LiBH_4$. Int. J. Hydrog. Energy **42**(9), 6189–6201 (2017)
103. M.K.N. Yenkie, G.S. Natarajan, Determination of specific surface area of granular activated carbon by aqueous phase adsorption of phenol and from pore size distribution measurements. Sep. Sci. Technol. **28**(5), 1177–1190 (1993)
104. A. Suslova et al., Recrystallization and grain growth induced by ELMs-like transient heat loads in deformed tungsten samples. Sci. Rep. **4**(1), 6845 (2014)
105. K. Alberi, M.A. Scarpulla, Suppression of compensating native defect formation during semiconductor processing via excess carriers. Sci. Rep. **6**(1), 27954 (2016)
106. A.S. Alshammari et al., Visible-light photocatalysis on C-doped ZnO derived from polymer-assisted pyrolysis. RSC Adv. **5**(35), 27690–27698 (2015)
107. Z. Jiang et al., Enhanced visible-light-driven photocatalytic activity of mesoporous TiO_2–xNx derived from the ethylenediamine-based complex. Nanoscale **5**(12), 5396–5402 (2013)

Part III
Cuprate Thin Film Materials and Nanocomposites

Potential of Copper Oxide High-Temperature Superconductors for Tailoring Ferromagnetic Spin Textures

Jordi Alcalà, Mercè Roig, Sergi Martín, Aleix Barrera, Alejandro Fernández-Rodríguez, Alberto Pomar, Lluís Balcells, Mariona Coll, Narcís Mestres, and Anna Palau

1 Introduction

Information technology revolution has powered the demand for large-scale computer facilities and data centres with extraordinarily high energy demands. Future of high-performance computing, with unprecedented society benefits, is completely unfeasible with today's technologies in light of their unsustainable power consumption [1]. The challenge ahead is a revolution with energy-efficient concepts and technologies able to reduce the related environmental and economic impacts.

Novel systems based on strongly correlated oxides are among the most outstanding challenges towards realizing high-performance energy-efficient functional devices for information storage and processing [2–4]. The interplay between charge and structural degrees of freedom in these materials can give rise to a numerous exotic ordering phenomenon, including colossal magnetoresistance in manganites, high-temperature superconductivity in copper oxides. Moreover, multiple transitions between distinct competing phases with drastically different electronic and magnetic order (ferromagnetic, paramagnetic, or antiferromagnetic ordering, superconductivity or Mott insulators), can be induced [5].

Superconducting technology, by virtue of its inherent low power dissipation, high speed, and lossless interconnect, emerges as a very promising and reliable approach to close the growing gap between required performance and sustainability in "big data" manipulation [6, 7]. Copper oxides, and particularly $YBa_2Cu_3O_{7-x}$

J. Alcalà · M. Roig · S. Martín · A. Barrera · A. Fernández-Rodríguez · A. Pomar · L. Balcells
M. Coll · N. Mestres · A. Palau (✉)
Institut de Ciència de Materials de Barcelona, ICMAB-CSIC, Bellaterra, Spain
e-mail: jalcala@icmab.es; smartin2@icmab.es; afernandez3@icmab.es; apomar@icmab.es; balcells@icmab.es; mcoll@icmab.es; narcis.mestres@icmab.es; palau@icmab.es

A. G. Roca et al. (eds.), *Surfaces and Interfaces of Metal Oxide Thin Films, Multilayers, Nanoparticles and Nano-composites*,
https://doi.org/10.1007/978-3-030-74073-3_7

(YBCO), have received significant attention for electronic applications since in addition to their superconducting behaviour they offer unique properties of non-conventional high-temperature superconductors. In particular, tuning of their carrier density to induce a reversible modulation of the insulator-superconductor quantum phase transition by an external electric field has attracted much attention due to their promise for transistors and memory devices with loss-free channels [8–10]. Moreover, the potentiality of nano-engineering the superconducting order parameter to create reconfigurable pinning landscapes offers the possibility to manipulate magnetic flux quanta in fluxtronics devices [11]. Exceptional behaviours have been found by combining superconductor (SC) and ferromagnetic (FM) materials in hybrid SC-FM systems. Interaction of these antagonistic order parameters has received enormous attention in condensed matter physics, in quality of unprecedented ways of designing system functionalities [12–14].

Here we introduce a new concept based on the combination of high-temperature cuprate oxides with soft FM structures as a promising approach to generate different magnetic states using superconducting stray fields or supercurrents, as a non-dissipative way to encode information. The manipulation of spin textures in magnetic materials opens up novel proposals for prospective high-performance magnetic memories [15, 16]. In these systems, the information can be stored in engineered non-trivial spin configurations, such as domain walls, vortices, or skyrmions [17, 18]. Key to the success of such devices is the precise generation and control of stable magnetic textures. The main bottleneck limiting the use of magnetic states as information bits remains on the valance between thermal stability and operation power. Highly stable states lead to large magnetic fields or currents for operation. The use of lossless supercurrents to stabilize the magnetic textures offers an attractive solution to reduce the power consumption of magnetic devices [14, 19, 20].

Direct evidence for the possibility to imprint non-trivial magnetic states in a FM layer by using superconducting stray fields was demonstrated in our previous work with SC-FM hybrid systems formed by micrometric high-temperature superconducting dots, covered by a soft FM thin film [19]. When an applied magnetic field is ramped up and down in those hybrid systems, regions with supercurrents circulating clockwise or counter-clockwise, following the shape of its external contour, are induced in alternating concentric layers, thus generating singular stray fields fully dependent on the SC dot geometry. By changing the shape of the dots and the magnetic history experienced by the hybrid system, a large manifold of magnetic patterns can be imprinted, stabilized, modified, and annihilate in the FM layer.

Figure 1 shows the magnetic domain structure obtained at 45K on SC-FM hybrid systems, based on YBCO dots covered with a Permalloy film, with different SC dot shapes. We obtained the magnetic images in remanence after saturating the SC dots. Different singular spin textures can be stabilized at zero external magnetic field, depending on the stray field generated by the SC structure below and sustained by lossless supercurrents. With a squared SC dot four tail-to-tail magnetic domains, separated by Bloch domain walls are imprinted. These domains are reversed to head-to-head spins in the negatively saturated state. In the case of triangular dots, the imprinted pattern is formed by three domains. Circular structures can be used to

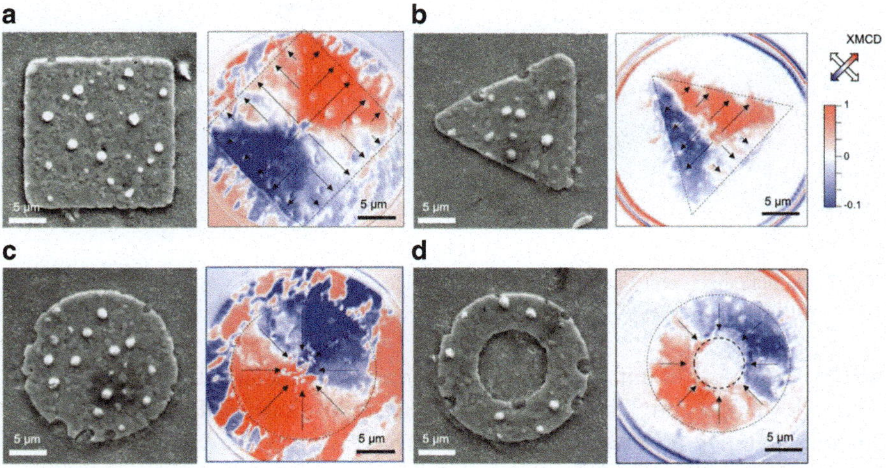

Fig. 1 (**a**) Scanning Electron Microscopy images (left) and the associated in-plane magnetic domain configurations imaged by X-ray magnetic circular dichroism (XMCD) of hybrid SC-FM structures with different SC dot geometry. (**a**) Square, (**b**) triangle, (**c**) disc, (**d**) ring. Arrows indicate the direction of the magnetization in the permalloy layer

create monopolar fields with magnetic charge easily switched from +1 to -1 with the sign of the applied field. This concept can be expanded to multiple geometries to provide a large manifold of magnetic topological interest for memory applications.

2 Experimental Details

2.1 Hybrid Superconducting-Ferromagnetic Devices

We fabricated hybrid SC-FM devices by combining $YBa_2Cu_3O_{7-x}$ (YBCO) superconducting strips with $Fe_{20}Ni_{80}$ (Permalloy) FM strips of different sizes. On one side, permalloy is a soft FM material which can be easily magnetized by an external magnetic field. Moreover, it has proved to have significant anisotropic magnetoresistance [21] that will be used to measure different magnetic configurations induced in the hybrid systems. On the other side, YBCO is a high-temperature copper oxide SC, able to carry large superconducting currents and thus trap high magnetic fields.

We used the chemical solution deposition approach to grow high-quality epitaxial YBCO films on 5 mm × 5 mm $LaAlO_3$ (LAO) single crystal substrates [22]. We used a photolithography process combined with chemical etching to pattern SC strips of two different widths, w(SC) = 50 μm and 100 μm. After patterning, YBCO structures were covered with a 15 nm thick film of Cerium Oxide grown by Atomic Layer Deposition, as an insulating layer to avoid current flowing through the SC when measuring the magnetoresistance of the ferromagnet. The following step was to pattern permalloy structures, also consisting of strips of different widths,

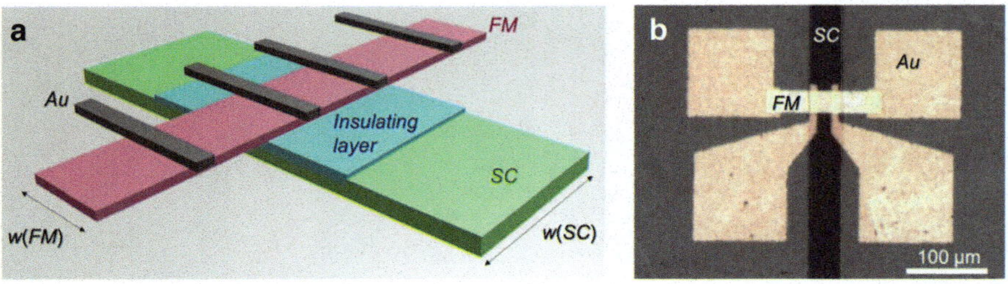

Fig. 2 (**a**) Schematic representation of SC-FM devices. (**b**) Optical image showing a SC_{50}-FM_{30} device with four Au contacts

w(FM) = 10 μm, 30 μm, 50 μm, 100 μm, arranged perpendicular to the SC strips. Permalloy strips were fabricated by sputtering deposition, combined with a lift-off photolithography process. Finally, we deposited Au metallic contacts by thermal evaporation and lift-off.

Figure 2a illustrates a schematic representation of the resulting devices denoted as $SC_{w(SC)}FM_{w(FM)}$, where the subscripts w(SC) and w(FM) indicate the width in μm of the SC and FM strips, respectively. An optical image of a representative hybrid device is shown in Fig. 2b.

2.2 *Anisotropic Magnetoresistance*

The anisotropic magnetoresistance (AMR) effect describes the change in the resistance of a magnetic material when the relative angle between the local magnetization and current direction changes. The AMR of a magnetic material can be described simply by the equation:

$$\rho(\theta) = \rho_{\perp} + \left(\rho_{\parallel} + \rho_{\perp}\right)\cos\theta \tag{1}$$

where $\rho(\theta)$ is the resistance as function of the angle θ between the local magnetization and current, $\rho_{\perp}$ and $\rho_{\parallel}$ are the resistances at $\theta = 90^{\circ}$ and a $\theta = 0^{\circ}$, respectively [23]. The AMR effect typically results in resistance changes of 0.1–2% and has been demonstrated to be a very sensitive probe for detection of sub-micron domain patterns, such us domain walls or vortex states, in Permalloy and Nickel nanostructures [24–27].

We used the AMR effect to detect different magnetic textures induced in the FM structures. Magnetotransport properties of permalloy structures were conducted in a four-point geometry using a Quantum Design Physical Properties Measurement System (PPMS).

We measured magnetoresistance hysteresis loops as a function of the external magnetic field, H, up to different maximum values, H_{max}. The magnetoresistance ratio was defined as:

$$dR = \frac{R_H - R_0}{R_0} \tag{2}$$

with R_H and R_0 being values of the sample resistance at an applied magnetic field H and zero magnetic field, respectively. All measurements were performed with the magnetic field applied out-of-plane and the current flow along the permalloy channel.

3 Magnetoresistance in Hybrid SC-FM Devices

3.1 *Effect of SC Stray Magnetic Fields on FM Magnetoresistance*

To determine the effect of SC trapped magnetic fields on the FM spin configuration of FM strips, we compare the magnetoresistance curves of permalloy above and below T_c. Figure 3 shows the magnetoresistance hysteresis loops measured for a hybrid structure (SC_{100}-FM_{50}) at 100K > T_c and 5K < T_c, sweeping up and down the magnetic field from -500 mT to 500 mT. Significant differences can be observed within the two loops. At 100K, with no influence of the superconducting trapped field, the magnetoresistance curves show a maximum at positive magnetic field for the sweep-up curve and at negative magnetic field for the sweep-down curve. These two peaks can be associated to the maximum disorder of the distribution of magnetic domains at the coercive field [23, 28].

The magnetoresistance ratio curves taken at different temperatures above T_c (100K, 200K, and 300K) present the same behaviour (Fig. 4). Below the

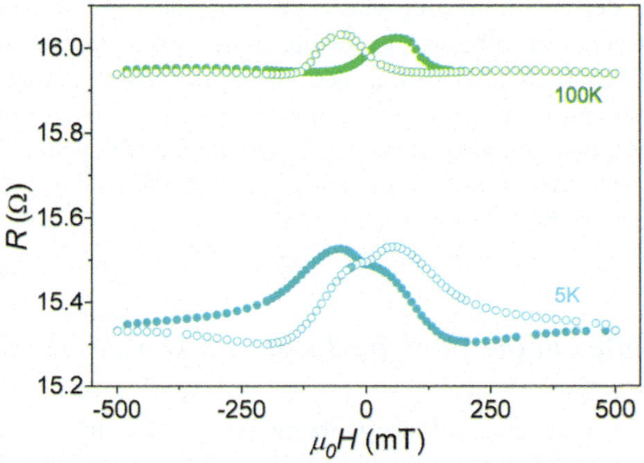

Fig. 3 Magnetoresistance curves for SC_{100}-FM_{50} measured sweeping-up (closed symbols) and sweeping-down (open symbols) the applied magnetic field at 100K and 5K

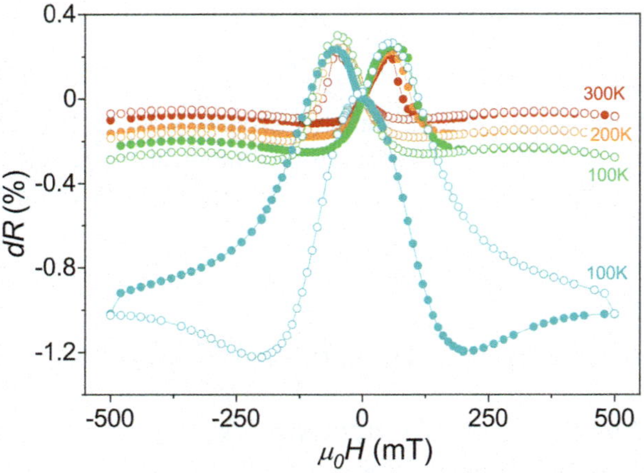

Fig. 4 Hysteresis curves of magnetoresistance ratio for SC_{100}-FM_{50}, measured sweeping-up (closed symbols) and sweeping-down (open symbols) the applied magnetic field at 300K, 200K, 100K, and 5K

superconducting critical temperature (5K), however, the hysteresis loop shows a different performance. In this case, the ascending branch, going from negative to positive magnetic field values, shows higher resistance than that of the descending branch and the magnetoresistance peak for the sweep-up curve appears at negative fields. Moreover, the normalized hysteresis loop is significantly enhanced with decreasing the temperature below T_c (see Fig. 4). These differences provide a direct indication that the magnetic structure of the permalloy strip is modulated by the SC stray fields.

We plot in Fig. 5 the magnetic field value at the magnetoresistance peak, appearing in the ascending branch of the hysteresis loops shown above ($H_{peak}{}^{up}$), as a function of the temperature. We observe a monotonic increase of the $H_{peak}{}^{up}$ value as the temperature is decreased, in agreement with an enhancement of the coercive field. This trend is not observed once the temperature is reduced below T_c, since as observed in the hysteresis loops the peak is shifted to negative fields when the magnetoresistance is measured at the superconducting state. The shift of $H_{peak}{}^{up}$ arises from the contribution of the magnetic field trapped in the SC strip that modifies the applied field necessary to reverse the magnetic domains on the FM structure, i.e. modify the coercive field, as observed in [19].

3.2 *Tunability of the Coercive Field with Device Geometry*

To better elucidate the effect of the SC trapped field on the magnetoresistive peak, associated with the FM coercive field, we analysed devices with different SC and FM widths. Figure 6 shows the magnetoresistance hysteresis loop obtained for the device shown above (SC_{100}-FM_{50}) compared with the one measured for a similar device with the same FM width but a narrower SC strip (SC_{50}-FM_{50}). Both devices show similar behaviours, with the magnetoresistance peaks of the ascending branch appearing at negative fields. However, the effect of SC stray field (shift of the peak to negative magnetic field values and magnetoresistance hysteresis) is significantly reduced by decreasing the SC strip width.

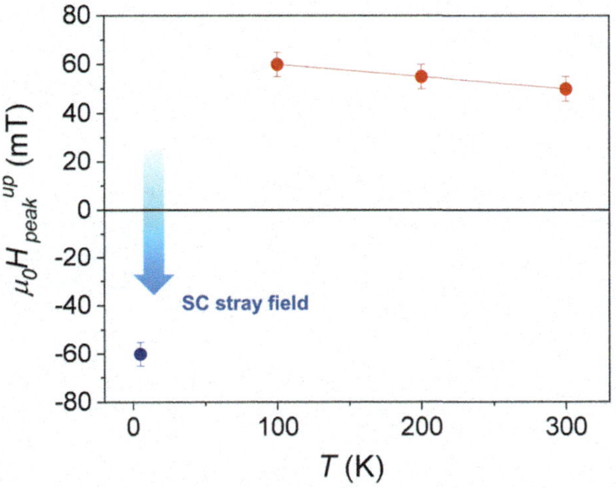

Fig. 5 Magnetic field at the magnetoresistance peak, obtained for SC_{100}-FM_{50} on the ascending branch of the hysteresis loop, at different temperatures

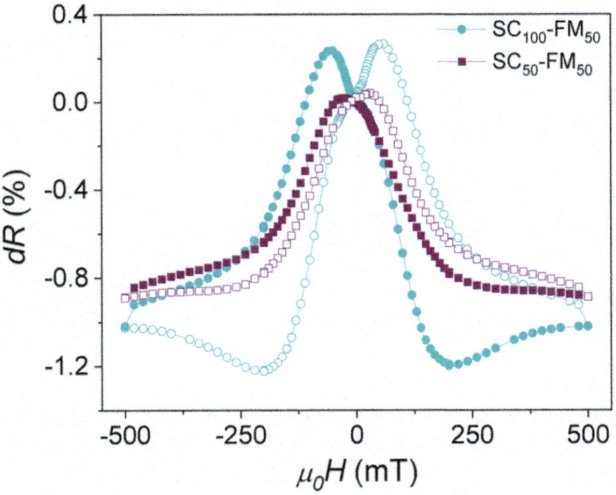

Fig. 6 Magnetoresistance ratio curves for SC_{100}-FM_{50} and SC_{50}-FM_{50} measured sweeping-up (closed symbols) and sweeping-down (open symbols) the applied magnetic field at 5K

We now turn to analyse the effect of the FM strip width on the magnetoresistance curves. Figure 7a shows the hysteresis loops obtained for two devices with a fixed SC strip of *w*(SC) = 50 μm and different FM width. By reducing the width of the FM strip, the effect of the SC trapped field on the shift of the magnetoresistance peak position becomes less notable. In this particular case, for the SC_{50}-FM_{10} device, the ascending magnetoresistance branch peak appears at positive applied field. Another notable feature is that by increasing the magnetic field from its minimum value, the

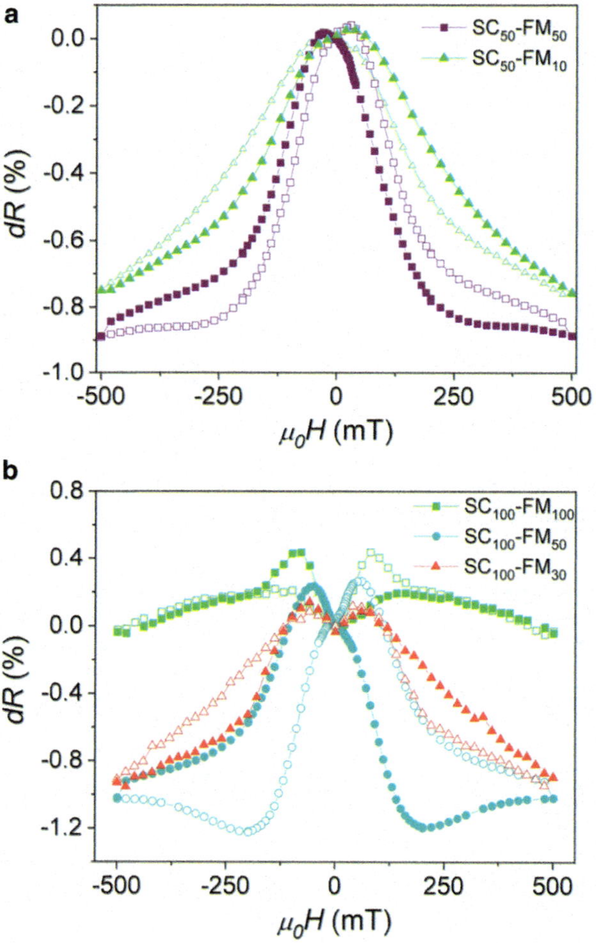

Fig. 7 Magnetoresistance ratio curves measured at 5K sweeping-up (closed symbols) and sweeping-down (open symbols) the applied magnetic field. (**a**) Devices with fixed *w*(SC) = 50 μm strip width and two different FM strip widths. (**b**) Devices with fixed *w*(SC) = 100 μm strip width and three different FM strip widths

magnetoresistance of SC_{50}-FM_{10} decreases (the ascending branch goes below the descending one) while it increases (the ascending branch goes above the descending one) for SC_{50}-FM_{50} and SC_{100}-FM_{50} (Fig. 6), where the effect of the SC was more evident. We show in Fig. 7b a series of hysteresis curves measured on devices with w(SC) = 100 μm. In this case, all samples show a magnetoresistance peak for the ascending branch at negative fields, although the peak shift is reduced as the FM strip width is decreased.

The evolution of the ascending branch magnetoresistance peak with the FM strip width is shown in Fig. 8, for devices with two different SC strip widths. As a general trend, we observe that the shift of H_{peak}^{up} at negative magnetic fields is enhanced for devices with wider SC and FM widths. A higher shift obtained by increasing w(SC) is expected since the trapped magnetic field in a SC structure increases by increasing its size [29]. In the case w(FM) the observed trend is not as straightforward though it is clear that the peak shift at negative fields decreases by reducing w(FM). It is worth mentioning that in some cases, two peaks are obtained for narrow FM strips as seen in Fig. 7b for the SC_{100}-FM_{30} device.

Concluding, magnetoresistance peaks associated to the maximum disorder of the distribution of magnetic domains at the coercive field are clearly modified by the presence of SC stray magnetic fields. The SC and FM strip dimensions can be used to modulate the FM spin configuration and the final behaviour of the magnetoresistance hysteresis loop thus providing large flexibility for the device design.

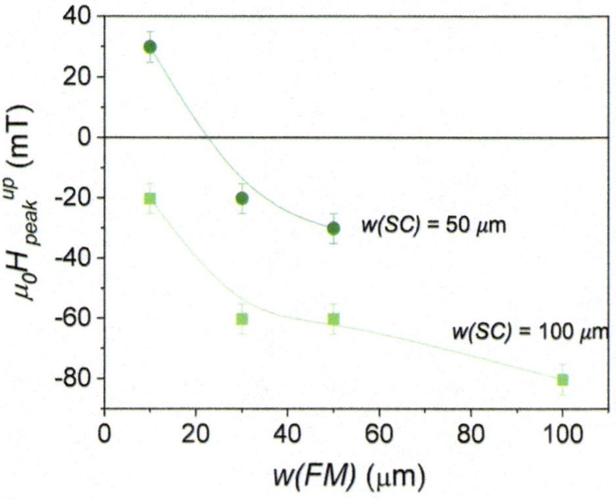

Fig. 8 Magnetic field at the magnetoresistance peak, obtained for devices with different values of w(SC), as a function of w(FM)

3.3 Magnetoresistance Modulation with SC Trapped Magnetic Fields

Given a particular SC-FM hybrid device, the reversal or magnetic domains can be modulated by performing minor hysteresis cycles and thus changing the magnetic history experienced by the SC strip. Figure 9 shows the magnetoresistance after a series of minor loops of increasing amplitude from $\mu_0 H = 50$ mT to $\mu_0 H = 500$ mT. The hysteresis is enhanced by increasing the maximum applied field in accordance with a higher trapped field within the SC structure. It is interesting to note that minor

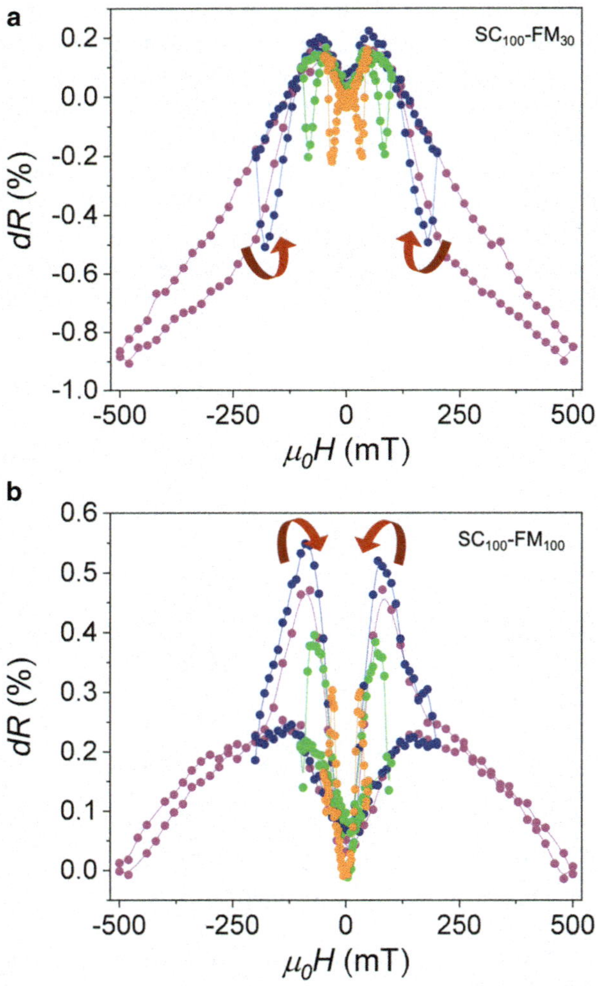

Fig. 9 Magnetoresistance hysteresis loops measured at different maximum applied fields (orange; 50 mT, green; 100 mT, blue; 200 mT and magenta; 500 mT), for devices with w(SC) = 100 μm and (**a**) w(FM) = 30 μm, (**b**) w(FM) = 100 μm

loops show a sharp switch in the magnetoresistance, which can be either enhanced or reduced depending on the FM strip width, as indicated with the arrows shown in Fig. 9.

By taking advantage of the different magnetic configurations that can be achieved in the SC-FM hybrid devices, multilevel magnetoresistance states may be induced at the remanent state, sustained by trapped SC stray fields. Figure 10 shows several minor loops illustrating the evolution of the magnetoresistance as a function of the maximum applied field obtained for a SC_{50}-FM_{30} device at 5K (closed symbols) and 100K (open symbols). In both cases, the first point has been measured in the demagnetizing state. The low values of applied magnetic field do not change the magnetoresistance of the FM structure at 100K. The obtained behaviour is very different at 5K where a clear magnetoresistance hysteresis is observed producing evident changes in the remanent magnetoresistance.

Figure 11 shows the remanent resistance state after each loop measured for different devices at 5K, compared with the values obtained at 100K for SC_{50}-FM_{30}. While no changes in the magnetoresistance ratio are obtained at 100K, a number of clearly distinguishable resistance levels can be induced at 5K. In this case, we obtained an increasing dependence of the magnetoresistance ratio with the applied magnetic field in agreement with a much larger trapped SC stray magnetic field. Further experimental study to determine the remanent structure obtained in each device would be necessary to identify the different remanent resistance values measured in each case.

As a whole, these results confirm that the remanent resistance of SC-FM devices may be modulated by changing the internal trapped magnetic field of SC structures. Resistance changes of 0.4% at remanence may be obtained by applying very small magnetic fields (tens of mT). This multilevel resistance behaviour, sustained at zero applied field, is completely stable as long as the trapped magnetic field is not

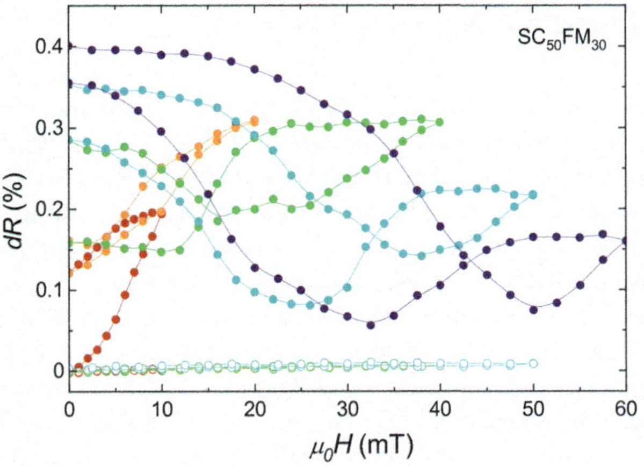

Fig. 10 Minor magnetoresistance loops measured after demagnetizing the sample SC_{50}-FM_{30} and applying a series of magnetic fields, red; 10 mT, orange: 20, green: 40, blue; 50 and purple; 60 mT, at 5K (closed symbols) and 100K (open symbols)

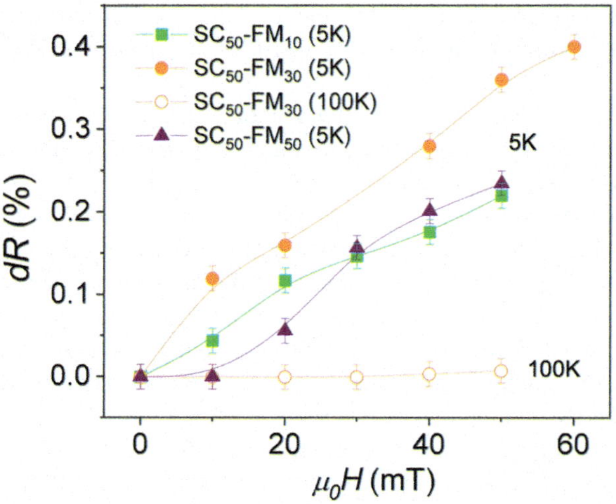

Fig. 11 Evolution of the remanent magnetoresistance ratio as a function of the applied magnetic field for different devices at 5K (closed symbols) and 100K (open symbols)

changed and may be exploited for energy-efficient non-volatile memory applications. Negative magnetic fields may be applied to demagnetize the SC structures and drive the FM resistance back to the initial resistive state or intermediate ones. Another possibility to erase the remanent magnetoresistance would be to locally increase the temperature of the required device above T_c and thus remove the trapped SC stray field.

3.4 *Magnetoresistance Modulation with Loss-Free Supercurrents*

Going one step further, we evaluated the possibility to modulate the FM resistance by applying a SC transport current, instead of generating the trapped supercurrents through an applied magnetic field. For that we apply a transport current through the SC strip while measuring the resistance of the FM structure. Results are shown in Fig. 12 where we compare the resistance measured at a certain applied current (blue dots) and after removing it (magenta squares).

Data shown in Fig. 12 clearly illustrate that the resistance of SC-FM devices may also be modulated with SC transport currents. Resistance changes of 0.1–0.15% may be produced by applying currents of several mA. It is worth to notice that the transport current is applied below T_c and thus these are lossless currents. The remanent state induced after applying a certain current and removing it maintains certain resistance change but the obtained ratio is notably reduced, as compared with the value measured with the current flowing through the SC strip. Moreover, contrary to what we found in the previous section, for the states induced by applying magnetic fields, there are no remanent trapped fields or currents and thus the induced magnetic states are not stable with time.

Figure 13 shows the time evolution of the remanent resistance measured at zero current, after applying different current intensities. A volatile behaviour is obtained, with resistance values that exhibit a decay to zero with 50–100 s, depending on the applied supercurrent. Fine tuning of this time decay may be used for volatile energy-efficient information processing.

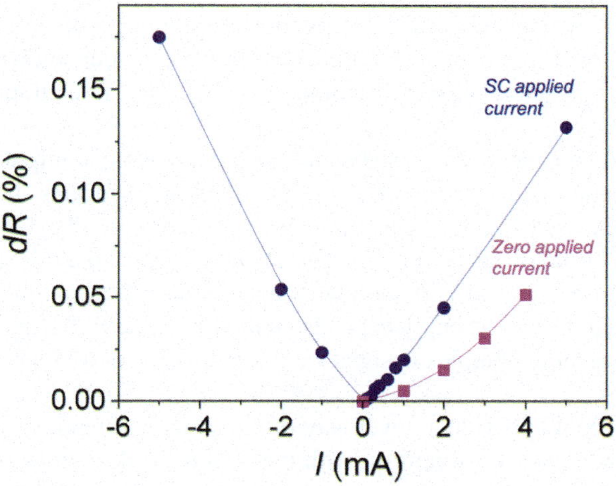

Fig. 12 Evolution of the resistance ratio obtained for $SC_{100}FM_{50}$ at 77K, as a function of the applied current. Blue dots show the resistance values obtained when current is applied through the SC strip and magenta squares the remanent values obtained when the applied current is removed

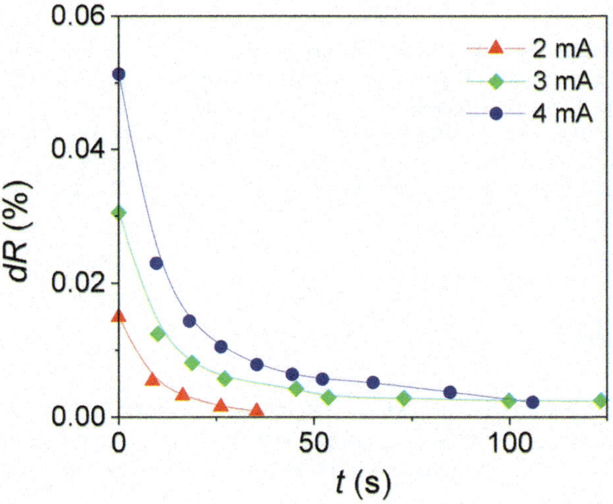

Fig. 13 Time evolution of the remanent resistance obtained after applying different values of transport current through the SC trip and going back to zero

3.5 Concluding Remarks and Outlook

We have presented here an original concept to manipulate magnetic states by exploiting the interplay of superconducting (SC) and ferromagnetic (FM) materials. Hybrid devices have been fabricated by combining SC and FM perpendicular strips of different sizes. We used copper oxide YBCO as a SC material, with unique potential benefits linked to the ability to sustain lossless high current densities. Trapped magnetic fields or transport current in the SC structures are able to manipulate and stabilize different magnetic states in Permalloy strips, clearly distinguished by magnetoresistance measurements. Therefore, we provide to our knowledge, the first example of magnetoresistance modulation with loss-free trapped fields or supercurrents.

Non-volatile magnetic states, stable at zero field, may be manipulated by changing the magnetic history experienced by the SC strips and thus their trapped SC stray field. Transport supercurrents may also be used to modify the magnetic configuration of permalloy strips but in this case, they evolve at the initial state once the current is removed, thus providing volatile data storage. Control of magnetic states in the presented devices strongly depends on their geometry offering much flexibility for realizing more complex designs with volatile or non-volatile interesting spin configurations.

This study could provide a novel way of encoding and reading magnetic states through superconducting structures, with the gain to work with loss-free magnetic fields and supercurrents. The presented effects could be used as an appealing approach to reduce the power consumption of magnetic recording devices, easily integrable with superconducting computing technologies. Further systematic research and theoretical efforts are needed to evaluate if the proposed systems fulfil all technology requirements essential to produce a significant impact in the field.

Acknowledgements We acknowledge financial support from Spanish Ministry of Science, Innovation and Universities (MICIU) through SUMATE project (RTI2018-095853-B-C21), co-financed by the European Regional Development Fund, and the SUPERSWITCH project (FUNMAT-FIP-2017). We also thank support from the European Union for NANOCOHYBRI project (Cost Action CA 16218), and from the Catalan Government with 2017-SGR-1519 and Xarmae. A.F.-R thanks the Spanish Ministry of Economy for the FPI Spanish grant (BES-2016-077310).

Bibliography

1. M. Mitchell Waldrop, More than Moore. Nature **530**, 145 (2016)
2. Y. Zhou, S. Ramanathan, Correlated electron materials and field effect transistors for logic: A review. Crit. Rev. Solid State Mater. Sci. **38**, 286–317 (2013). https://doi.org/10.1080/10408436.2012.719131
3. M. Bibes, J.E. Villegas, A. Barthélèmy, Ultrathin oxide films and interfaces for electronics and spintronics. Adv. Phys. **60**, 5–84 (2011). https://doi.org/10.1080/00018732.2010.534865

4. M. Coll, J. Fontcuberta, M. Althammer, et al., Towards oxide electronics: A roadmap. Appl. Surf. Sci. **482**, 1–93 (2019). https://doi.org/10.1016/j.apsusc.2019.03.312
5. M. Imada, A. Fujimori, Y. Tokura, Metal-insulator transitions. Rev. Mod. Phys. **70**, 1039–1263 (1998). https://doi.org/10.1103/RevModPhys.70.1039
6. D.S. Holmes, S. Member, A.L. Ripple, M.A. Manheimer, Energy-efficient superconducting computing – Power budgets and requirements. IEEE Trans. Appl. Supercond. **23**, 1701610 (2013)
7. I.I. Soloviev, N.V. Klenov, S.V. Bakurskiy, et al., Beyond Moore's technologies: Operation principles of a superconductor alternative. Beilstein J. Nanotechnol. **8**, 2689–2710 (2017). https://doi.org/10.3762/bjnano.8.269
8. C.H. Ahn, J.-M. Triscone, J. Mannhart, Electric field effect in correlated oxide systems. Nature **424**, 1015–1018 (2003). https://doi.org/10.1038/nature01878
9. J. Mannhart, High- Tc transistors. Supercond. Sci. Technol. **9**, 49–67 (1996)
10. A. Palau, A. Fernandez-rodriguez, J.C. Gonzalez-rosillo, et al., Electrochemical tuning of metal insulator transition and nonvolatile resistive switching in superconducting films. ACS Appl. Mater. Interfaces **10**, 30531 (2018). https://doi.org/10.1021/acsami.8b08042
11. A. Crassous, R. Bernard, S. Fusil, et al., Nanoscale electrostatic manipulation of magnetic flux quanta in ferroelectric/superconductor $BiFeO_3/YBa_2Cu_3O_7$-d heterostructures. Phys. Rev. Lett. **107**, 1–5 (2011). https://doi.org/10.1103/PhysRevLett.107.247002
12. J. Linder, J.W.A. Robinson, Superconducting spintronics. Nat. Phys. **11**, 307–315 (2015). https://doi.org/10.1038/nphys3242
13. M.G. Blamire, J.W.A. Robinson, The interface between superconductivity and magnetism: Understanding and device prospects. J. Phys. Condens. Matter **26**, 453201 (2014). https://doi.org/10.1088/0953-8984/26/45/453201
14. G. Shaw, S.B. Alvarez, J. Brisbois et al., Magnetic recording of superconducting states. 1–18 (2019). https://doi.org/10.3390/met9101022
15. H.-B. Braun, Topological effects in nanomagnetism: From superparamagnetism to chiral quantum solitons. Adv. Phys. **61**, 1–116 (2012). https://doi.org/10.1080/00018732.2012.663070
16. R. Wiesendanger, Nanoscale magnetic skyrmions in metallic films and multilayers: A new twist for spintronics. Nat. Rev. Mater **1**, 16044 (2016). https://doi.org/10.1038/natrevmats.2016.44
17. M.S. Lee, T.A. Wynn, E. Folven, et al., Tailoring Spin textures in complex oxide micromagnets. ACS Nano **10**, 8545 (2016). https://doi.org/10.1021/acsnano.6b03770
18. S.S.P. Parkin, M. Hayashi, L. Thomas, Magnetic racetrack memory. Science **320**, 190 (2014). https://doi.org/10.1126/science.1145799
19. A. Palau, S. Valencia, N. Del-Valle, et al., Encoding magnetic states in monopole-like configurations using superconducting dots. Adv. Sci **3**, 1600207 (2016). https://doi.org/10.1002/advs.201600207
20. V. Rouco, R. Córdoba, J.M. De Teresa, et al., Competition between superconductor – Ferromagnetic stray magnetic fields in $YBa_2Cu_3O_7$−x films pierced with Co nano-rods. Sci. Rep. **7**, 5663 (2017). https://doi.org/10.1038/s41598-017-05909-6
21. G. Nahrwold, J.M. Scholtyssek, S. Motl-Ziegler, et al., Structural, magnetic, and transport properties of Permalloy for spintronic experiments. J. Appl. Phys. **108**, 013907 (2010). https://doi.org/10.1063/1.3431384
22. X. Obradors, T. Puig, A. Pomar, et al., Progress towards all-chemical superconducting $YBa_2Cu_3O_7$-coated conductors. Supercond. Sci. Technol. **19**, S13–S26 (2006). https://doi.org/10.1088/0953-2048/19/3/003
23. M. Viret, D. Vignoles, D. Cole, et al., Spin scattering in ferromagnetic thin films. Phys. Rev. B **53**, 8464–8468 (1996)
24. D.W. Wong, I. Purnama, G.J. Lim, et al., Current-induced three-dimensional domain wall propagation in cylindrical NiFe nanowires. J. Appl. Phys. **119**, 153902 (2016). https://doi.org/10.1063/1.4946753

25. T. Wren, O. Kazakova, Anisotropic magnetoresistance effect in sub-micron nickel disks. J. Appl. Phys. **117**, 17E134 (2015). https://doi.org/10.1063/1.4918967
26. M. Bolte, M. Steiner, C. Pels, et al., Magnetotransport through magnetic domain patterns in permalloy rectangles. Phys. Rev. B Condens. Matter Mater. Phys **72**, 1–8 (2005). https://doi.org/10.1103/PhysRevB.72.224436
27. M. Hayashi, L. Thomas, C. Rettner, et al., Dependence of current and field driven depinning of domain walls on their structure and chirality in permalloy nanowires. Phys. Rev. Lett. **97**, 1–4 (2006). https://doi.org/10.1103/PhysRevLett.97.207205
28. W. Gil, D. Görlitz, M. Horisberger, J. Kötzler, Magnetoresistance anisotropy of polycrystalline cobalt films: Geometrical-size and domain effects. Phys. Rev. B Condens. Matter Mater. Phys **72**, 1–10 (2005). https://doi.org/10.1103/PhysRevB.72.134401
29. C. Jooss, J. Albrecht, H. Kuhn, et al., Magneto-optical studies of current distributions in high- Tc superconductors. Rep. Prog. Phys. **65**, 651–788 (2002). https://doi.org/10.1088/0034-4885/65/5/202

Effect of the Neutron and Heavy-ion Irradiation on the Pinning Properties of Cuprate Superconductors

Noriko Chikumoto

1 Introduction

Critical current density (J_c), the maximum current density we can pass with zero resistance, is one of the most important properties for application of superconductors. Because the resistivity in superconductor is caused by vortex motion due to the Lorentz force from passed electric current, it is necessary to introduce so-called pinning centers that prevent the vortex motion. So a lot of efforts have been made to explore pinning centers that provide strong pinning force (F_p) to improve J_c. It is widely known that imperfections in superconducting matrix, such as normal precipitates, grain boundaries, dislocations, stacking faults, etc., can be pinning centers. At such imperfections, the depression of superconducting order parameter occurs. When the vortex passes through the imperfections, the energy of vortices changes, resulting in a vortex pinning interaction.

Irradiation of a material with particles, such as electrons, protons, neutrons, and ions, has known to be a useful way to introduce defects in a controlled way; the sizes and the densities of the irradiation defects can be changed by irradiation conditions such as kind, energy, and fluence of radiated particles. The greatest advantage of this method is that it can be used independently of the material synthesis process.

Among various kinds of irradiation, the high-energy heavy-ion irradiation is attracted much attention, since it produces columnar defects that is expected to give strongest pinning force if the flux line is aligned with their length direction. On the other hand, the neutron irradiation has an advantage of a large penetration range. It

N. Chikumoto (✉)
Center of Applied Superconductivity and Sustainable Energy Research, Chubu University, Kasugai, Japan
e-mail: nchiku@isc.chubu.ac.jp

A. G. Roca et al. (eds.), *Surfaces and Interfaces of Metal Oxide Thin Films, Multilayers, Nanoparticles and Nano-composites*,
https://doi.org/10.1007/978-3-030-74073-3_8

is also expected to introduce effective pinning centers, because the size of neutron damage is comparable to the size of the vortex core.

A great deal of irradiation experiment has been performed since the discovery of the high T_c superconductors and there are a plenty of reports on it [1]. So in this chapter, I would like to show some data that are not much studied. This chapter is organized as follows. First the basic understandings regarding the introduction of defects by irradiation will be explained. Then we show some of our experimental result of neutron and the heavy-ion irradiation effects on the pinning properties of two kinds of cuprate high T_c superconductors (HTSC), $(La_{1-x}Sr_x)_2CuO_{4-\delta}$ (La214) and $REBa_2Cu_3O_{7-\delta}$ (RE123, RE:Y, Nd, Gd). In La214, the neutron and heavy-ion irradiation effect on the J_c-B characteristics was investigated using single crystal samples. On the other hand, in RE123, we performed irradiation studies using coated conductors. We only performed heavy-ion irradiation, but the irradiation conditions, irradiation species, etc., were changed in order to change the size, length, and the arrangement of irradiation defects. In addition, we performed oxygen annealing to examine the stability of irradiation defects.

2 Defects Introduced by Irradiation

The effects of radiation on crystalline solids are highly variable depending on the type of material and the type of incident radiation. However, there are roughly two effects. One is the ionization and electronic excitation often results in chemical reactions and finally disappears. The other is the knocking-on the atom to cause the atomic displacement. This effect is very important because it results in the radiation damages that remain in the crystal and often causes the change of the material properties. These two effects are derived from nuclear energy loss and electric energy loss of radiation in condensed matter, and the contribution of each phenomenon differs among the kind and the energies of the incident particles. In the case of electron and proton irradiation, electric energy loss is dominant. They may produce point-like defects in the form of interstitial ions and vacancies (Frenkel pairs). On the other hands, nuclear energy loss is also important in the case of ions and neutrons and larger defects are created, as follows.

Neutron irradiation is generally carried out by exposing a material in a fission reactor that provides a wide spectrum of neutrons. Defect cascades are mainly produced by fast neutrons with energies above 0.1 MeV. According to the transmission electron microscope (TEM) observation reported in Ref. [2], the major observable defects introduced by the neutron irradiation are spherical collision cascades, with sizes ranging from below 1 nm up to 5 nm in diameter, surrounded by the strain field. Since the size of cascade defects is comparable to the coherence length of HTSC's, it yields a preferable condition for core pinning. Indeed, large improvements of pinning properties had been reported for various HTSC by several groups [3]. On the other hand, the lower energy neutrons may create small size defects, such as point defects (Frenkel pairs) and defect clusters. It mainly results in the oxygen disordering and causes the depression of T_c [4].

In case of high-energy ion irradiation, the incident ions give energy to the target material through the electric energy loss procedure and slow down themselves. So the extent of the electronic energy loss is often called to be the electron stopping power, S_e. It is known that the damage morphology strongly depends on the electronic stopping power S_e of the incident in the target. For instance, in the case of Y123 polycrystalline samples, discontinuous extended defects are introduced when the S_e exceeds the threshold value of » 8 keV/nm [5]. As S_e increases, the defects are more elongated and finally the continuous columnar defects are created at S_e > 20 keV/nm [5].

Figure 1 shows typical columnar defects in Nd123 single crystal observed by TEM. The irradiation was carried out at Grand Accélérateur National d'Ions Lourds (GANIL) (Caen, France) using 5.8 GeV Pb-ion. Direction of the incident beam was parallel to the crystal *c*-axis. Figure 1a is the plan-view of the columnar defects. The defects are seen as white contrasts. The amorphous regions with diameter » 7 nm are randomly distributed in the sample. The incident ion beam crossed the whole thickness of the sample < 100 μm and create parallel continuous columnar defects (Fig. 1b).

It is noteworthy that such irradiation defects are also influenced strongly by the irradiation temperature. At very low temperature, the defects cannot move, but above a particular temperature, rearrangement or recombination of defects occurs through thermal diffusion process. It causes the formation of some kinds of complex defect clusters such as dislocation loops and stacking faults instead of annihilation.

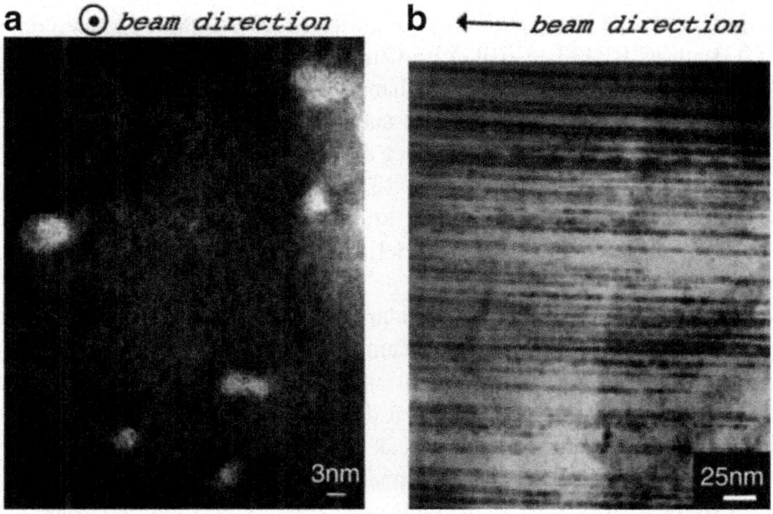

Fig. 1 Transmission electron micrographs of columnar defects in the 5.8 GeV Pb-ion irradiated Nd123 single crystal. (**a**) TEM image of the track cross-section. (**b**) Images taken along the columnar defects

Radiochemical effects during irradiation make the situation more complicated. Local heating effects by thermal spikes, etc., are expected to accelerate the moving of interstitial atoms and vacancies. Important point is that the radiation damages might be modified or even recovered by annealing the irradiated material at a certain temperature.

3 Irradiation Effects in La214 Single Crystals

3.1 Neutron Irradiation of La214 Single Crystal

In HTSC, it has been recognized that the electronic anisotropy and the interlayer coupling between superconducting CuO_2 layers affect their magnetization behaviors in the mixed state. Among various kinds of HTSC, $(La_{1-x}Sr_x)_2CuO_{4-\delta}$ (La214) is known to have the simplest crystal structure with only one CuO_2 plane. In this system, the anisotropy can be varied by changing the Sr content, x. According to resistivity measurement reported by Kimura et al. [6], the anisotropy ratio ρ_c/ρ_{ab} at 50 K decreased drastically from nearly 4000 to 160 with increasing x. Kobayashi et al. [7] reported that the magnetization behavior of the crystal also changes with x. In the under-doped crystal, $x < 0.075$, hysteresis width, ΔM, decreases exponentially with the applied magnetic field. On the other hand, in the over-doped crystal, $x > 0.075$, an anomaly in magnetization curves, in particular the depression of the magnetization at intermediate field range followed by a secondary maximum ("the second peak effect") is observed, when the magnetic field was applied perpendicular to the CuO_2 plane.

Similar second peak effect has been reported in various high T_c superconductors (HTSC), such as RE123 [8–10], $YBa_2Cu_4O_{8-\delta}$ (Y124) [11], Bi2212 [12–15], and Tl-based compounds [16]. Several mechanisms have been proposed to explain "the second peak effect": (1) second phase such as oxygen-deficient region is driven normal by applied field and start to act as additional pinning centers [8]. (2) A change in flux creep dynamics [9, 12]. (3) 3D-2D transition in the vortex system [15] (4) Matching of vortex spacing to pinning center periodicity ("matching effect") [14], (5) the field-driven disordering transition from Bragg glass phase to vortex glass phase [10].

Considering the facts that the secondary peak is observed only for over-doped compositions in La214 and a substantial amount of the oxygen deficiency is reported to exist only in the over-doped composition of La214 [17]; this behavior is likely to be attributed to the existence of oxygen defects. So we investigated the effect of neutron irradiation on the secondary peak in the over-doped La214 crystals, in an attempt to reveal the role of defects in this phenomenon.

In this study, we used the La214 single crystal with $x = 0.11$ grown by a travelling-solvent floating-zone (TSFZ) method. The details of crystal growth are described in details in Ref. [6]. The sample was sequentially irradiated with neutrons to the fluences of 2 and 4 × 10^{20} m^{-2}. The irradiations were carried out in a nuclear research reactor (JRR-4) at Japan Atomic Energy Research Institute. The flux density of fast neutron was about 5.5 × 10^{16} $m^{-2}s^{-1}$. During the irradiation, the sample was wrapped with quartz wool and enclosed in an evacuated quartz tube. The number of the displacement of the atoms due to the irradiation is calculated to be less than 10^{-4} d.p.a. (displacement per atom) at the fluence of 4 × 10^{20} m^{-2}, and there was no observable change in T_c (~ 29.4 K) after each irradiation. Magnetization measurements were performed by a Vibrating Sample Magnetometer (VSM, EG&G Princeton Applied Research, Model 4500), with applied magnetic fields parallel to the c-axis. The field sweep rate during the magnetization measurement had been kept at 20 mT/s.

Typical magnetization curves recorded at $T = 15$ K for La214 single crystal with $x = 0.11$ before and after the neutron irradiations are shown in Fig. 2. The height of the secondary peak was strongly enhanced and the secondary peak field, B_{pk}, was shifted downward, after the neutron irradiation. On the other hand, there was no observable change in the hysteresis width for $B_a > B_{pk}$.

The temperature dependence of the magnetization curves of the irradiated sample with the fluence of 2 × 10^{20} m^{-2} is shown in Fig. 3. The secondary peak appears over a wide temperature range from 4.2 K to near T_c. The peak position shifts to lower fields as the temperature is increased, similar to that observed in the unirradiated sample [7]. Similar behavior was also observed in the irradiated sample with the fluence of 4 × 10^{20} m^{-2}.

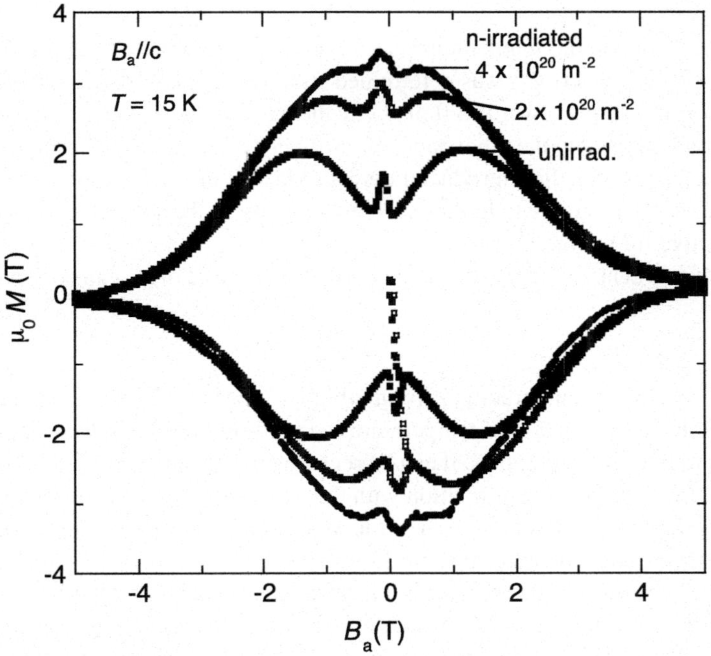

Fig. 2 Magnetization curve of $(La_{0.89}Sr_{0.11})_2CuO_{4-\delta}$ recorded at T=15 K before and after neutron irradiation. The neutron fluences were 2 and 4 × 10^{20} m^{-2}

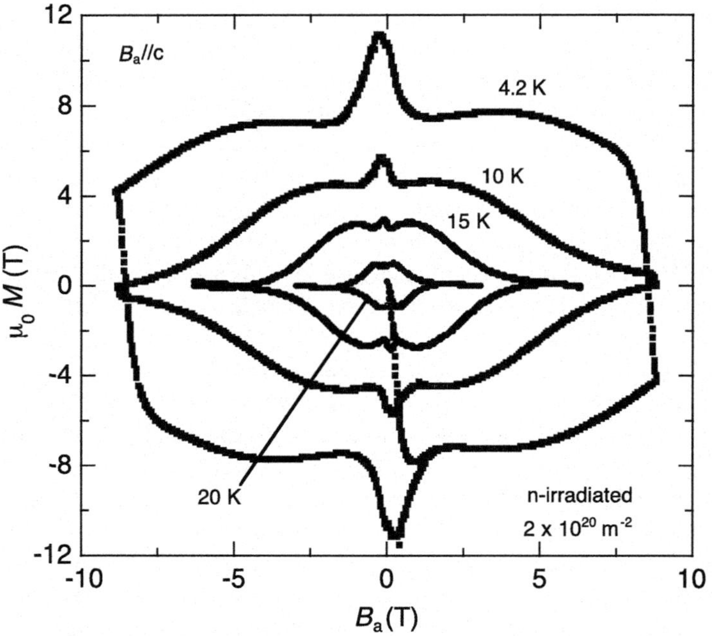

Fig. 3 Temperature dependence of magnetization curves for neutron-irradiated La214 ($x = 0.11$) crystal (fluence: 2×10^{20} m^{-2})

Figure 4 shows the temperature variation of the observed B_{pk} and the irreversibility field, B_{irr}, where B_{irr} was determined from the locus of the vanishing of hysteresis width. The downward shift of B_{pk} by the irradiation was observed over a wide temperature range. On the other hand, B_{irr} was unchanged after each irradiation step. This result indicates that introduced defect cascades play an important role in the low and the intermediate field range where we observe the secondary peak, but it is not effective for raising B_{irr}.

Kobayashi et al. found that the temperature dependence of B_{pk} can be expressed by,

$$B_{pk}^{1/2} = A\left(T - T_c\right) \tag{1}$$

where T_c is critical temperature, A is an arbitrary constant [7]. So the square root of B_{pk} is plotted as a function of temperature in Fig. 5, in order to see the applicability of (1) in the irradiated samples. It can be seen that the square root of B_{pk} of irradiated samples also exhibit a linear relation with the temperature. This result indicates that the secondary peaks observed in the irradiated samples are due to the same mechanism as that in the unirradiated sample.

Further insight into the efficiency of defect cascades introduced by the neutron irradiation is given by analyzing the pinning force density, F_p, by the irradiation. Figure 6 shows the magnetic field variation of F_p at $T = 15$ K calculated from the data shown in Fig. 3. While F_p increases at low fields with increasing the irradiation

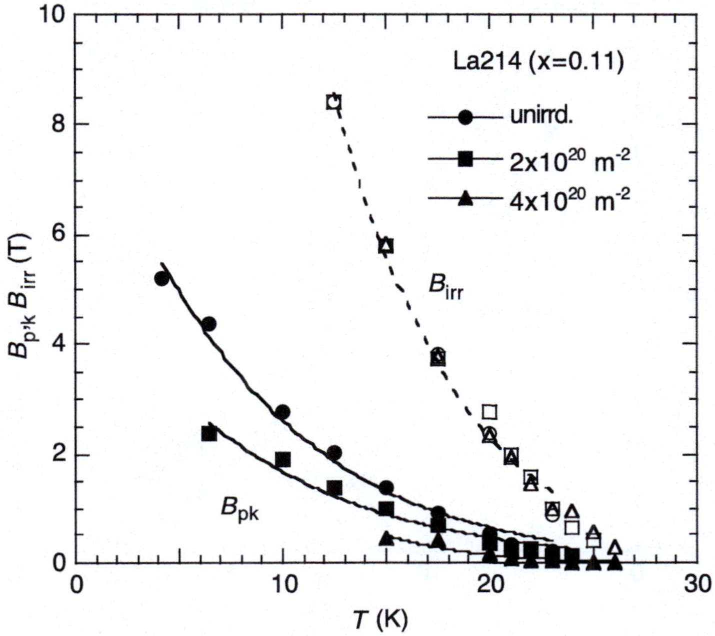

Fig. 4 The temperature variation of B_{pk} and B_{irr} of unirradiated and neutron-irradiated La214 crystal (x=0.11)

fluence, it saturated at high fields. At high fields, F_p decreases as $1 - B/B_{irr}$ with the increasing field. Such a saturation behavior has been often observed in the conventional superconductors with rather weak pinning centers [18]. Matsushita et al. [19] explained such a behavior in terms of avalanching depinning model. The flux lattice is considered to contain many defects, and local plastic deformations may take place under the driving force. If the pinning is not strong enough, such a local instability might develop to a catastrophic avalanching flux flow. The probability of occurrence of a plastic deformation is considered to be proportional to defects of flux lattice. As for the intermediate field region, just below the peak field, deformation of the flux lattice allows the vortices to be pinned most effectively ("synchronization effect") and increase in pinning strength is expected. This may be the possible origin of the secondary peak phenomenon. However, further discussion is needed.

3.2 Heavy-Ion Irradiation of La214 Single Crystals

For heavy-ion irradiation experiment, we used two La214 single crystals with different Sr contents, x = 0.05 (under-doped) and 0.11 (over-doped). Both of them were grown by a travelling-solvent floating-zone (TSFZ) method [5]. The irradiation with 5.8 GeV Pb-ions was performed at GANIL at room temperature. Direction

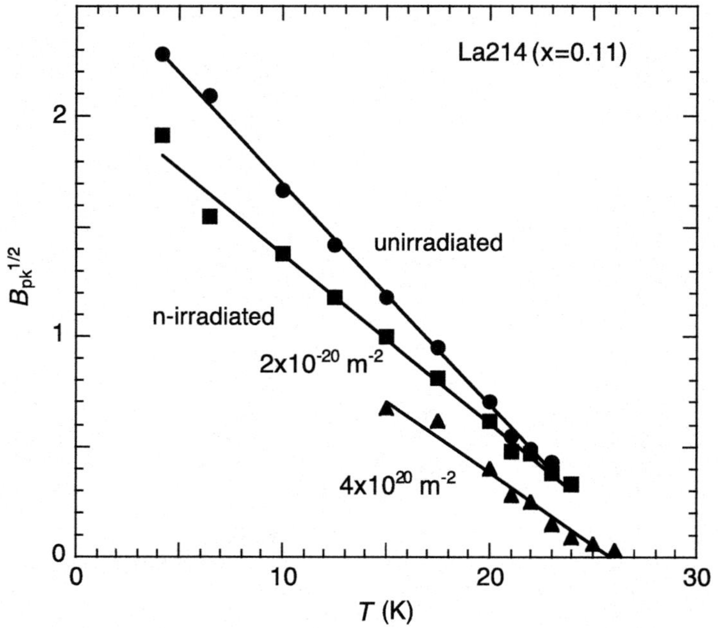

Fig. 5 Temperature dependence of the square root of B_{pk}

of the incident beam was almost parallel to the c-axis. The samples were sequentially irradiated to the fluences of 10^{10} and 1.1×10^{11} ions/cm^2, corresponding to a dose-equivalent flux density of $B_\Phi = 0.2$ and 2.2 T, respectively.

Firstly, we show the irradiation effect in under-doped La214. Figure 7 shows typical magnetization curves of Pb-ion irradiated La214 crystals ($x = 0.05$) measured at $T = 10$ K. Total irradiation fluences were $B_\Phi = 0.2$ and 2.2 T. In the $B_\Phi = 0.2$ T sample, a double-peak structure appeared at low field and it shifted toward higher field after the higher dose irradiation ($B_\Phi = 2.2$ T). Owing to this peak structure, the lightly irradiated sample showed a larger hysteresis at low fields.

The temperature dependence of the magnetization curves of the irradiated samples ($B_\Phi = 0.2$ and 2.2 T) is shown in Fig. 8. In the $B_\Phi = 0.2$ T sample, the double-peak structure appears at temperature above 10 K, while the $B_\Phi = 2$ T sample exhibits double-peak structure at whole temperature range from 4.2 K to T_c. It was found that the peak field is almost independent of temperature.

Next, we show the irradiation effect in over-doped La214. Figure 9 presents typical magnetization curves of unirradiated and Pb-ion irradiated La214 sample ($x = 0.11$), recorded at 10 K. Total fluences were $B_\Phi = 0.2$ and 2.2 T. After the irradiation, the hysteresis width is dramatically enhanced at low field range, while only a slight change was observed in the high fields.

In the magnetization curves of the heavily Pb-ion irradiated crystal ($B_\Phi = 2.2$ T), a new peak structure appeared at lower field. This new peak is very sharp and its position was found to be temperature independent, similar to that observed in the

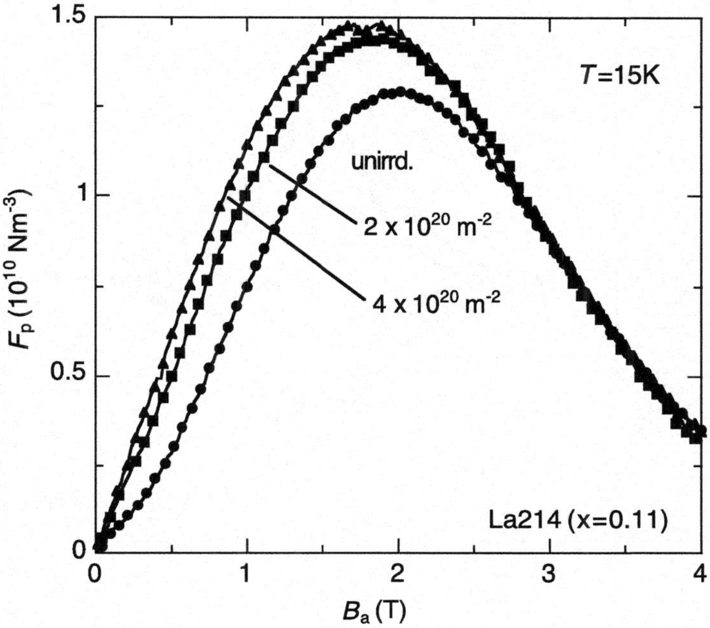

Fig. 6 Magnetic field variation of pinning force density, F_p, at $T = 15$ K before and after the neutron irradiation

irradiated La214 crystals with under-doped composition ($x = 0.05$). Moreover, the peak field was same as that of under-doped sample with the same irradiation dose ($B_\Phi = 2.2$ T). As far as the temperature dependence is concerned, the origin of this new peak is different from that of the secondary peak, which exhibits strong temperature dependence, as shown previously.

Figure 10 compares the effect of the irradiation on the irreversibility field, B_{irr} for $x = 0.05$ and 0.11 samples. The irreversibility field was determined from the closing point of hysteresis loop. In $x = 0.05$ sample, a slight increase of B_{irr} with irradiation fluence was observed at an intermediate temperature range. Decrease of B_{irr} above 20 K observed in $B_\Phi = 2.2$ T sample is due to the degradation of T_c, from 27.5 K to 24.8 K. On the other hand, irreversibility line was almost unchanged in case of $x =$ 0.11 sample. The difference in irradiation effect on B_{irr} between the two samples can be explained by the difference of pre-existing defects. In $x = 0.05$, pinning strength of the pre-existing defects is very weak and B_{irr} is much lower than $x = 0.11$. In this case, introduction of strong pinning center by irradiation effectively increases. On the other hand, in $x = 0.11$ the effect of introducing irradiation defects hardly appeared, since there were already sufficiently strong pinning centers.

Now we discuss the origin of the peak in magnetization curves observed in heavy-ion irradiated La214 samples. The peak observed in the Pb-ion irradiated sample exhibits four distinct features:

1. The peak has a very sharp structure,

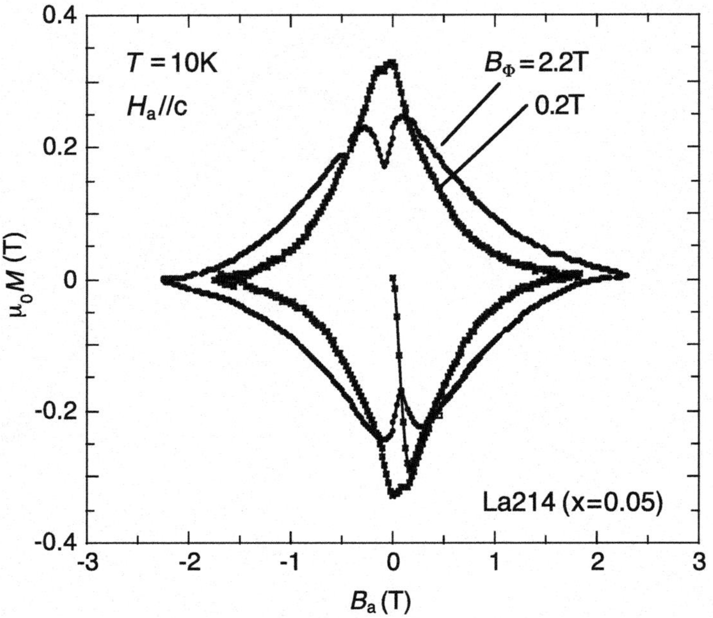

Fig. 7 Magnetization curves of Pb-ion irradiated La214 crystals (x = 0.05) (B_Φ = 0.2, 2.2 T) recorded at T =10 K

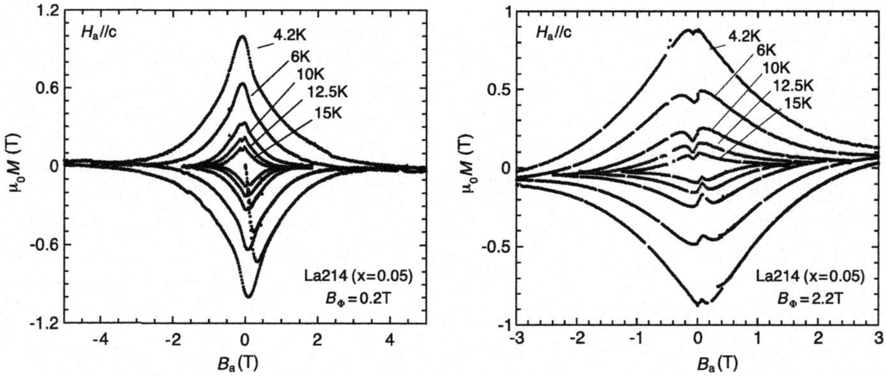

Fig. 8 Temperature dependence of magnetization curves of Pb-ion irradiated under-doped La214 crystals (x = 0.05) (B_Φ = 0.2, 2.2 T)

2. The peak field is independent of temperature,
3. The samples with the same irradiation fluence have the same peak field, independent of sample composition,
4. The peak moves to higher fields with increasing irradiation fluence.

The peak was also appeared in the unirradiated sample with the over-doped composition. However, it showed completely different characteristics: The peak appears

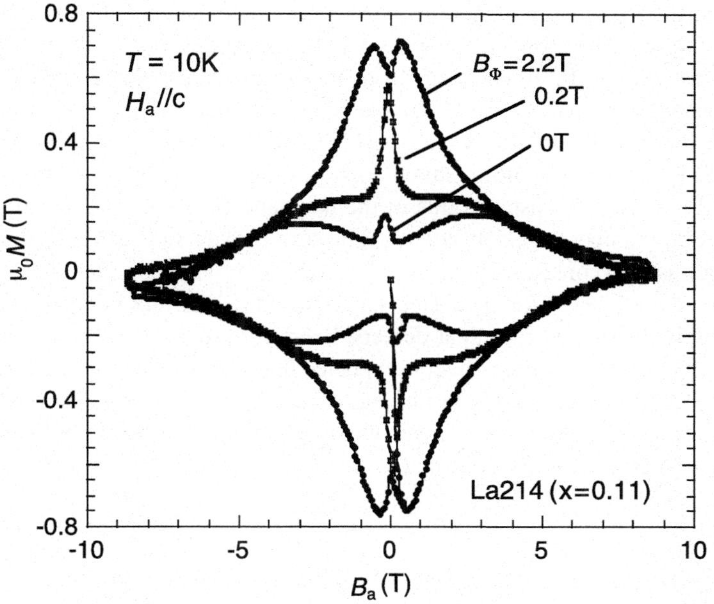

Fig. 9 Magnetization curves of unirradiated and Pb-ion irradiated ($B_{\Phi} = 0.2$, 2.2 T) La214 crystals ($x = 0.11$) recorded at $T = 10$ K

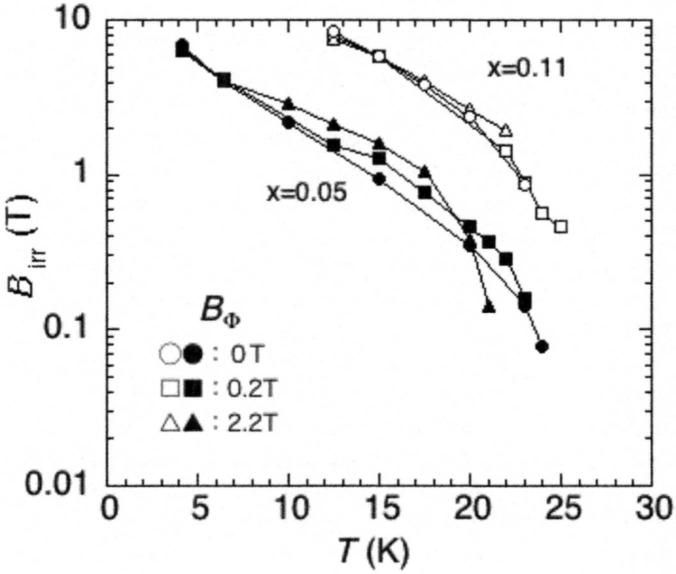

Fig. 10 The irreversibility line of unirradiated and Pb-ion irradiated La214 crystals (x=0.05 (closed symbols) and 0.11(open symbols))

at an intermediate field which is strongly temperature dependent. And it has rather broad structure. Therefore, we suppose that the peak appeared in the Pb-ion irradiated sample has a different origin from that in the unirradiated sample. Several mechanisms have been proposed to explain peak effect as shown previously. Among them, the present observation seems to be consistent with the matching effect mechanism. Since a matching of flux lattice to a periodic distribution of pinning center is expected to occur at constant value of the magnetic field, the peak position should not show any temperature dependence. The shift of peak field after the higher dose irradiation is also consistent with such a picture.

In the case of heavy-ion irradiated samples, pinning of flux lattice is mainly due to randomly distributed columnar defects. Even if the array of pinning center is not perfectly periodic, matching effect is expected to occur when the mean vortex spacing matches the mean spacing of pinning center, in the case of rather concentrated system of pinning centers. It is reasonable then that such matching peak appears at lower field than calculated value of B_{Φ}.

4 Irradiation Effects in RE123 Coated Conductors

4.1 Effect of Heavy-Ion Irradiation in RE123 Coated Conductors

Since RE123 coated conductors have relatively excellent high magnetic field characteristics among various HTSCs, it is expected to be applied to magnets. So a number of studies have been conducted to explore a new method to introduce effective pinning centers that can improve J_c at high magnetic field. Most of the introduction methods are based on chemical methods. The most common way is adding impurities to a raw material and dispersing it as a precipitate in superconducting RE123 phase during the manufacturing process. In the pulsed laser deposition method (PLD method), it has been reported that the $BaMO_3$ (M=Hf, Zr, Sn) inclusion form long columnar inclusion with its diameter ~10 nm within RE123 matrix and they provide considerable J_c enhancement at high magnetic fields [20–22]. For the further enhancement of J_c, we need to optimize many parameters, such as size, shape, arrangement, inclement, density of pinning centers. However, it is difficult to control these parameters in case of chemical method.

On the other hand, the irradiation method has the advantage that these parameters can be controlled independently of the manufacturing process of the coated conductors. In this method, the damage morphology strongly depends on the electronic stopping power, S_e. In the case of Y123 polycrystalline samples, the continuous columnar defects are created at $S_e > 20$ keV/nm [5]. It has also been reported that the diameter of columnar defect increases with S_e [23]. Since the size and shape of columnar defects are very similar to $BaMO_3$ nano-rods, this technique can be used to create a model system for pinning optimization.

Therefore, in this study, we used the heavy-ion irradiation technique to explore an optimum condition for improving J_c in RE123 tape.

In the present study, we used Y123 tape prepared by a pulsed laser deposition (PLD) technique. The dimension of each layer of the tape used in the present study is as follows: Ag (5 μm), Y123 (0.5 μm), CeO_2 (0.55 μm), $Gd_2Zr_2O_7$ (0.8 μm), and Hastelloy (100 μm). One-cm-width tape was cut into the size of about 2.0 mm × 2.0 mm, and they were irradiated either by 450 MeV $^{129}Xe^{23+}$, 500 MeV $^{197}Au^{31+}$, or 400 MeV $^{84}Kr^{18+}$ ions to the Y123 layer from Ag layer side by using the AVF cyclotron of Japan Atomic Energy Agency. In order to change the energy of heavy ions incident to the Y123 layer, Ag layer of some sample was removed by a chemical etching method. The irradiation fluence was varied from 1.0×10^{10} to 1.0×10^{12} ions/cm^2, while the direction of the incident beam was kept for perpendicular to the tape surface (parallel to the *c*-axis of Y123 layer.

We also calculated S_e in Y123 layer by using TRIM code (TRansport of Ion in Matter). For calculations, we assumed that the composition of the Y123 layer is $YBa_2Cu_3O_{6.9}$, density is 6.5 g/cm^3 and the displacement energy E_d = 20 eV [24]. Calculated results are shown in Fig. 11. We can say from the calculation that introduced ions have enough energy to penetrate Y123 layer, and S_e is almost constant within the Y123 layer in each condition.

We also estimated the diameter of columnar defects from Ref. 21. Irradiation conditions and corresponding S_e values and estimated diameter of columnar defects are summarized in Table 1.

Before and after the irradiation, magnetic hysteresis measurement was performed with physical property measuring system (PPMS), Quantum design, and J_c was

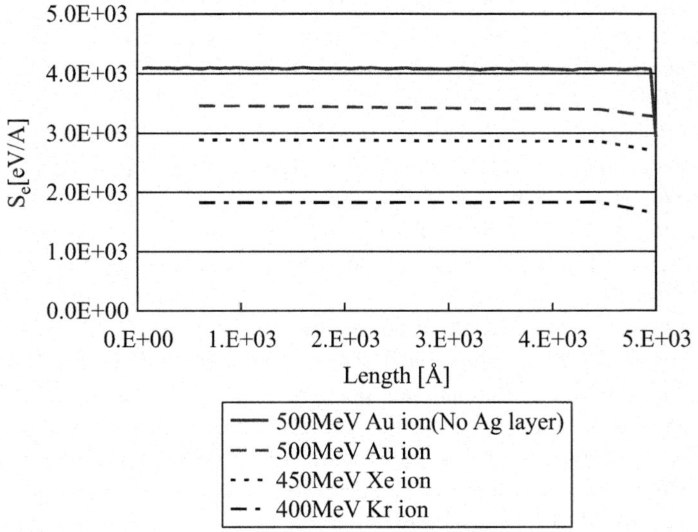

Fig. 11 S_e in Y123 layer for 500 MeV Au-ion, 450 MeV Xe-ion, and 400 MeV Kr ion irradiation. The calculation was carried out using TRIM code (TRansport of Ion in Matter), assuming Y123 is covered by 5 μm of Ag layer

Table 1 Irradiation conditions, calculated S_e values and diameter of columnar defects

	Thickness of Ag on top of Y123 (μm)	S_e (keV/nm)	Estimated diameter of columnar defects (nm)
400 MeV $^{84}Kr^{18+}$	5.0	18	4
450 MeV $^{129}Xe^{23+}$	5.0	29	8
500 MeV $^{197}Au^{31+}$	0	41	13
	5.0	34	11

calculated using modified Bean model [25]. T_c was determined from AC susceptibility measurement.

At first, we show typical columnar defects observed by TEM for Xe-ion and Au-ion irradiated Y123 tapes in Fig. 12. In both cases, the continuous columnar defects (shown by arrows) are successively created in the Y123 layer.

Figure 13 shows the fluence dependence of J_c at $T = 77$ K and for applied magnetic field $B_a = 0.2$ and 2 T for Kr, Xe and Au-ion irradiation. At $B_a = 0.2$ T, J_c was most improved at 1×10^{11} ions/cm^2 in all irradiations except Kr, and the degree of increase was almost constant regardless of the ion species. On the other hand, at $B_a = 2$ T, the largest improvement in J_c was observed in the higher fluence region (3 to 5×10^{11} ions/cm^2), and the maximum value of J_c varies with ion species. We can also see that the Xe-irradiation is most effective for enhancement of high field J_c. It is noteworthy that the defect diameter of Xe-irradiation is not largest among three species.

Here, we would like to define the fluence at which the maximum improvement in J_c is obtained at each applied magnetic field B_a as "optimum fluence (ϕ_{op} (B_a))." Then we consider the relationship between the density of the columnar defects and the applied magnetic field at ϕ_{op} (B_a). At $B_a = 0.2$ T, the flux line density (d_{fl}) is calculated to be 1×10^{10}/cm^2. Assuming that each incident ion creates one columnar defect, and each columnar defect pins one flux line, the irradiation to a fluence of 1×10^{10} ions/cm^2 is considered to be sufficient. However, in reality, there is not much improvement in J_c at that fluence, and ϕ_{op} (0.2 T) is ~1×10^{11} ion/cm^2, ten times larger than d_{fl}. At $B_a = 2$ T, $d_{fl} = 1 \times 10^{11}$/cm^2, but ϕ_{op} (2 T) = 3 ~ 5×10^{11} ions /cm^2, which is three times larger than d_{fl}. This can be understood by follows using the schematic illustration shown in Fig. 14. While the defects introduced by heavy-ion irradiation have a random distribution, the flux lines try to form a triangular lattice as much as possible due to the repulsive force between the flux lines. So, it may be easier for the flux line to find an ideal position to form triangular lattice for higher pinning density and there for can obtain larger pinning force. However, when the fluence becomes higher (for example, 1×10^{12} ion/cm^2), the portion of damage area Y123 becomes large, resulting in the decrease of the superconducting volume and thus large decrease of J_c is observed.

The reason why Au-ion irradiation did not give a large improvement in J_c at B_a = 2 T may be understood by considering the decrease of the superconducting volume due to the irradiation defects. For example, for Au (Ag: 0 mm), Au (Ag: 5 mm), and

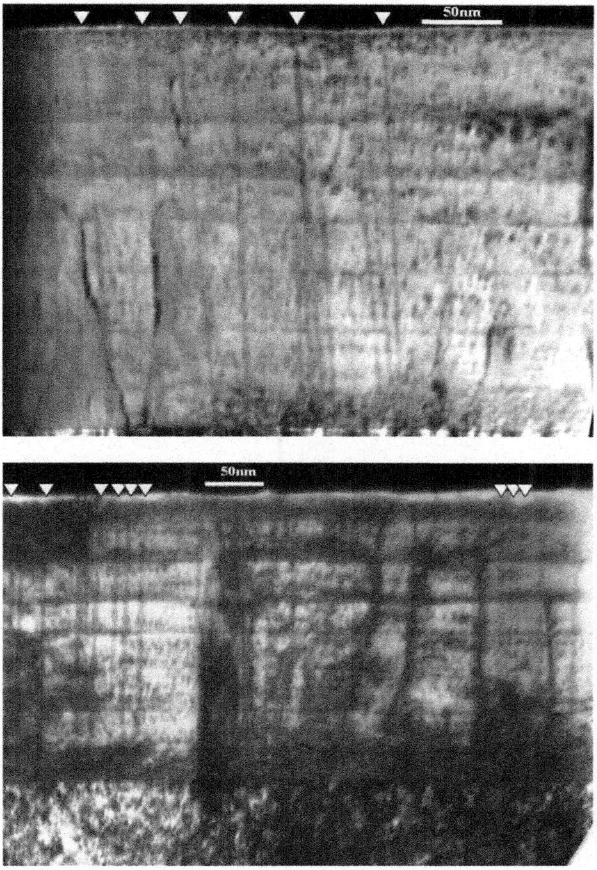

Fig. 12 TEM images of columnar defects in heavy-ion irradiated Y123 tapes. The locations of columnar defects are shown by white arrows. (**a**) 450 MeV $^{129}Xe^{23+}$ irradiation, fluence 1×10^{11} [ions/cm^2] (**b**) 500 MeV $^{197}Au^{31+}$ irradiation, fluence 1×10^{11} [ions/cm^2], Ag layer removed

Xe (Ag: 5 mm), the volume % of defects at irradiation fluence 5×10^{11} ions/ cm^2, assuming no the overlap of defects, is ~ 66%, ~ 47%, and ~ 25%, respectively, suggesting that the superconducting current path is significantly impeded.

In addition, large disturbance of Y123 crystal lattice is observed in Au-ion irradiated sample. Figure 15 shows the irradiation fluence dependence of normalized c-axis length, c/c_0, in which c_0 is the c-axis length before irradiation, and increase of FWHM (Full Width at Half Maximum) of (006) peak, ΔFWHM, of Y123 measured by X-ray diffraction measurement (XRD). We can see a remarkable extension of c-axis length as well as the increase of FWHM by Au-ion irradiation.

Figure 16 compares the Raman spectroscopy data for before and after irradiation with various species to the fluence of 1×10^{12} ions/ cm^2. We can see that the peaks around 335 and 500 cm^{-1} shifted to lower value by irradiation. It is known that these peaks are sensitive to oxygen defects, and result indicates the formation of oxygen

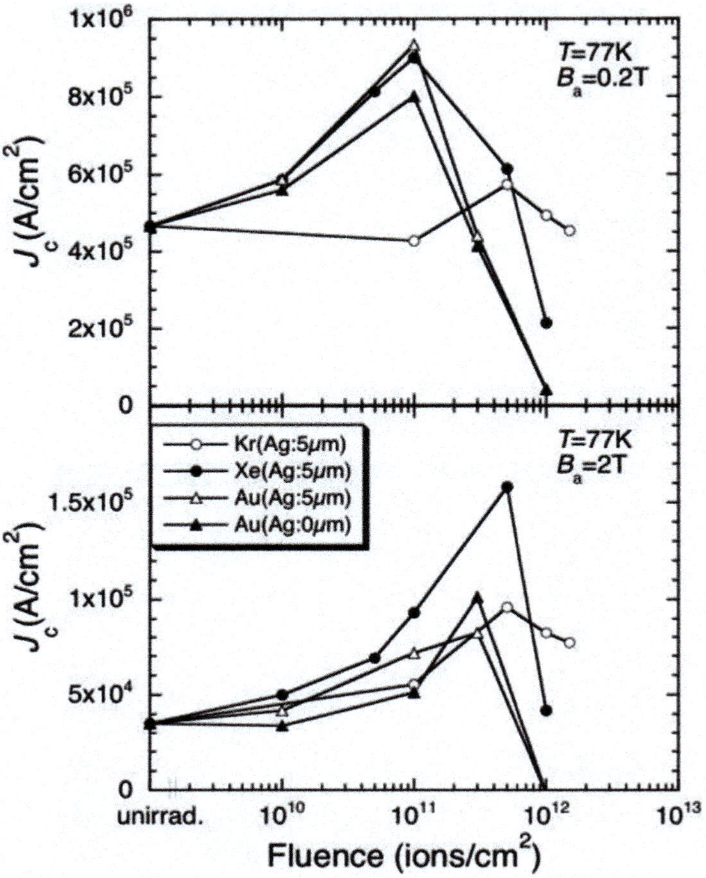

Fig. 13 Fluence dependence of J_c at 77 K for 400 MeV Kr-, 450 MeV Xe-, and 500 MeV Au-ion irradiations

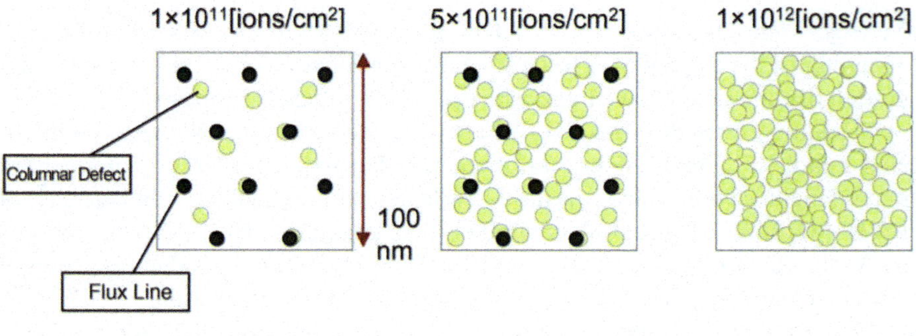

Fig. 14 Schematic illustration of the distribution of columnar defects and flux lines for B_a=2 T

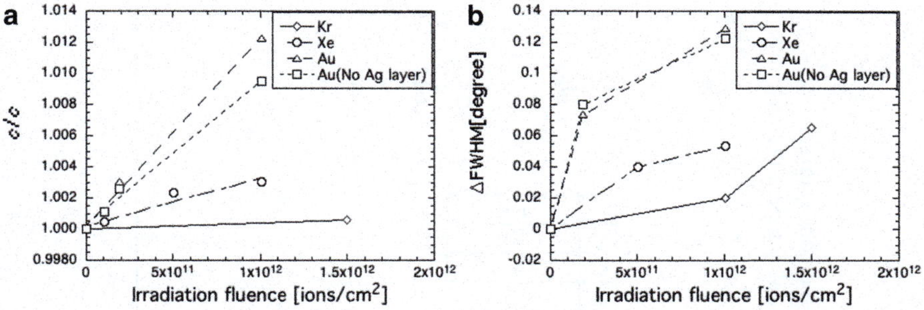

Fig. 15 Irradiation fluence dependence of c/c_0 (**a**) and ΔFWHM of (006) peak in XRD data (**b**)

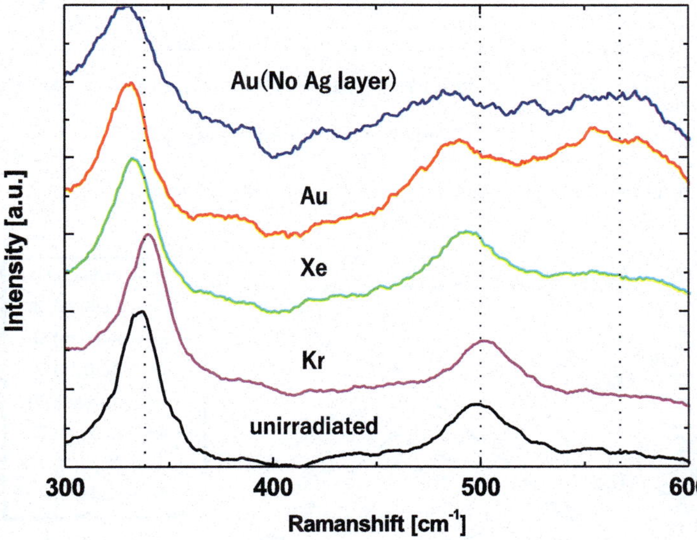

Fig. 16 Raman spectroscopy data for Y123 before and after irradiation. All the irradiation was performed at the same fluence 1×10^{12} ions/cm^2

vacancies by irradiation [26]. In addition, in Au-ion irradiated samples we observed an appearance of broad peak around 570 cm^{-1}. According the refs. [27, 28], this peak suggests the presence of cation disorder in Y123 layer.

We also measured T_c before and after irradiation. Figure 18 shows the fluence dependence of T_c for three different irradiation conditions. T_c are found to be almost constant at low fluences below 1.0×10^{11} ions/cm^2, while it decreases rapidly at a certain fluence, namely critical fluence, ϕ_{cr}, depending on the irradiated species. We can see that the smaller ϕ_{cr} for larger S_e, such as Au-ion irradiation.

From these results, we consider that not only the reduction of superconducting volume but also the reduction of superconducting condensation energy due to the introduction of the disorder inside the Y123 crystal lattice causes the reduction of J_c, in particular for Au-ion irradiation.

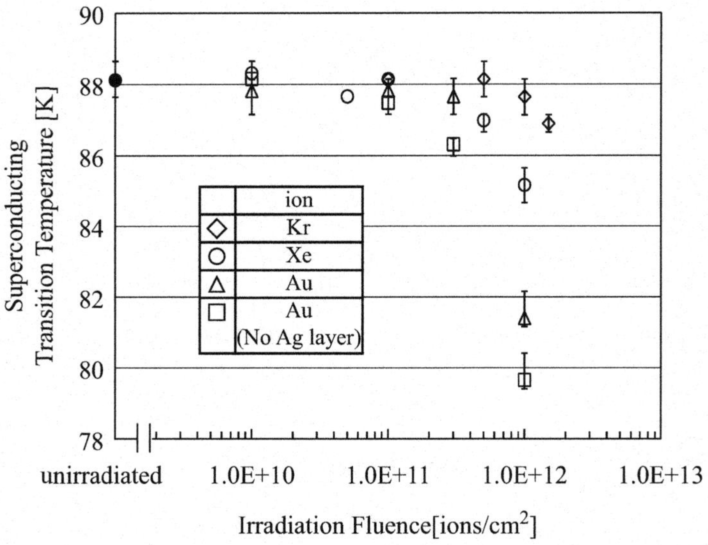

Fig. 17 Effect of heavy-ion irradiation on T_c in Y123 coated conductor

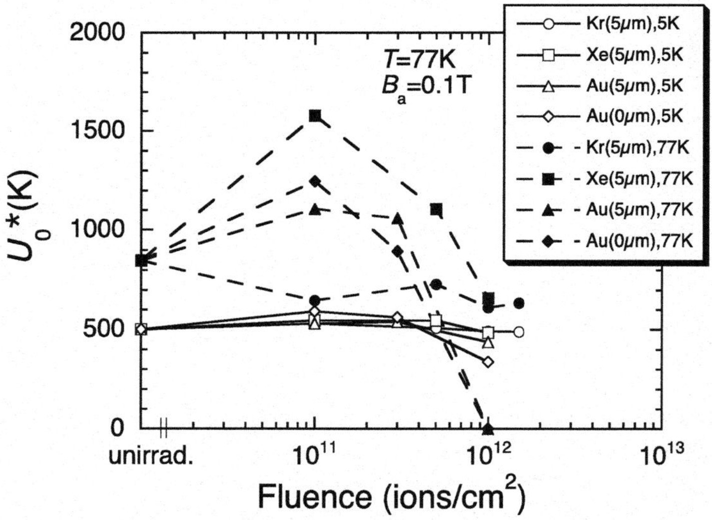

Fig. 18 Fluence dependence of U_0* at 77 K and 0.1 T for 400 MeV Kr-, 450 MeV Xe-, and 500 MeV Au-ion irradiations

Regarding Kr ion irradiation, there was almost no improvement in J_c at B_a = 0.2 T, whereas at B_a = 2 T, improvement in J_c was observed at a high fluence ~ 5 × 10^{11} ions/cm^2. In case of Kr ion irradiation, estimated size of columnar defect is about 4 nm, which is much smaller than the superconducting coherence length and the pinning force provided by the columnar defect is thought be the same order as existing defects, such as spiral dislocations. In such situation, introduction of pinning center by irradiation is thought to be not effective for improving J_c in a low magnetic field because d_{fl} is much smaller than the total number of pinning centers. On the other hand, in a high magnetic field where d_{fl} exceeds the density of existing defects, the magnetic flux lines that were not pinned before irradiation will be pinned by the introduction of defects, leading to the improvement of J_c.

In order to verify this idea, magnetic flux creep measurement was performed and the change in apparent pin potential U_0^* due to heavy-ion irradiation was investigated. Flux creep measurement was performed as follows. At first, we applied a sufficiently high magnetic field (5 T) to give full penetration of flux line in the sample. Then we reduced the magnetic field to 0.1 T and then measured the time dependence of magnetization. U_0^* were obtained by using Anderson-Kim model [29, 30] from the magnetic relaxation results.

Figure 18 shows the fluence dependence of U_0^* at 5 K and 77 K of Kr, Xe, and Au-ion irradiated samples. At 5 K, there is no significant change in U_0^* due to irradiation, suggesting the pinning strength is almost the same as that of the existing defect.

On the other hand, at 77 K, U_0^* tends to increase at 10^{11} ions/cm^2 and decrease at larger fluences, except for Kr ion irradiation. Regarding Kr ion irradiation, there was almost no change in U_0^* due to irradiation. These results suggest that the columnar defect introduced by Au and Xe-ion irradiation has larger pinning potential in this temperature range, while the pinning potential of columnar defect introduced by Kr ion irradiation seems be comparable to existing defects.

4.2 *Effect of Post-Annealing in Heavy-Ion Irradiated RE123 Coated Conductors*

As discussed in the previous section, we consider that the reason for reduction of J_c in Au-ion irradiation is due to the reduction of superconducting condensation energy caused by the introduction of the disorder inside the Y123 crystal lattice.

In the present study, we performed post-annealing to anneal out introduced cation disorder and oxygen defects and to see if it is efficient for restoring superconducting properties. For this post-annealing process, selection of annealing temperature is very important. If the annealing temperature is too high, large deficiency of oxygen occurs in Y123 and the superconductivity will be lost. So, we chose two moderate temperature, 473 K and 673 K for the treatment. The annealing was performed in the flowing oxygen atmosphere for 2 h and then quenched.

Figure 19 shows the J_c values measured at $B_a = 2.0$ T before irradiation, after irradiation, and after annealing. We observed an increase of J_c after annealing at 673 K, while it was reduced at 473 K for 3.0×10^{11} ions/cm^2. On the other hand, in case of 1.0×10^{12} ions/cm^2, the decreased J_c by the irradiation did not recovered by the annealing. We also measured change in T_c after irradiation and annealing. As

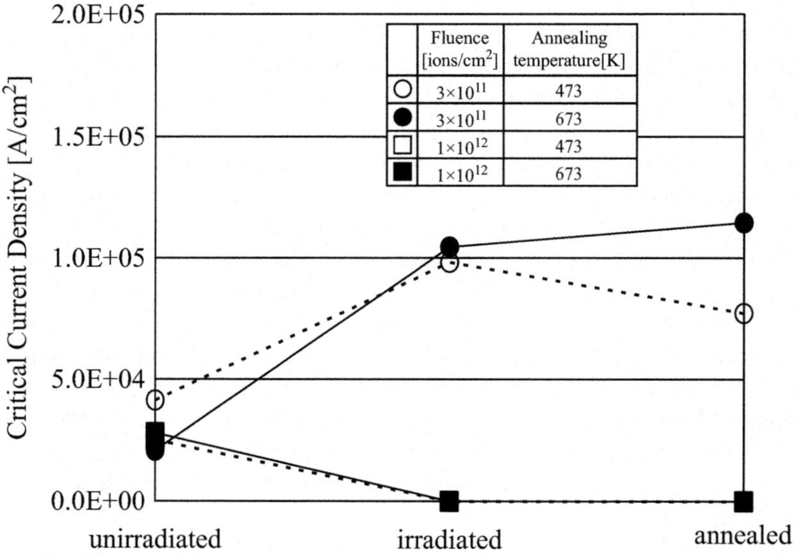

Fig. 19 Change in J_c at $B_a = 2.0$ T after Au-ion irradiation and post-annealing for Y123 (removed silver layer)

shown in Fig. 18, T_c was decreased by about 8 K after Au-ion irradiation (no Ag layer) with 1.0×10^{12} ions/cm^2. After annealing, we observed slight recovery of T_c, $\Delta T_c \sim 0.5$ K for 473 K and ~1 K for 673 K.

These results suggest that the post-annealing was a good way to recover T_c without decreasing J_c. However, optimization of annealing condition is required.

5 Summary

In this chapter, irradiation studies on La214 single crystals and Y123 coated conductor are reviewed. All of the results shown here lead to the conclusion that both neutron and heavy-ion irradiation are powerful tools for improving the pinning properties. Besides that, irradiation is also very effective in studying the pinning mechanism. In La214, there is big difference in J_c-B characteristics between neutron-irradiated sample and heavy-ion irradiated sample. The former causes so-called "fish-tail" type 2nd peak, while latter causes "matching effect" type 2nd peak in J_c-B curve.

In Y123 coated conductors, we examined the effect columnar defect size for the improvement of J_c. We modified the defect size by changing S_e. S_e is changed either by changing the ion species to be irradiated or by changing the thickness of Ag layer on top of Y123 layer. The enhancement of J_c was observed especially at the high magnetic field in 77 K, but the degree of the enhancement was different for the irradiation condition. The largest enhancement of J_c at 77 K has been observed in case of the Xe-ion irradiation, whereas TRIM code calculation gives largest S_e value for Au-ion irradiation without Ag layer. XRD and Raman microscopy data suggested the introduction of the disorder inside the Y123 crystal lattice in Au-ion irradiated sample, which likely causes the depression of superconducting properties. This result suggests that we must carefully choose ion species that can create columnar defects without damaging crystallinity of Y123 layer. However, we show that post-annealing may be effective to recover the damage inside Y123 for further J_c improvement.

Acknowledgments The author would like to thank Dr. T. Kimura for La214 single crystals, Dr. A. Ibi for Y123 coated conductors, and Dr. M. Konczykowski for heavy- ion irradiation in GANIL. The irradiation and post-annealing of Y123 had been studied in collaboration with K. Nakashima and T. Terai.

References

1. *For example,* Wei-Kan Chu, Jia Rui Liu, Zu Hua Zhang, Nuclear instruments and methods in physics research section B: Beam interactions with materials and atoms 59–60, Part 2, 1447 (1991); L. Civale, Supercond. Sci. Technol. **10**, A11 (1997).
2. M.C. Frischherz, M.A. Kirk, J. Farmer, L.R. Greenwood, H.W. Weber, Physica C **232**, 309 (1994)
3. H.W. Weber, in *Proceedings of the 10th Anniversary HTS Workshop on Physics, Materials and Applications*, ed. by B. Batlogg, C.W. Chu et al., (World Scientific, Singapore, 1996) pp163 and references there in.
4. F.M. Sauerzopf, Phys. Rev. B **57**, 10959 (1998)
5. V. Hardy, D. Groult, M. Hervieu, J. Provost, B. Raveau, Nucl. Instrum. Methods B **54**, 472 (1991)
6. T. Kimura, K. Kishio, T. Kobayashi, Y. Nakayama, N. Motohira, K. Kitazawa, K. Yamafuji, Physica C **192**, 247 (1992)
7. T. Kobayashi, Y. Nakayama, T. Kimura, K. Kishio, K. Kitazawa, K. Yamafuji, Appl. Phys. Lett. **62**, 1830 (1993)
8. M. Daeumling, J.M. Seuntjens, D.C. Labalestier, Nature **346**, 332 (1990)
9. L. Krusin-Elbaum, L. Civale, V.M. Vinokur, F. Holtzberg, Phys. Rev. Lett. **69**, 2280 (1992)
10. T. Nishizaki, T. Naito, N. Kobayashi, Phys. Rev. B **58**, 11169 (1998) Physica C282-287, 2117(1997)
11. M. Xu, D.K. Finnemore, G.W. Crabtree, V.M. Vinokur, B. Dabrowski, D.G. Hinks, K. Zhang, Phys. Rev. B **48**, 10630 (1993)
12. N. Chikumoto, M. Konczykowski, N. Motohira, A.P. Malozemoff, Phys. Rev. Lett. **69**, 1260 (1992)
13. T. Tamegai, Y. Iye, I. Oguro, K. Kishio, Physica C **213**, 33 (1993)
14. G. Yang, P. Shang, S.D. Sutton, I.P. Jones, J.S. Abell, C.E. Gough, Phys. Rev. B **48**, 4054 (1993)
15. E. Zeldov, D. Majer, M. Konczykowski, A.I. Larkin, V.M. Vinokur, V.B. Geshkenbein, N. Chikumoto, H. Shtrikman, Europhys. Lett. **30**, 367 (1995)
16. V. Hardy, A. Wahl, A. Ruyter, M. Maignan, C. Martin, L. Coudrier, J. Provost, C. Simon, Physica C **232**, 347 (1994)
17. J.D. Jorgensen, P. Lightfoot, S. Pei, B. Dabrowski, D.R. Richards, D.G. Hinks, *Advances in Superconductivity III* (Springer-Verlag, Tokyo, 1990), p. 337

18. For review, T. Matsushita, in *Composite Superconductors* (Marcel Dekker Inc., New York, 1994), p. 21.
19. T. Matsushita, H. Kuepfer, J. Appl. Phys. **63**, 5048 (1988)
20. H. Tobita, K. Notoh, K. Higashikawa, M. Inoue, T. Kiss, T. Kato, T. Hirayama, M. Yoshizumi, T. Izumi, Y. Shiohara, Supercond. Sci. Technol. **25**, 062002 (2012)
21. J.L. Macmanus-Driscoll, S.R. Foltyn, Q.X. Jia, H. Wang, A. Serquis, L. Civale, B. Maiorov, M.E. Hawley, M.P. Maley, D.E. Peterson, Nat. Mater. **3**, 439 (2004)
22. S. Kang, A. Goyal, J. Li, A.A. Gapud, P.M. Martin, L. Heatherly, J.R. Thompson, D.K. Christen, F.A. List, M. Paranthaman, D.F. Lee, Science **311**, 1911 (2006)
23. Y. Zhu, Z.X. Cai, R.C. Budhani, M. Suenaga, D.O. Welch, Phys. Rev. B **48**, 6436 (1993)
24. T.E. Mitchell et al., Appl. Phys. Lett. **55**, 283 (1989)
25. E.M. Gyorgy, R.B. van Dover, K.A. Jackson, L.F. Schneemeyer, J.V. Waszczak, Appl. Phys. Lett. **55**, 283 (1989)
26. R.M. Macfarlane, H.J. Rosen, E.M. Engler, R.D. Jacowitz, V.Y. Lee, Phys. Rev. B **38**, 284 (1988)
27. G. Gibson, J.L. MacManus-Driscoll, L.F. Cohen, IEEE Trans. Appl. Supercond. **7**, 2130 (1997)
28. G. Gibson, L.F. Cohen, R.G. Humphreys, J.L. MacManus-Driscoll, Physica C **333**, 139 (2000)
29. P.W. Anderson, Y.B. Kim, Rev. Mod. Phys. **36**, 39 (1964)
30. R.A. Doyle, W.S. Seow, J.D. Johnson, A.M. Campbell, P. Berghuis, R.E. Somekh, J.E. Evetts, G. Wirth, J. Wiesner, Phys. Rev. B **51**, 12763 (1995)

Interfaces in *RE*BCO-Based Nanocomposite Thin Films and their Contribution to Vortex Pinning

Alok K. Jha and Kaname Matsumoto

1 Interfaces and their Structures

An interface may be termed as a small number of atomic layers which separate two different crystalline solids in contact with each other [1]. For example, a thin film deposited on a crystalline substrate is separated by the substrate-thin film interface. Another example could be the interfaces inside a superconducting film which separate the superconducting matrix and non-superconducting inclusions. The property at the so-called interface is very much different from the crystalline solids it separates [2].

Interfaces are broadly classified into two types: (i) homophase interfaces and (ii) heterophase interfaces. An interface separating two regions of the same phase is called a homophase interface which is often termed as a grain boundary. An interface separating two different crystal phases is called a heterophase interface [3].

Interfaces are also classified in other three classes based on their structural features. These are called (i) coherent, (ii) semi-coherent, and (iii) incoherent interfaces. Starting with a references structure and making a series of operations, these interfaces can be constructed exhibiting different structural features [4].

Figure 1 shows the construction of a heterophase interface between two different crystal phases: phase 1 and phase 2. The reference structure is the single crystal of phase 1. The dashed line in Fig. 1a indicates the location of the interface to be constructed. The first operation transforms the upper portion of the crystal 1 into phase 2 while maintaining registry along the interface as shown in Fig. 1b. As the two phases (1 and 2) adjoining the interface are maintained in the registry, the resulting

A. K. Jha (✉) · K. Matsumoto
Department of Materials Science and Engineering, Kyushu Institute of Technology, Kitakyushu, Japan
e-mail: akjha@post.matsc.kyutech.ac.jp

A. G. Roca et al. (eds.), *Surfaces and Interfaces of Metal Oxide Thin Films, Multilayers, Nanoparticles and Nano-composites*,
https://doi.org/10.1007/978-3-030-74073-3_9

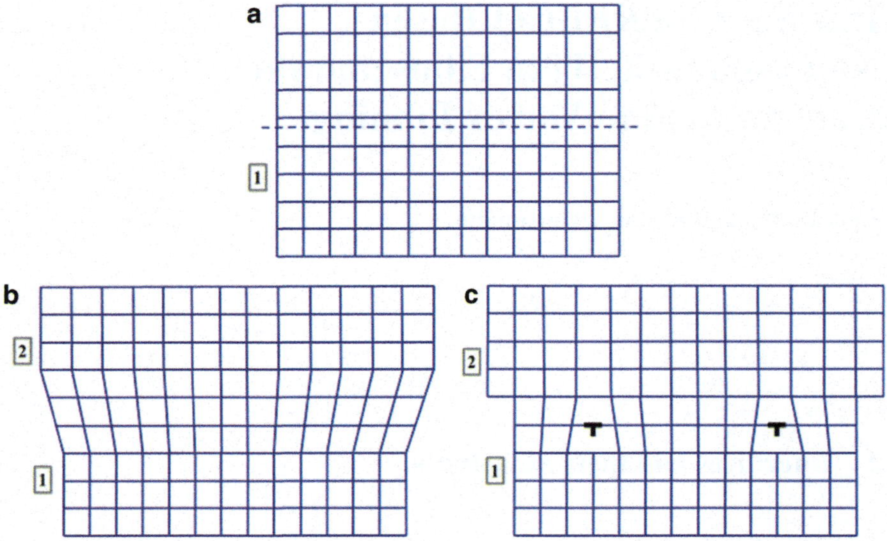

Fig. 1 Schematic diagram of a heterophase interface. (**a**) Phase 1 is the reference crystal. (**b**) Transformation of the upper region of the interface into phase 2 while maintaining coherence. (**c**) Relaxation of long-range stresses by the introduction of dislocations at the interface between the phases

interface is coherent. In order to maintain the interface registry, long-range coherency stresses are required.

The formation of a semi-coherent interface requires another operation in which these stresses are eliminated by introducing some dislocations in the interface as represented in Fig. 1c. Thus the resulting interface consists of patches of coherent interface separated by dislocations. In this process, the overall coherence of the interface is destroyed by the cuts and displacements resulting in dislocations, and such an interface is, therefore, considered to be semi-coherent with respect to the reference structure in Fig. 1a.

If the density of the dislocation is so large that their cores overlap, then the coherence characteristics of the reference structure will be destroyed, and it will result in an incoherent interface. The cores of incoherent interfaces, therefore, consist of continuous slabs of poor material, and the interfaces lack long-range order.

In *RE*BCO-based nanocomposite thin films, vortex pinning and subsequently critical current densities tend to be sensitive to not only the geometry and concentration of nanoscale inclusions but also to the interfacial composition and the resulting nanoscale lattice defects.

2 *RE*BCO Superconductors and their Critical Parameters

Superconductivity in $YBa_2Cu_3O_{7-\delta}$ (YBCO) was first discovered in its bulk polycrystalline form and the superconducting transition was observed above the liquid nitrogen temperature (77 K) [5]. The discovery of YBCO accelerated the search for higher transition temperature (T_c) materials and in the quest of discovering higher T_c superconductor, other rare-elements were tried to replace Y in YBCO [6–8]. Apart from the critical temperature (T_c) and critical magnetic field (H_c), the critical current density (J_c) is the most relevant parameter of a superconductor which determines the practical utility of a superconductor [9–11].

Soon after the discovery of *RE*BCO superconductors in their polycrystalline forms, it became possible to make their thin films on single crystal substrates such as $SrTiO_3$, Al_2O_3, MgO, *etc.* [12–14]. The current carrying capability of *RE*BCO superconducting thin films significantly enhanced over the years due to the development of the fabrication and processing techniques. Many fabrication techniques such as pulsed laser deposition (PLD) [15], chemical solution deposition (CSD) [16], metal organic chemical vapor deposition (MOCVD) [17] have been developed over the years to deposit high quality oriented *RE*BCO thin film which exhibit high J_c of 1–5 MA/cm^2 at 77 K, self-field [18, 19].

3 Vortex Pinning by APCs in *RE*BCO Thin Films

In the critical state of a type-II superconductor (such as *RE*BCO), the flow of electrical current is limited by temperature and applied magnetic field. When an electrical current is passed across a superconducting sample subjected to an applied magnetic field, the vortices experience a Lorentz force in a direction perpendicular to both the current and the applied magnetic field as depicted in Fig. 2. The movement of the vortices due to Lorentz force would generate an electric field leading to energy dissipation. The vortices, therefore, need to be pinned at their respective locations inside the superconducting matrix [20].

Some microscopic defects are naturally generated during the growth of thin films, such as grain boundaries, oxygen vacancies, point defects, dislocations, *etc.* Some of these defects act as pinning centers for vortices against the Lorentz force. Figure 3 shows the schematic diagram of such naturally generated defects in a typical superconducting *RE*BCO thin film. These naturally generated pinning centers, however, are not strong enough to prevent the vortex motion under applied magnetic fields at elevated temperatures [11, 21, 22].

An alternative strategy employed to prevent the motion of the vortices is the generation of artificial defects inside superconducting matrix. Such intentionally generated pinning centers are termed as artificial pinning centers (APCs). The APCs are introduced into *RE*BCO films through different methods which include addition and/or substitution of rare-earth atoms [23, 24] or incorporation of nanoscale

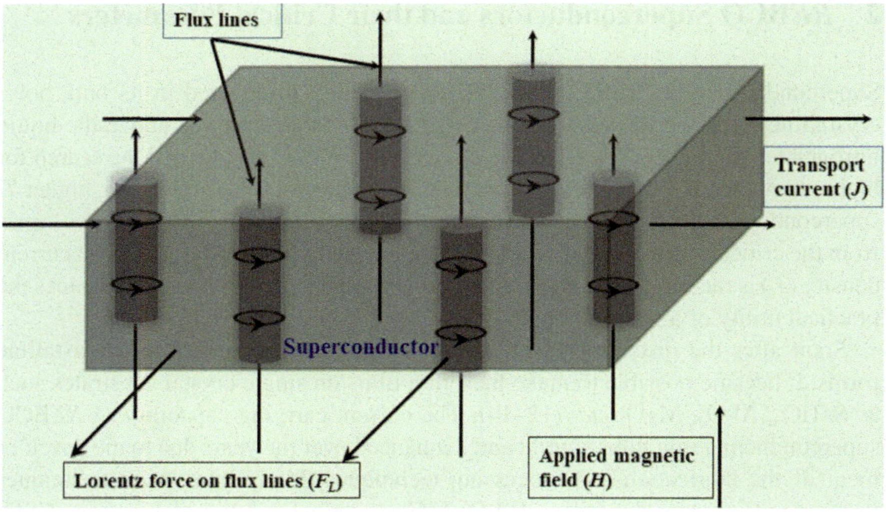

Fig. 2 The mixed state in a type-II superconductor. Magnetic flux starts penetrating the sample in the form of vortices, with quantized flux $\Phi_0 = h/(2e)$. (Reproduced from Jha and Matsumoto [20])

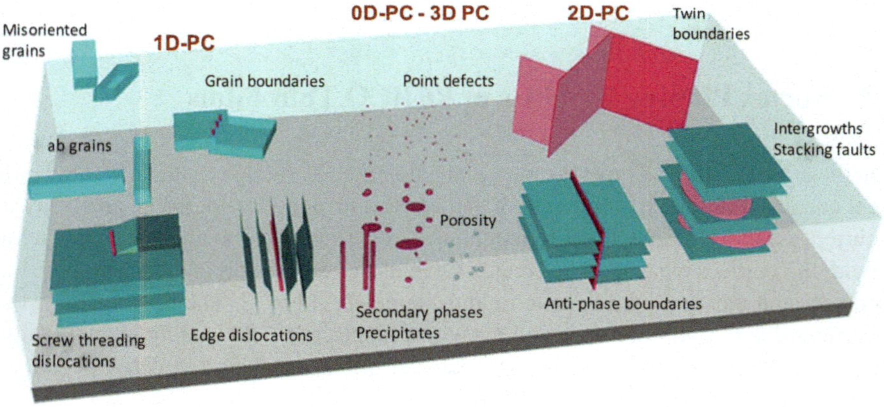

Fig. 3 Schematic diagram of various defects generated during the growth of *RE*BCO thin films. The ones shown in pink are useful for vortex pinning whereas others are detrimental to the superconducting properties. (Reproduced from Palau et al. [60])

secondary phases into the *RE*BCO film matrix [25–37] or decoration/modification of the substrate surface [38, 39]. Improved vortex pinning in *RE*BCO films through incorporation of secondary phase nanoinclusions has been frequently observed. These secondary phases are self-assembled either in the form of nanocolumns or nanoparticles inside the *RE*BCO matrix as represented in the schematics in Fig. 4. The nanoscale inclusions of secondary phases such as Y_2BaCuO_5 [25, 26], Y_2O_3 [27], $BaTiO_3$ [28], $BaZrO_3$ [29–31], $BaSnO_3$ [32, 33], $BaHfO_3$ [34], YBa_2NbO_6 [35, 36], and YBa_2TaO_6 [37] have been successfully observed to enhance the in-field J_c of *RE*BCO thin films over a wide range of temperature and applied magnetic field.

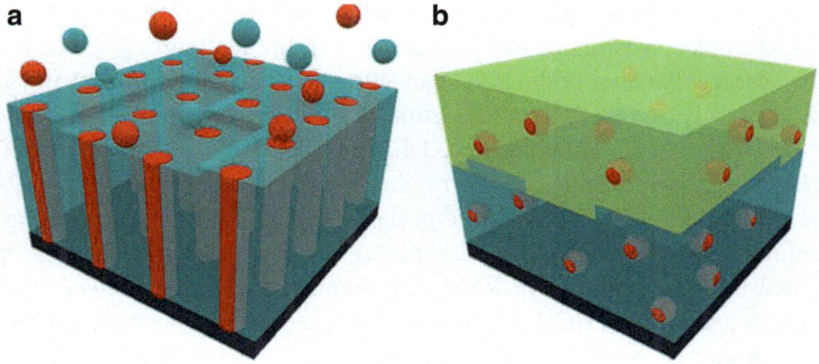

Fig. 4 The schematic diagram representing the formation of (**a**) columnar and (**b**) spherical nanostructures with simultaneous and sequential nucleation of secondary phases and the *RE*BCO matrix, respectively. (Reproduced from Palau et al. [60])

4 Interfaces in *RE*BCO Thin Films Consisting of Columnar Nanostructures (1D APCs)

Several secondary phase materials, when incorporated into the *RE*BCO thin films, tend to self-assemble in the form of nanocolumnar structures running through the thickness of the thin films. Because of their orientation along a particular direction (along the *c*-axis of *RE*BCO), they are also termed as one-dimensional (1D) APCs. Several perovskite and double-perovskite materials such as $BaZrO_3$ [29–31], $BaSnO_3$ [32, 33], $BaHfO_3$ [34], *etc.*, have been observed to self-assemble in the form of nanocolumnar structures inside *RE*BCO films. These nanocolumns have been successfully observed to enhance the in-field J_c of *RE*BCO thin films over a wide range of temperature and applied magnetic field.

In analogy to H_c of a superconductor, irreversibility field (H_{irr}) is another very important parameter which is defined as the magnetic field at which the vortex pinning strength become negligibly small due to large resistivity values caused by the vortex motion [40, 41]. The irreversibility line separates *H-T* phase diagram of a superconductor into reversible and irreversible (or hysteretic) portions. In the reversible portion of this *H-T* phase diagram, the vortex movements due to thermal fluctuation are so large that the ordered vortex lattice state is destroyed [42, 43]. It is highly desired to shift this irreversibility line towards higher *H-T* regime by artificial pinning center (APC) technology [9, 11, 44, 45] and many reports have successfully addressed this issue by introducing nanocolumnar structures inside *RE*BCO thin films.

In a recent work, Horii et al. [46] have successfully introduced columnar nanostructures of Ba-Nb-O (BNO) phase inside YBCO thin films. In this study, it has been estimated that a lower T_c matrix annulus with a lateral dimension of ~12–14 nm exists around each BNO nanocolumns resulting from the strong interface strains. The generation of lower T_c region around the BNO nanocolumns enhances vortex pinning significantly.

Figure 5 shows the schematic diagram of the interfaces (cross-sectional view) in the YBCO films consisting of BNO nanocolumns. In this figure, *d*, *l*, and *D* represent the average diameter of the BNO nanocolumns, distance from the YBCO/BNO interface of a relatively lower T_c region, and average inter-columnar distance. According to their model, it is suggested that stronger vortex pinning is expected for $D > 2l+d$, and a significant reduction of vortex pinning is apparent for $D \leq 2l+d$.

The generation of the lattice strain in the YBCO matrix by incorporating BNO nanocolumns was ascertained by calculating the *c*-axis lengths from reciprocal space measurement (RSM). The RSM measurements confirmed the coexistence of the *a*- and *b*-axes of the YBCO film, which is induced by twin microstructures parallel to the surface of the STO substrate. Whereas the same could not be observed for the YBCO film incorporating BNO nanocolumns which suggested that the broadening of the diffraction spots in the YBCO matrix is apparently induced by the introduction of the BNO nanocolumns.

Figure 6 shows the comparison of the irreversibility field of pristine YBCO and YBCO+BNO nanocomposite films with varying concentrations of BNO phase. It can be clearly observed that the irreversibility lines for YBCO films with BNO nanocolumns were clearly shifted toward higher *H*-*T* regions by BNO incorporation, and the vortex-glass region became qualitatively larger with the increase in the nominal BNO concentration [47–49]. It is well known that the H_{irr} value is correlated with the magnitude of vortex pinning energy through the change in electromagnetic anisotropy in high-T_c cuprate superconductors [50, 51]. It was suggested in their study that all the YBCO matrix regions are dominated by the lower T_c regions, which is induced by the lattice strain at the nanorod/matrix interface.

The incorporation of BZO and BSO nanocolumns into ErBCO and SmBCO thin films has also been reported due to which strong enhancement in the in-field J_c has been observed [52, 53]. The role of interfaces although was not highlighted in these studies.

The incorporation of a double-perovskite phase YBNO has been reported to achieve extremely high pinning force density (F_p) values in YBCO thin films [54]. The YBNO nanocolumns were observed to grow along the *c*-axis of the YBCO film. The continuous nanocolumns also exhibited some segmentation, and the distortions arising from stacking faults in the YBCO matrix near the YBCO/YBNO interface

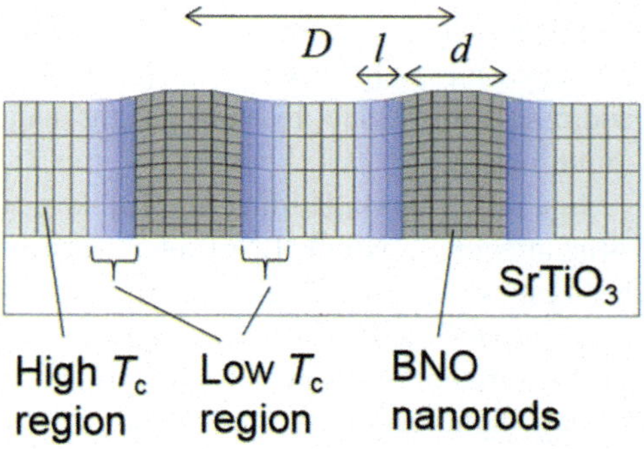

Fig. 5 Schematic diagram of the cross-sectional view of an YBCO film with BNO nanocolumns. The strained regions near the interfaces have their T_cs reduced. (Reprinted from Horii et al. [46], with the permission of AIP Publishing)

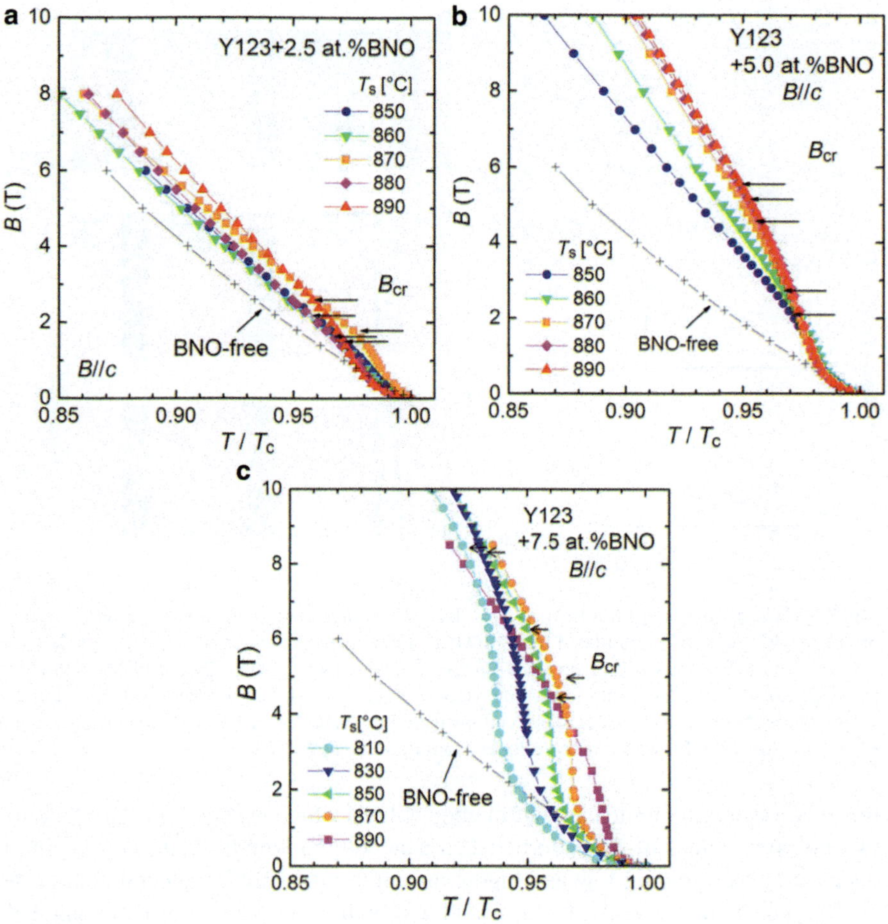

Fig. 6 Variation of irreversibility field for YBCO films with varying concentrations of BNO (**a**) 2.5%, (**b**) 5%, and (**c**) 7.5% deposited at different substrate temperatures. (Reprinted from Horii et al. [46], with the permission of AIP Publishing)

were also observed. A thin layer of amorphous material was observed at these points which are important as they provide a mechanism to release the lattice strain resulting from the large lattice misfit between the phases.

Recently, Cantoni et al. [55] have reported that the strain-induced oxygen-deficient matrix regions exist at YBCO/BZO interfaces, and the distance of such regions from the interface is ~ 10 nm in an YBCO film with BZO nanocolumns. Figure 7a shows the cross-sectional view of the YBCO/BZO (4 vol%) nanocomposite film. BZO phase can be observed in the form of *c*-axis oriented nanocolumns which is common in YCBO/BZO films grown by PLD technique. The lattice misfit occurs between BZO and YBCO phases along the vertical interfaces. The distribution of the out-of-plane strain has been calculated in this study which results from

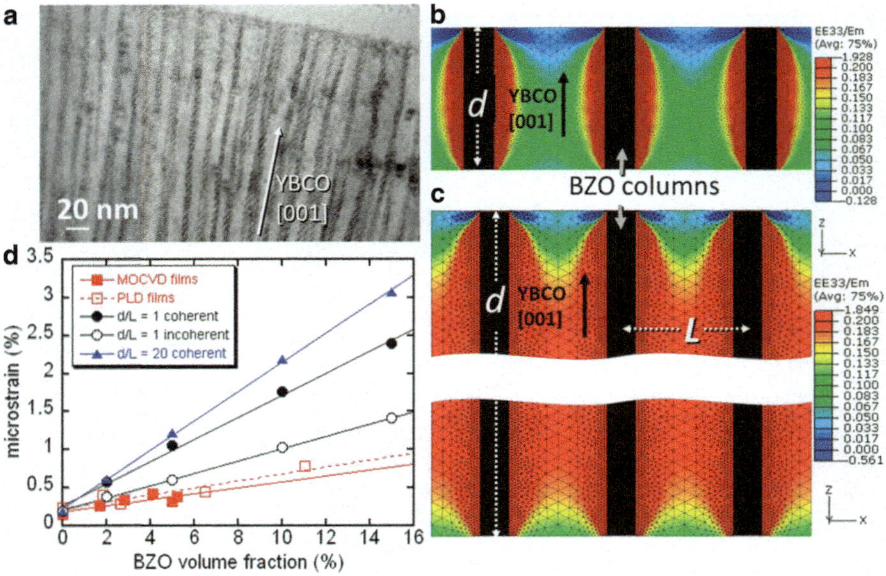

Fig. 7 (**a**) Cross-sectional TEM image of an YBCO/BZO nanocomposite film. Volumetric average of the out-of-plane strain calculated from FEM calculations for the cases *d*/*L* = 1 (**b**) and *d*/*L* = 20 (**c**). (**d**) Variation of microstrain versus BZO concentration as calculated by the FEM method for the cases *d*/*L* = 20 coherent interface (blue triangles), *d*/*L* = 1 coherent interface (black dots), *d*/*L* = 1 incoherent interface (black circles), and measured by XRD for PLD films (open squares) and MOCVD films (filled squares). (Reproduced from Cantoni et al. [55])

this misfit using the finite element method (FEM) which is shown in Fig. 7b, c for two different values of the ratio between nanocolumn length, *d*, and inter-column distance *L*, representing low and high concentrations of BZO nanocolumns in the YBCO matrix, respectively. It was found that high concentration of BZO nanocolumns results in a larger overall positive strain of the YBCO along the *c*-axis causing elongation of the *c*-axis. Figure 7d shows the variation of microstrain as a function BZO concentration as calculated from the Williamson-Hall plot of the XRD data. It can be observed that the microstrain also increases with increasing BZO concentration. The comparative results indicated that the experimental value is significantly reduced in comparison to the simulated value for the case of *d*/*L* = 20 and coherent interface, which indicate that much of the lattice strain has been relieved through misfit dislocations that are generated at the YBCO/BZO interface, giving rise to a semi-coherent YBCO/BZO interface.

The HRTEM image of such an interface is shown in Fig. 8. In Fig. 8b, large number of defects, such as dislocations, can be observed which are marked with white arrows. The high-resolution Z-STEM micrograph of the BZO nanocolumn in Fig. 8a is shown in Fig. 8b which reveals that the BZO *c*-axis and the *RE*BCO *c*-axis are tilted with respect to each other by nearly 3°. This commonly observed tilt is accommodated by tilt dislocations as marked by circles. The misfit dislocations are

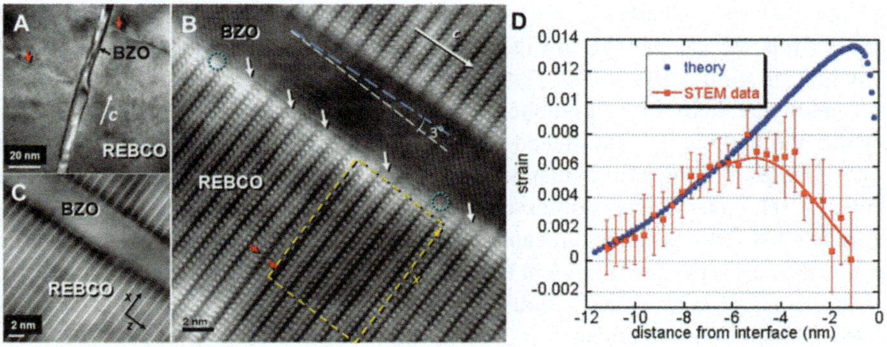

Fig. 8 (**a**) TEM image of a BZO nanocolumn inside *RE*BCO matrix. (**b**) High-resolution Z-STEM image of the same nanocolumn in (**a**). Light blue circles indicate pure tilt dislocations and white arrows indicate misfit dislocations. (**c**) Bright field image acquired simultaneously to image (**b**). (**d**) Comparison between theoretically calculated and experimentally measured strain. (Reproduced from Cantoni et al. [55])

also revealed in this image which are effective in relieving the lattice strain (arrows in Fig. 8b).

The local strain has been quantified through the Z-STEM images by measuring the variation of the *RE*BCO *c*-axis lattice parameter. Figure 8d shows the comparison between theoretically calculated strain (blue dots) and experimentally measured strain (red squares). It has been observed that the strain increases when moving from the matrix interior toward the BZO nanocolumn until it reaches a peak value at a distance of ~5 nm from the interface. However, near the BZO interface, the strain decreases rapidly over a distance of ~3 nm. In the narrow *RE*BCO strip, where the tilt and misfit dislocations cause strong lattice distortions, both compressive and tensile strain components result in atomic displacements along the columns. The analytical prediction of the local strain distribution near the YBCO/BZO boundary is also shown in Fig. 8d. The observed peak of the measured strain separates two concentric YBCO columns with widely varying T_cs. The defective YBCO column (closer to the boundary) has been found to be highly oxygen deficient, which is expected to have a much lower T_c than the T_c of the YBCO phase which exist on the another side of the strain peak.

The role of *RE*BCO/APC interface in improving vortex pinning has further been highlighted in a recent work by Gautam et al. [56]. In this study, they have made two different kinds of nanocomposites: YBCO+BZO and YBCO+BHO. Both BZO and BHO phases tend to form nanocolumns inside *RE*BCO matrix prepared by PLD and MOCVD techniques. These *c*-axis oriented nanocolumns or the so-called 1D APCs have proven to provide strong correlated pinning of vortices resulting in enhanced in-field J_c in superconducting YBCO films. Although, there has always been discussions on the issue as to what would fundamentally determine the pinning efficiency of different 1D APCs. Both BHO and BZO phases were observed to have similar lateral dimensions of 5–6 nm. Figure 9a, b, c shows the cross-sectional TEM image of YBCO+BZO and YBCO+BHO nanocomposite films in which the formation of nanocolumns can clearly be observed. The corresponding Fourier filtered images of these films are also shown in Fig. 9d, e, f. While comparing the Fourier filtered images of both YBCO+BZO and YBCO+BHO nanocomposite films, it was observed that there is larger number of dislocations along the YBCO/BZO interface which the authors have attributed to the larger lattice misfit between YBCO and BZO phases as compared to that between YBCO and BHO phases. In contrast, the YBCO/BHO interface was observed to be coherent, although a few randomly distributed dislocations and significant *ab*-plane buckling was also observed. The high interfacial defect concentration in the YBCO+BZO films is attributed to the release of strain in a short range of 1–2 nm from the interface, where oxygen deficiency was earlier found to be significant [55].

Figure 10 shows the variation of J_c and F_p with respect to applied magnetic field for YBCO+BZO and YBCO+BHO nanocomposite films measured at 77 K and 65 K. The in-field J_c values of the YBCO+BHO nanocomposite films are much higher than their BZO counterpart which is reflected in much higher $F_{p\max}$ values of YBCO+BHO films (182 GNm^{-3} at 65 K). The authors have attributed the highly enhanced vortex pinning properties of YBCO+BHO films to a highly coherent YBCO/BHO interface. The coherent nature of the YBCO/BHO interface is attributed to the smaller lattice misfit between YBCO and BHO phases which would allow BHO and YBCO lattices to compromise to each other due to low rigidity of the BHO nanocolumns. In contrast, the BZO nanocolumns are prone to the interface defects, resulting in a semi-coherent YBCO/BZO interface.

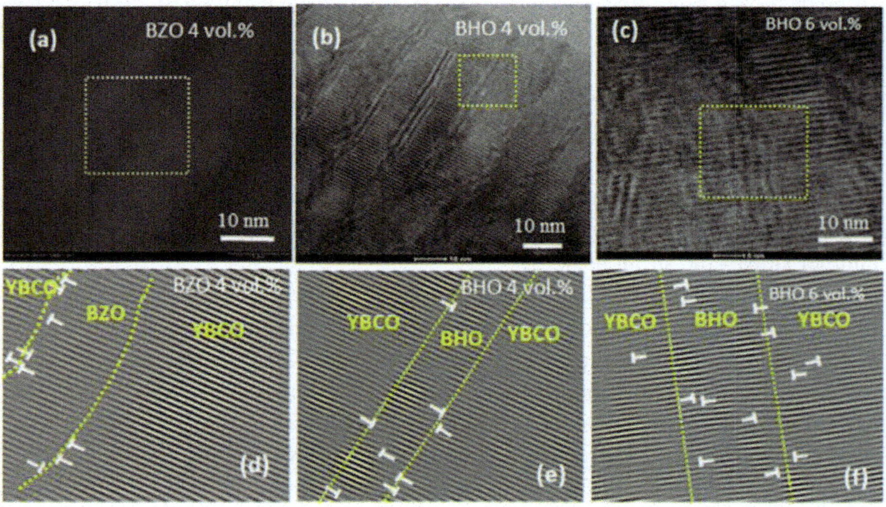

Fig. 9 Cross-sectional HRTEM images of YBCO nanocomposite films consisting of (**a**) 4 vol.% BZO, (**b**) 4 vol.% BHO and (**c**) 6 vol.% BHO. The corresponding Fourier filtered images of the marked areas (rectangles) in (**a**), (**b**), and (**c**) are shown in (**d**), (**e**), and (**f**) respectively. (Reprinted from Gautam et al. [56], with the permission of AIP Publishing)

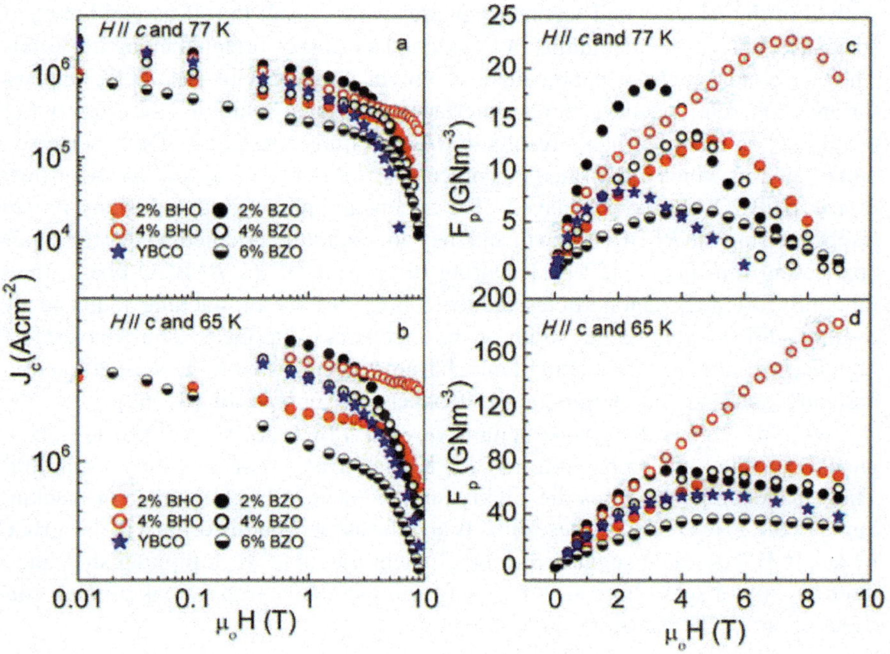

Fig. 10 Variation of J_c and F_p with applied magnetic field for pristine YBCO film in comparison to YBCO films with varying concentrations of BZO and BHO nanocolumns. These data are obtained for (*H*//*c*-axis) at 77 K (**a**) and (**c**) and at 65 K (**b**) and (**d**). (Reprinted from Gautam et al. [56], with the permission of AIP Publishing)

5 Interfaces in *RE*BCO Thin Films Consisting of Spherical Nanostructures (3D APCs)

The secondary phase materials when incorporated inside *RE*BCO film matrix exhibit different geometries which depend upon many parameters. Lattice misfit between the phases and surface diffusion co-efficient of adatoms are crucial in determining the interface between the two phases which may be coherent, semi-coherent, or incoherent [57]. The surface energies of coherent and semi-coherent interfaces are much lower in comparison to that of the incoherent interfaces and the large surface energies of the incoherent interfaces result in coarsening of the grains in polycrystalline films [58] and faceting of precipitates and grain boundaries [1]. It has also been reported that the morphology of the secondary phase nanoinclusions inside *RE*BCO matrix is determined by the combined effect of lattice misfit and elastic properties of *RE*BCO and secondary phase materials [59].

The morphology or geometry of the secondary phase nanoinclusions is also determined by the growth or deposition techniques [60]. For example, in-situ growth techniques such as pulsed laser deposition (PLD), chemical vapor deposition (CVD), or hybrid liquid phase epitaxy (HLPE) allow simultaneous nucleation and growth of the *RE*BCO and secondary phases which result in self-assembled nanostructures, such as nanocolumns of the secondary phases oriented along the *c*-axis. These secondary phase nanoinclusions are in contact with the *RE*BCO phase through coherent or semi-coherent interfaces. Chemical solution deposition (CSD) technique, on the other hand, results in *RE*BCO films consisting of nanoscale secondary phases which are formed in a precursor film in which nucleation and growth of the *RE*BCO matrix is delayed. The secondary phase nanoinclusions remain trapped during the *RE*BCO growth and they are randomly oriented within the superconducting film matrix [57, 61]. During the growth of the *RE*BCO phase, these randomly oriented nanoparticles induce a high density of stacking faults in the resulting film to relieve the strain at the incoherent interfaces. The nanostrained regions surrounding the stacking faults have been suggested to be very efficient in inducing isotropic core pinning due to Cooper pair suppression [61, 62].

Figure 11 shows the nanostructural view of YBCO+BZO and YBCO+YBTO nanocomposite films prepared from CSD technique as reported by Palau et al. [60]. These nanocomposite films are highly strained with a large number of stacking faults in the YBCO phase. These defects and the associated distortions in the YBCO matrix [63] have been suggested to be efficient isotropic weak pinning sites and a large density of such short stacking faults improve the isotropic weak pinning contribution in YBCO nanocomposite thin films.

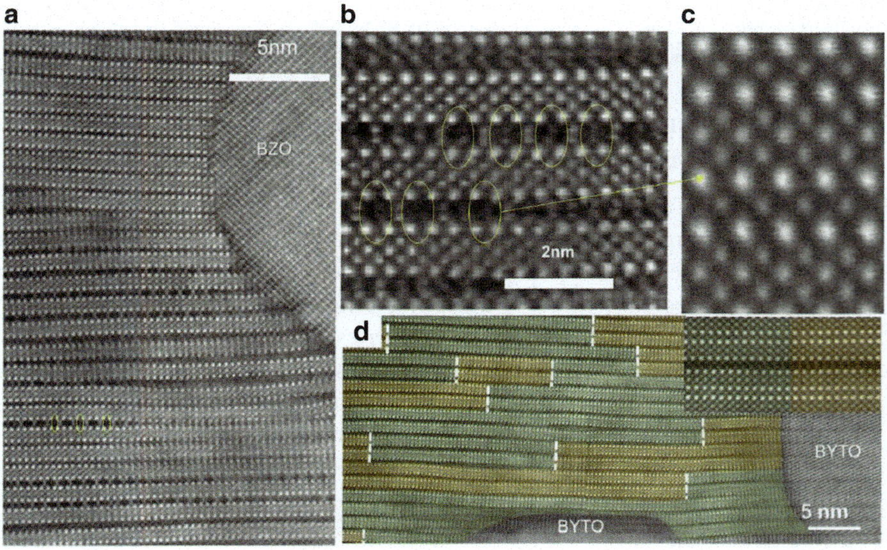

Fig. 11 High-resolution cross-sectional HAADF-TEM images of an YBCO+BZO10% nanocomposite film (**a**)–(**c**) with a large density of stacking faults, at different magnifications. (**d**) Twin boundary domains in a highly distorted region of an YBCO+YBTO10% nanocomposite film. Green and brown colored regions mark different twin domains, corresponding to [100] and [010] YBCO zone axes, respectively. The inset shows the magnified view of a TB. (Reproduced from Palau et al. [60])

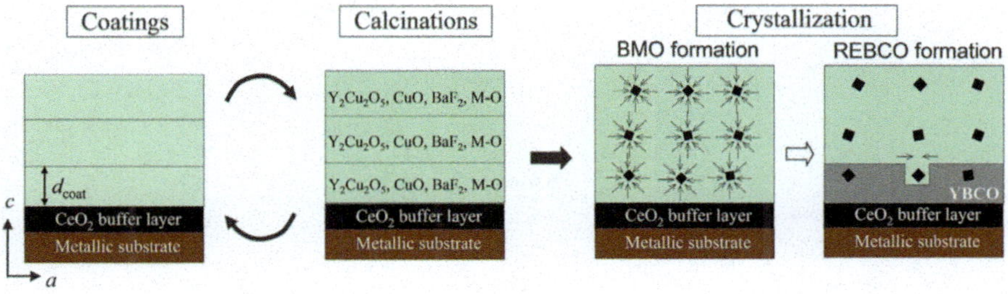

Fig. 12 Schematic diagram of the fabrication steps in Metal organic deposition (MOD) technique for the deposition of (Y,Gd)BCO films consisting of $BaMO_3$ (M = Zr, Hf) nanoparticles. (1) Coating a chemical solution, (2) calcination to obtain a precursor containing several amorphous phases and (3) crystallization of the superconducting phase. (Reproduced from Miura et al. [67])

In recent years, metal organic deposition (MOD) has also been evolved as a promising technique for depositing high performance *RE*BCO superconducting thin films [64–66]. In MOD technique, the formation of secondary phase nanoinclusions takes place before host superconducting matrix formation. Figure 12 shows the schematic diagram depicting the steps of *RE*BCO nanocomposite film growth. MOD technique consists of coatings, calcinations, and crystallization steps. The nanoparticles are formed earlier in the temperature ramp of the crystallization before the crystalline growth of the *RE*BCO matrix, resulting in randomly distributed and oriented nanoparticles of secondary phases. This method has been demonstrated to be capable of introducing large concentrations of secondary phases which is otherwise difficult in in-situ deposition techniques. Since the crystallization process of *RE*BCO phase in MOD technique is liquid mediated, the *RE*BCO/APC interface is incoherent and the matrix crystalline quality is less affected as compared to in-situ techniques which allow larger concentration of secondary phase nanoinclusions.

Figure 13a shows the variation of *c*-axis length with the concentration of secondary phase nanoinclusions for PLD-grown *RE*BCO films consisting nanocolumns, PLD-grown *RE*BCO film consisting of nanoparticles, and MOD grown *RE*BCO film consisting of nanoparticles as reported by Miura et al. [67]. It can be observed that while PLD-grown films exhibit elongation of the *c*-axis length for larger concentration of secondary phase nanoinclusions, MOD films keep their *c*-axis almost unaltered even for larger concentration of secondary phase nanoinclusions. The influence of intact crystalline quality of MOD *RE*BCO films is visible in the electrical transport properties which are shown in Fig. 13b, c. The T_c and self-field J_c of the MOD grown *RE*BCO films also do not degrade even for higher concentration of secondary phase nanoinclusions.

In order to understand the difference between the microstructures of PLD-grown and MOD grown *RE*BCO films, high-resolution plan-view TEM images and geometrical phase analysis revealing the strain induced by secondary phase nanoinclusions in the PLD and MOD grown films are shown in Fig. 13d and e. The strain component maps reveal that in PLD-grown films, the compressive and tensile stains

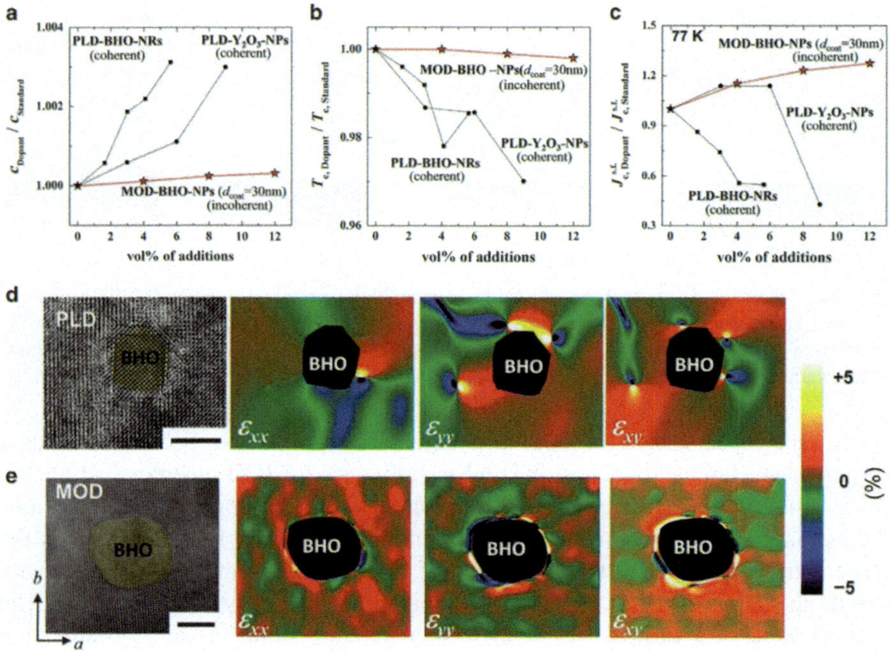

Fig. 13 Comparison of the structural and superconducting properties as a function of the concentration of secondary phase nanoinclusions and strain maps for PLD grown and MOD grown films. Variation of (**a**) normalized *c*-axis length, (**b**) T_c and (**c**) self-field J_c at 77 K as a function the concentration of the nanoinclusions. (**d**) and (**e**) show high-resolution plan-view images, where strain (ε_{xx}, ε_{yy}, and ε_{xy}) maps were determined by geometrical phase analysis for PLD grown GdBCO+BHO (nanocolumns) and MOD grown (Y,Gd)BCO +BHO (nanoparticles) films. The horizontal scale bars in (**d**) and (**e**) indicate 5 nm. (Reproduced from Miura et al. [67])

are localized near the *RE*BCO/nanocolumn coherent interfaces and spread over a distance of ~10 nm whereas, for MOD grown thin films, the strain is localized over far shorter distance (~2 nm) around the incoherent *RE*BCO/nanoparticle interfaces. The much smaller strained regions in the *RE*BCO matrix results in much less degradation of the superconducting properties. It has also been shown in this study that by selecting suitable secondary phase, it is possible to control the size of the nanoparticles which would result in large number of pinning sites with less deterioration of the superconducting matrix and subsequently much higher critical current performance could be realized.

6 Conclusions

In order to improve the *in-field* J_c of *RE*BCO superconducting thin films, many methods have been tried to incorporate APCs inside *RE*BCO film matrix. These methods have been developed over the years to intentionally introduce nanoscale secondary phases into *RE*BCO superconducting thin films. Different secondary phases exhibit different morphologies when incorporated in *RE*BCO film matrix which determine the interface structures between the phases. The artificial pinning centers induced by secondary phase nanoscale inclusions and the associated interfacial strain influence the vortex pinning properties of *RE*BCO thin films. The interfaces in *RE*BCO superconducting thin films are governed by many parameters which include lattice misfit between the phases, surface diffusion co-efficient of adatoms, and the processing techniques. While *in-situ* fabrication techniques rely on the coherent interfaces, *ex-situ* techniques rely on incoherent interfaces for vortex pinning enhancement.

Acknowledgments We thank T. Horide, S. Saini, P. Mele, A. Ichinose, Y. Yoshida, and S. Awaji for cooperative research on vortex pinning studies in *RE*BCO nanocomposite films.

References

1. A.P. Sutton, R.W. Balluffi, *Interfaces in Crystalline Materials* (Oxford University Press, Oxford, 1996)
2. H. Oda, S. Kiyohara, K. Tsuda, T. Mizoguchi, *J. Phys. Soc. Jpn.* **86**, 123601 (2017)
3. J.M. Howe, *Interfaces in Materials* (Wiley, New York, 1997)
4. R.W. Balluffi, S.M. Allen, W.C. Carter, *Kinetics of Materials* (Wiley, Hoboken, 2005)
5. M.K. Wu, J.R. Ashburn, C.J. Torng, P.H. Hor, R.L. Meng, L. Gao, Z.J. Huang, Y.Q. Wang, C.W. Chu, *Phys. Rev. Lett.* **58**, 908 (1987)
6. G.V.M. Williams, J.L. Tallon, *Physica C* **258**, 41 (1996)
7. M. Nakamura, H. Kutami, Y. Shiohara, *Physica C* **260**, 297 (1996)
8. J.G. Lin, C.Y. Huang, Y.Y. Xue, C.W. Chu, X.W. Cao, J.C. Ho, *Phys. Rev. B* **51**, 12900 (1995)
9. D. Larbalestier, A. Gurevich, D.M. Feldmann, A. Polyanskii, *Nature* **414**, 368 (2001)
10. A.P. Malozemoff, *Nat. Mater.* **6**, 617 (2007)
11. S.R. Foltyn, L. Civale, J.L. Macmanus-Driscoll, Q.X. Jia, B. Maiorov, H. Wang, M. Maley, *Nat. Mater.* **6**, 631 (2007)
12. D. Dijkkamp, T. Venkatesan, X.D. Wu, S.A. Shaheen, N. Jisrawi, Y.H. Min-Lee, W.L. McLean, M. Croft, *Appl. Phys. Lett.* **51**, 619 (1987)
13. X.W. Cao, X.J. Xu, Z.H. Wang, J. Fang, R.L. Wang, H.C. Li, *Physica C* **282**, 1993 (1997)
14. M. Badaye, F. Wang, Y. Kanke, K. Fukushima, T. Morishita, *Appl. Phys. Lett.* **66**, 2131 (1995)
15. B. Schey, in *Chapter 14 in "Pulsed Laser Deposition of Thin Films"*, ed. by R. Eason, (Wiley, Hoboken, 2007), pp. 313–331
16. M. Miura, in *Chapter 5 in "Oxide Thin Films, Multilayers, and Nanocomposites"*, ed. by P. Mele, T. Endo, S. Arisawa, C. Li, T. Tsuchiya, (Springer, Cham, Heidelberg, 2015), pp. 3–26
17. A. Ignatiev, in *Chapter 15 in "Second-Generation HTS Conductors"*, ed. by A. Goyal, (Kluwer Academic Publishers, Dordrecht, 2005), pp. 245–259
18. D. B. Chrisey, G. K. Hubler (eds.), *Chapter 8 in "Pulsed Laser Deposition of Thin Films"* (Wiley, New York, 1994)
19. R.K. Singh, D. Kumar, *Mater. Sci. Eng. R* **22**, 113 (1998)
20. A.K. Jha, K. Matsumoto, *Front. Phys.* **7**, 82 (2019)
21. B. Dam, J.M. Huijbregtse, F.C. Klaassen, R.C.F. Van der Geest, G. Doornbos, J.H. Rector, A.M. Testa, S. Freisem, J.C. Martinez, B. Stauble-Pumpin, R. Griessen, *Nature* **399**, 439 (1999)
22. J.M. Huijbregtse, F.C. Klaassen, A. Szepielow, J.H. Rector, B. Dam, R. Griessen, B.J. Kooi, J.T.M. de Hosson, *Supercond. Sci. Technol.* **15**, 395 (2002)
23. S.H. Wee, A. Goyal, P.M. Martin, L. Heatherly, *Supercond. Sci. Technol.* **19**, 865 (2006)

24. C. Cai, B. Holzapfel, J. Hanishch, L. Fernandez, L. Schultz, *Phys. Rev. B* **69**, 104531 (2004)
25. T. Haugan, P.N. Barnes, R. Wheeler, F. Meisenkothen, M. Sumption, *Nature* **430**, 867 (2004)
26. A.K. Jha, K. Matsumoto, T. Horide, S. Saini, P. Mele, A. Ichinose, Y. Yoshida, S. Awaji, *J. Appl. Phys.* **122**, 093905 (2017)
27. S.K. Viswanathan, A.A. Gapud, M. Varela, J.T. Abiade, D.K. Christen, S.J. Pennycook, D. Kumar, *Thin Solid Films* **515**, 6452 (2007)
28. A.K. Jha, N. Khare, R. Pinto, *Solid State Commun.* **151**, 1447 (2011)
29. J.L. MacManus Driscoll, S.R. Foltyn, Q.X. Jia, H. Wang, A. Serquis, L. Civale, B. Maiorov, M.E. Hawley, M.P. Maley, D.E. Peterson, *Nat. Mater.* **3**, 439 (2004)
30. A. Goyal, S. Kang, K.J. Leonard, P.M. Martin, A.A. Gapud, M. Varela, M. Paranthaman, A.O. Ijaduola, E.D. Specht, J.R. Thompson, D.K. Christen, S.J. Pennycook, F.A. List, *Supercond. Sci. Technol.* **18**, 1533 (2005)
31. A.K. Jha, N. Khare, R. Pinto, *J. Supercond. Nov. Magn.* **25**, 377 (2012)
32. C.V. Varanasi, J. Burke, H. Wang, J.H. Lee, P.N. Barnes, *Appl. Phys. Lett.* **93**, 092501 (2008)
33. P. Mele, K. Matsumoto, A. Ichinose, M. Mukaida, Y. Yoshida, S. Horii, R. Kita, *Supercond. Sci. Technol.* **21**, 125017 (2008)
34. S. Miura, Y. Yoshida, Y. Ichino, A. Tsuruta, K. Matsumoto, A. Ichinose, S. Awaji, Jpn. J. Appl. Phys. **53**, 090304 (2014)
35. S.H. Wee, A. Goyal, Y.L. Zuev, C. Cantoni, V. Selvamanickam, E.D. Specht, *Appl. Phys. Express* **3**, 023101 (2010)
36. A.K. Jha, K. Matsumoto, T. Horide, S. Saini, P. Mele, Y. Yoshida, S. Awaji, *Supercond. Sci. Technol.* **27**, 025009 (2014)
37. S.H. Wee, A. Goyal, E.D. Specht, C. Cantoni, Y.L. Zuev, V. Selvamanickam, S. Cook, *Phys. Rev. B* **81**, 140503 (2010)
38. T. Aytug, M. Paranthaman, A.A. Gapud, S. Kang, H.M. Christen, K.J. Leonard, P.M. Martin, J.R. Thompson, D.K. Christen, R. Meng, I. Rusakova, C.W. Chu, T.H. Johansen, *J. Appl. Phys.* **98**, 114309 (2005)
39. A.K. Jha, N. Khare, R. Pinto, *J. Appl. Phys.* **110**, 113920 (2011)
40. A. Gurevich, *Nat. Mater.* ***10***, 255 (2011)
41. A.P. Malozemoff, T.K. Worthington, Y. Yeshurun, F. Holtzberg, P.H. Kes, *Phys. Rev. B* **38**, 7203 (1988)
42. A. Schilling, R. Jin, J.D. Guo, H.R. Ott, *Phys. Rev. Lett.* **71**, 1899 (1993)
43. H. Pastoriza, M.F. Goffman, A. Arribere, F. de la Cruz, *Phys. Rev. Lett.* **72**, 2951 (1994)
44. K. Matsumoto, P. Mele, *Supercond. Sci. Technol.* **23**, 014001 (2010)
45. X. Obradors, T. Puig, *Supercond. Sci. Technol.* **27**, 044003 (2014)
46. S. Horii, M. Haruta, A. Ichinose, T. Doi, *J. Appl. Phys.* **118**, 133907 (2015)
47. S. Horii, K. Yamada, H. Kai, A. Ichinose, M. Mukaida, R. Teranishi, R. Kita, K. Matsumoto, Y. Yoshida, J. Shimoyama, K. Kishio, *Supercond. Sci. Technol.* **20**, 1115 (2007)
48. S. Horii, H. Kai, M. Mukaida, K. Yamada, R. Teranishi, A. Ichinose, K. Matsumoto, Y. Yoshida, R. Kita, J. Shimoyama, K. Kshio, *Appl. Phys. Lett.* **93**, 152506 (2008)
49. A. Ichinose, K. Naoe, T. Horide, K. Matsumoto, R. Kita, M. Mukaida, Y. Yoshida, S. Horii, *Supercond. Sci. Technol.* **20**, 1144 (2007)
50. J. Shimoyama, K. Kitazawa, K. Shimizu, S. Ueda, S. Horii, N. Chikumoto, K. Kishio, *J. Low Temp. Phys.* **131**, 1043 (2003)
51. K. Takenaka, K. Mizuhashi, H. Takagi, S. Uchida, *Phys. Rev. B* **50**, 6534 (1994)
52. M. Namba, S. Awaji, K. Watanabe, S. Ito, E. Aoyagi, H. Kai, M. Mukaida, R. Kita, *Appl. Phys. Express* **2**, 073001 (2009)
53. T. Ozaki, Y. Yoshida, Y. Ichino, Y. Takai, A. Ichinose, K. Matsumoto, S. Horii, M. Mukaida, Y. Takano, *J. Appl. Phys.* **108**, 093905 (2010)
54. D.M. Feldmann, T.G. Holesinger, B. Maiorov, S.R. Foltyn, J.Y. Coulter, I. Apodaca, *Supercond. Sci. Technol.* **23**, 095004 (2010)
55. C. Cantoni, Y. Gao, S.H. Wee, E.D. Specht, J. Gazquez, J. Meng, S.J. Pennycook, A. Goyal, *ACS Nano* **5**, 4783 (2011)

56. B. Gautam, M.A. Sebastian, S. Chen, S. Misra, J. Huang, F.J. Baca, R. Emergo, T. Haugan, Z. Xing, H. Wang, J.Z. Wu, *Appl. Phys. Lett.* **113**, 212602 (2018)
57. J. Gutierrez, A. Llordes, J. Gazquez, M. Gibert, N. Roma, S. Ricart, A. Pomar, F. Sandiumenge, N. Mestres, T. Puig, X. Obradors, *Nat. Mater.* **6**, 367 (2007)
58. C.V. Thomson, *Annu. Rev. Mater. Sci.* **30**, 159–190 (2000)
59. J. Wu, J. Shi, *Supercond. Sci. Technol.* **30**, 103002 (2017)
60. A. Palau, F. Vallès, V. Rouco, M. Coll, Z. Li, C. Pop, B. Mundet, J. Gàzquez, R. Guzman, J. Gutierrez, X. Obradors, T. Puig, *Supercond. Sci. Technol.* **31**, 034004 (2018)
61. A. Llordes, A. Palau, J. Gazquez, M. Coll, R. Vlad, A. Pomar, J. Arbiol, R. Guzman, S. Ye, V. Rouco, F. Sandiumenge, S. Ricart, T. Puig, M. Varela, D. Chateigner, J. Vanacken, J. Gutierrez, V. Moshchalkov, G. Deutscher, C. Magen, X. Obradors, *Nat. Mater.* **11**, 329 (2012)
62. G. Deutscher, *Appl. Phys. Lett.* **96**, 122502 (2010)
63. R. Guzman, J. Gazquez, B. Mundet, M. Coll, X. Obradors, T. Puig, *Phys. Rev. Mater.* **1**, 024801 (2017)
64. M. Miura, B. Maiorov, J.O. Willis, T. Kato, M. Sato, T. Izumi, Y. Shiohara, L. Civale, *Supercond. Sci. Technol.* **26**, 035008 (2013)
65. M. Weigand, N.A. Rutter, J.H. Durrell, *Supercond. Sci. Technol.* **26**, 105012 (2013)
66. M. Miura, B. Maiorov, F.F. Balakirev, T. Kato, M. Sato, Y. Takagi, T. Izumi, L. Civale, *Sci. Rep.* **6**, 20436 (2016)
67. M. Miura, B. Maiorov, M. Sato, M. Kanai, T. Kato, T. Kato, T. Izumi, S. Awaji, P. Mele, M. Kiuchi, T. Matsushita, *NPG Asia Mater.* **9**, e447 (2017)

Fabrication of $(Nd_{0.33}Eu_{0.33}Gd_{0.33})Ba_2Cu_3O_y$ Superconducting Thin Films by Pulsed Laser Ablation of Melt-Textured Targets

Shiv Jee Singh, Muralidhar Miryala, and Paolo Mele

1 Introduction

In order to be suitable for practical applications (frictionless transportation, winding of magnets, etc.), superconducting materials should possess not only a high transition temperature T_c, but also high critical current density J_c which is associated with the pinning force (F_p). Introduction of nanosized artificial pinning centers (APCs) was widely used to enhance J_c and F_p for high-T_c superconductors (HTSCs) like $YBa_2Cu_3O_x$ (YBCO, $T_c = 92$ K). In the past 10 years, the nano-engineering approach to control microstructure, distribution, concentration, and dimensionality of APCs has been developed as a powerful tool to produce high performance superconductors. For example, ablation of mixed $BaSnO_3$ BSO (4 wt%) -YBCO targets by pulsed laser deposition (PLD) [1] has enabled fabrication of high-quality YBCO thin films with incorporated BSO in the form of nanorods, classified as one-dimensional APCs (1D-APCs). YBCO films added with 4 wt% BSO have J_c = 0.3 MA/cm^2 and $F_p^{MAX} = 28.3$ GN/m^3 (77 K, 3 T, *B*//*c*). However, J_c is intrinsically anisotropic with respect to the magnetic field direction and generally has maximum value for *B*//*c*-axis. This feature is critical for those applications requiring J_c independent magnetic field direction. An attempt has been made to solve this issue by introducing isotropic (or 3D) pinning centers to the YBCO films in the form of Y_2O_3

S. J. Singh
Institute of high pressure Physics (UNIPRESS), Polish Academy of Sciences, Warsaw, Poland

M. Miryala
Graduate School of Engineering and Science, Shibaura Institute of Technology, Tokyo, Japan

P. Mele (✉)
SIT Research Laboratories, Shibaura Institute of Technology, Saitama, Japan
e-mail: pmele@shibaura-it.ac.jp

A. G. Roca et al. (eds.), *Surfaces and Interfaces of Metal Oxide Thin Films, Multilayers, Nanoparticles and Nano-composites*,
https://doi.org/10.1007/978-3-030-74073-3_10

nanoscopic islands: for example, 5.44 at% Y_2O_3 addition resulted in F_p^{MAX} =14.3 GN/m^3 (77 K, 3 T, *B*//*c*)) [2]. Combination of 1D- and 3D-APCs in the form of BSO nanorods and Y_2O_3 nanoparticles in multilayered YBCO film brought an encouraging result F_p^{MAX} =17.6 GN/m^3 (77 K, 2.2 T, *B*//*c*) in the [(90 nm YBCO+BSO)/(30 nm YBCO+Y_2O_3)]×3 multilayer [3].

From the practical application point of view, performance of superconducting thin films requires a further progress. In particular, irreversibility field (H_{irr}), i.e., the maximum field at which the superconductor still carries a nonzero electric current, needs to be improved. Typical value of H_{irr} at liquid nitrogen temperature (77 K) for YBCO-BSO prepared by PLD is about 11 T [1]. Further enhancements of F_p and H_{irr} are necessary for production of really applicable thin films. Besides the addition of APCs to YBCO, another successful route has been demonstrated to be the replacement of Y with RE (RE = light rare earths like Nd, Sm, Gd). Due to the larger ionic radius of RE^{3+} respect to Y^{3+}, T_c of REBCO is higher than T_c of YBCO and so is the H_{irr} as it was shown for RE = Gd [4], RE = Sm [5], and RE = Nd [6]. Furthermore, partial substitution of Y with RE has been suggested by Driscoll et al. [7] as an effective way to improve the superconducting properties of YBCO. This approach has been independently confirmed by several research groups worldwide: for example, in an early report Haugan et al. [8] described twofold increase of low field J_c (*H*//*c*, B = 0.2 T) in (Y, Eu)BCO PLD films respect to the case of pure YBCO. Later Xu et al. reported H_{irr} = 14.8 T (77 K) for MOCVD (Y,Gd)BCO films doped with 15 vol% Zr, though F_p^{MAX} = 14 GN/m^3 at 6 T was only about half of that of YBCO-BSO [9] .

A bit earlier, extensive studies have been made on bulk triple REs mixed $(Nd_{0.33}Eu_{0.33}Gd_{0.33})Ba_2Cu_3O_x$ (NEG-123), which has been proved as a superconducting material with marvelous performance [10]. Its J_c value at 77 K reaches of the order of 10^5 A/cm^2 (*H*//*c*-axis) at the self-field and 5 T, and irreversibility field approaches to 15 T [11, 12]. A very important aspect was the possibility to reduce the pinning defects' size up to nanoscale level and to bring it close to the coherence length. Nanoscale particles of secondary phases, Zn, Mo, Ti, Nb, etc., are enhanced the flux pinning properties of these materials up to 3 times compared to that of YBCO bulk compounds. As a result, pinning in these materials is strong up to liquid oxygen temperature (90.2 K), which enables the levitation up to vicinity of critical temperature [12]. So, these materials have extended the application temperature range by up to 13 K.

Very few studies based on the thin film preparation of NEG-123 are available in the literature. Cai et al [13, 14], Inoue et al [15], and Ko et al [16] have reported NEG-123 thin films with excellent transport properties by laser ablation of conventional sintered NEG-123 targets.

Pulsed laser deposition (PLD) is a conservative thin film deposition technique, allowing the transfer of stoichiometry and microstructure from target to thin film. Therefore, we decided to use PLD with the purpose to fabricate NEG-123 thin films with exceptional superconducting properties. In this study, we propose for the first time the use of melt-textured $(Nd_{0.33}Eu_{0.33}Gd_{0.33})Ba_2Cu_3O_y$ (NEG-123) + Gd_2BaCuO_5 (Gd-211) pellets, instead of sintered materials, as targets for the PLD laser ablation.

In addition, we shall discuss the optimization of growth of thin film by varying the deposition temperature inside vacuum chamber in a broad range, from 700 to 900 °C, and studying the effect of O_2 chamber pressure during the deposition. The post-annealing effects of these thin films will also discussed at different annealing time. Finally, we have concluded the best conditions to grow high-quality superconducting thin film of NEG-123+Gd-211 which shows the highest T_c of 92.6 K and H_{c2} of 173 T.

2 Materials and Methods

Dense pellets of $(Nd_{0.33}Eu_{0.33}Gd_{0.33})Ba_2Cu_3O_y$ (NEG-123) + Gd_2BaCuO_5 (Gd-211) were prepared by oxygen controlled melt growth (OCMG) procedure, according with a procedure described in [11].

NEG-123+Gd-211 films were prepared on $SrTiO_3$ (STO) (100) single crystalline substrates by pulsed laser deposition (PLD) using Nd:YAG laser (fourth harmonic, $\lambda = 266$ nm) focused on a dense OCMG NEG-123+Gd211 target with an energy density of about 4.2 J/cm^2. The target was rotated during the irradiation of laser beam. The substrates were glued with silver paste on Inconel disk customized for ultrahigh vacuum applications. Laser frequency of 10 Hz, number of laser pulses 10000, distance target-substrate 24 mm, and rotation speed of the target 30% rpm were kept unchanged as experimental parameters during all the deposition routes. On the contrary, all other parameters were varied in order to determine the optimal growth window: deposition time 2–30 min; substrate temperature 700–900 °C; partial O_2 pressure 200–1000 mTorr. Additionally, selected samples were submitted to post-annealing in 200 sccm of flowing oxygen for time variable between 0 and 60 h.

The structural characterization was done by XRD (Rigaku Ultima V), and morphology was checked by SEM (JEOL, FESEM). The electrical resistivity vs temperature (ρ-T) characteristics were measured by conventional four-probe technique from 77 to 300 K by means of Quantum Design PPMS. Magnetic field dependence of resistivity was evaluated at 65–100 K in magnetic field parallel to the c-axis in the interval of 0–9 T.

3 Results and Discussion

Typical films obtained by ablation of melt-textured target are highly dense with presence of minor outgrowths and few pores on a flat surface as shown in Fig. 1. Inset shows that thickness of the film is about 300 nm for a growth time of 30 min: the resulting growth rate of 0.17 nm/s is about 5 times slower than typical growth rates of YBCO or NEG-123 films produced by ablation of conventional sintered targets. To justify this, it may be invoked the higher hardness of our MTG targets in comparison with the sintered ones: in fact after several deposition routes, it was not

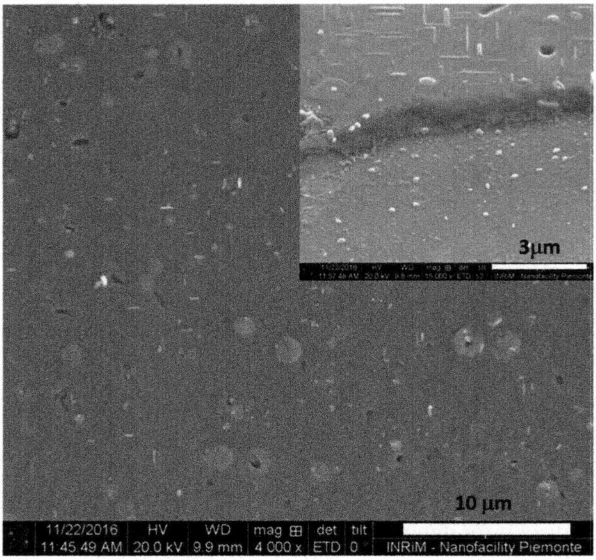

Fig. 1 SEM image of the surface of $(Nd_{0.33}Eu_{0.33}Gd_{0.33})Ba_2Cu_3O_y$ film deposited under following conditions: t_{dep} = 30 min, T_{dep} = 800 °C, P_{dep} = 600 mTorr, and post-annealed at 500 °C for 15 h in 200 sccm O_2. The inset shows tilted image of the same film, with evidence of a thickness of about 300 nm

visible the typical "ring" which usually appears quickly on sintered targets. According with representative in-plane XRD θ-2θ scan (Fig. 2) of same sample shown in Fig. 1, films grew with c-axis orientation. The out-of-plane XRD (Φ scans) showed no peaks which means that films were not epitaxial.

We have measured the temperature dependence of the in-plane resistivity of these grown films and by considering the transition temperature behaviors, the optimization of growth conditions for the NEG-123 thin films has been concluded.

In the first step, the deposition temperature was fixed at 800 °C and oxygen pressure at 600 mTorr, both as reported in literature [15, 16], and laser deposition time was varied in the range 15–30 min. The resistivity behaviors of these films are shown in Fig. 3. As the deposition time is increased, the resistivity is decreased with the enhanced T_c from 90 to 92.6 K. The highest T_c of 92.6 K is obtained for 30 min of deposition time which indicated the good quality of this film.

In next step, we have optimized the deposition temperature by fixing the deposition time of 30 min. Hence, the various thin films are grown by fixing the constant deposition time of 30 min and oxygen pressure to 600 mtorr, while deposition temperature is changed from 700 to 800 °C. The temperature dependence of the resistivity of these grown films is presented in Fig. 4.

Again the obtained transition temperature T_c is improved with the deposition time, and best temperature for deposition is 800 °C with T_c = 92.6 K. This T_c value is almost same as reported for OCMG bulk material of the same composition.

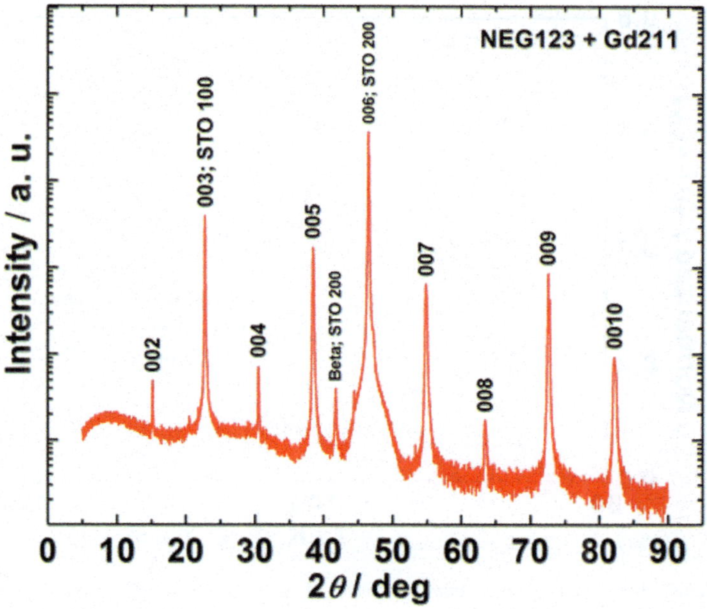

Fig. 2 In-plane XRD (θ–2θ) scan of $(Nd_{0.33}Eu_{0.33}Gd_{0.33})Ba_2Cu_3O_y$ film deposited under following conditions: $t_{dep} = 30$ min, $T_{dep} = 800$ °C, $P_{dep} = 600$ mTorr, and post-annealed at 500 °C for 15 h in 200 sccm O_2

To find the best optimal oxygen pressure during the thin film growth, we have fixed the best deposition temperature of 800 °C and deposition time of 30 min and then grown the films with various chamber pressure from 200 to 1000 mTorr. Again, these films were characterized by the resistivity measurements as shown in Fig. 5. The film prepared at 200 mTorr has T_c of 90 K with very broad transition. With increasing pressure, the resistivity reached to lowest value and sharp transition with highest T_c value for the deposition pressure of 600 mTorr. Further increased the pressure start to decrease the quality of thin film and obtained very broad transition as almost similar to that for 200 mTorr. It is worth mentioning that optimized pure YBCO films are fabricated under 200 mTorr of O_2 [1].

From the analysis of Figs. 3, 4 and 5, we can say, the best conditions for preparing NEG-123 thin films by ablation of MTG targets are determined as follows: $t_{dep} = 30$ min, $T_{dep} = 800$ °C, $P_{dep} = 600$ mTorr. It is worth to inform that before measuring the transport measurements, all the films described above were submitted to post-annealing in oxygen under the following conditions: $T = 500$ °C for 15 h in 200 sccm O_2. The conclusive optimization step was exploring different times of post-annealing conditions for films deposited under the best conditions: 0, 15 h, and 60 h. The film without post-annealing showed semiconducting behavior without any sign of superconducting transition. As presented in Fig. 6, the long annealing of 60 h depresses the superconducting transition, so that 15 h is the ideal post-annealing duration time.

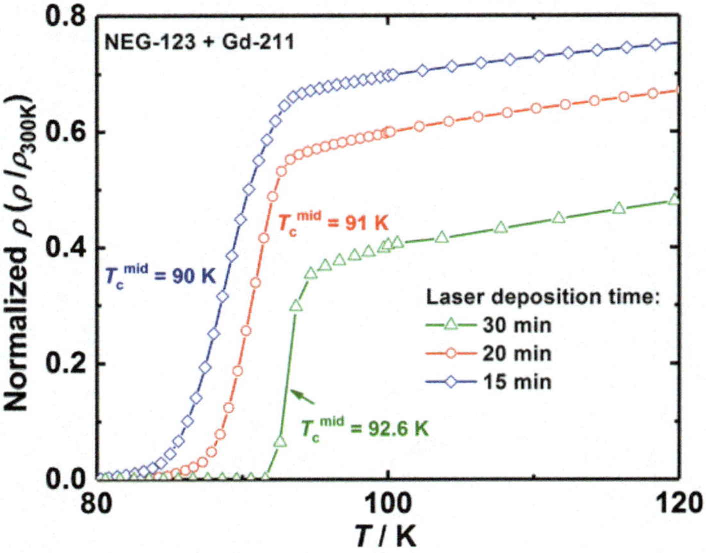

Fig. 3 Resistivity vs. temperature curves of $(Nd_{0.33}Eu_{0.33}Gd_{0.33})Ba_2Cu_3O_y$ films deposited under following conditions: T_{dep} = 800 °C, P_{dep} = 600 mTorr, and post-annealed at 500 °C for 15 h in 200 sccm O_2, and laser deposition time 15, 20 and 30 min

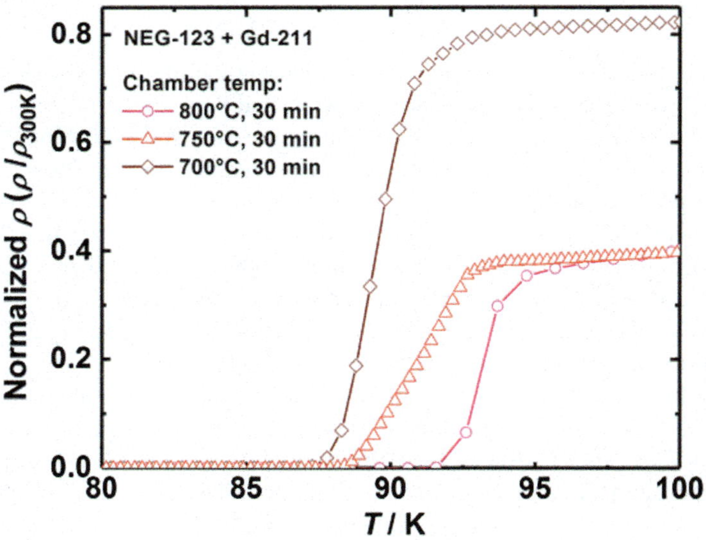

Fig. 4 Resistivity vs. temperature curves of $(Nd_{0.33}Eu_{0.33}Gd_{0.33})Ba_2Cu_3O_y$ films deposited under following conditions: t_{dep} = 30 min, P_{dep} = 600 mTorr, and post-annealed at 500 °C for 15 h in 200 sccm O_2, and T_{dep} = 700, 750, 800 °C

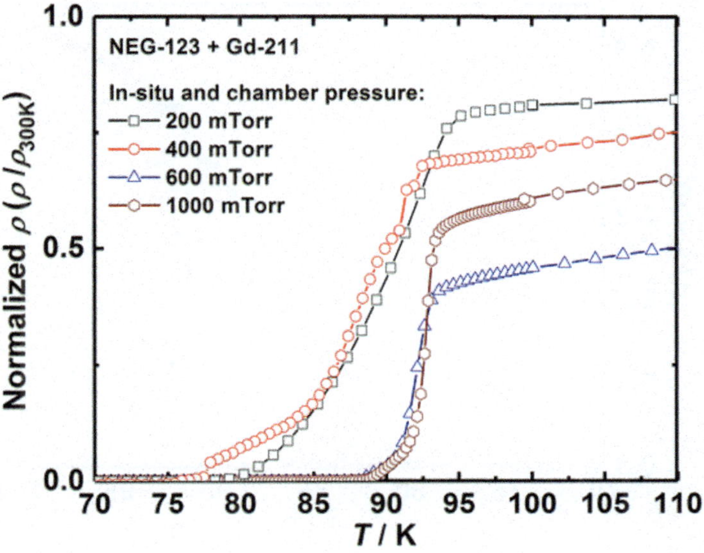

Fig. 5 Resistivity vs. temperature curves of $(Nd_{0.33}Eu_{0.33}Gd_{0.33})Ba_2Cu_3O_y$ films deposited under following conditions: $t_{dep} = 30$ min, $T_{dep} = 800$ °C, and post-annealed at 500 °C for 15 h in 200 sccm O_2, and $P_{dep} = 200, 400, 600, 1000$ mTorr

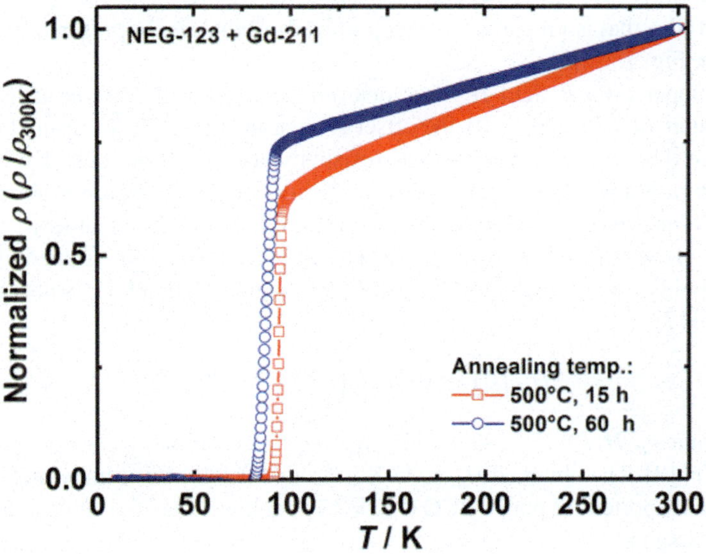

Fig. 6 The temperature dependence of resistivity of $(Nd_{0.33}Eu_{0.33}Gd_{0.33})Ba_2Cu_3O_y$ films deposited under following conditions: $t_{dep} = 30$ min, $T_{dep} = 800$ °C, $P_{dep} = 600$ mTorr and post-annealed at 500 °C for 0, 15 and 60 h in 200 sccm O_2

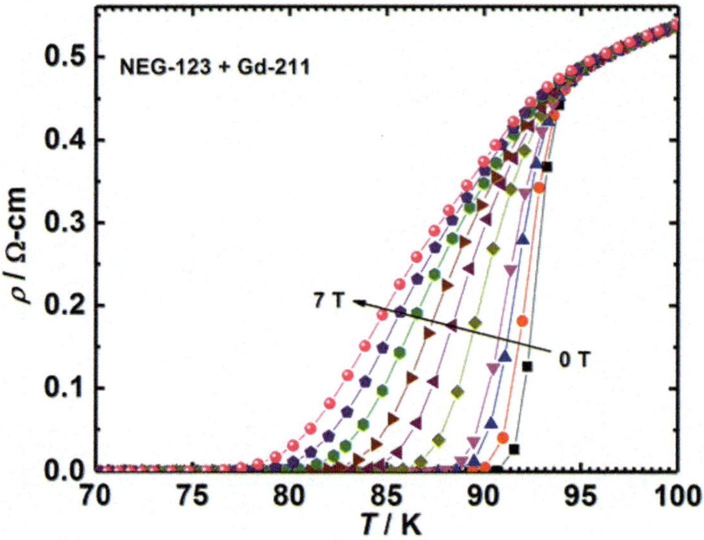

Fig. 7 Resistivity vs. temperature curves under external magnetic field up to 9T (*H*//c) of $(Nd_{0.33}Eu_{0.33}Gd_{0.33})Ba_2Cu_3O_y$ films deposited under following conditions: t_{dep} = 30 min, T_{dep} = 800 °C, P_{dep} = 600 mTorr and post-annealed at 500 °C for 0, 15 h in 200 sccm O_2

Furthermore, we have also study the magneto-resistive behavior up to 9 T of the best thin film that is prepared by the optimized conditions. Representative data are shown in Fig. 7.

The upper critical field H_{c2} was determined from such measurements. At an applied field of 9 Tesla (Fig. 7), the offset of the transition shifts by around 10 K. We have used the 90% of ρ_n (normal state resistivity at $T = T_c$) criteria to define H_{c2} and 10% criteria for the corresponding irreversibility field (H_{irr}). Figure 8 depict H_{c2} and H_{irr} as a function of temperature obtained magnetic field dependent resistivity studies. The slope dH_{c2}/dT is estimated to be −2.7 T/K. The zero-temperature upper critical field $H_{c2}(0)$ is extrapolated using Werthamer–Helfand–Hohenberg (WHH) formula [17]:

$$H_{c2}(0) = -0.693T_c\left(\mathrm{d}H_{c2}/\mathrm{d}T\right)T = T_c.$$

The obtained $\mu_0 H_{c2}(0)$ is ~ 173 T which is lower than that of typical values of YBCO (~250 T) [18]. The obtained H_{irr} at 77 K is around 9 T for this film, which is higher than value reported for pure YBCO (H_{irr} = 7 T) but lower than the value of best MTG NEG-123 sample: 15 T [12]. H_{irr} (77 K) is 33 T, about 1/3 of reported values of pure YBCO.

Our study suggests that more study will be needed in this direction to improve the quality of these best NEG-123 films obtained by ablation of MTG targets. Since H_{irr} and H_{c2} must be higher before evaluating their critical currents. We will explore the feasibility of different substrates like MgO and $LaAlO_3$ and the effect of artificial pinning centers on the critical fields.

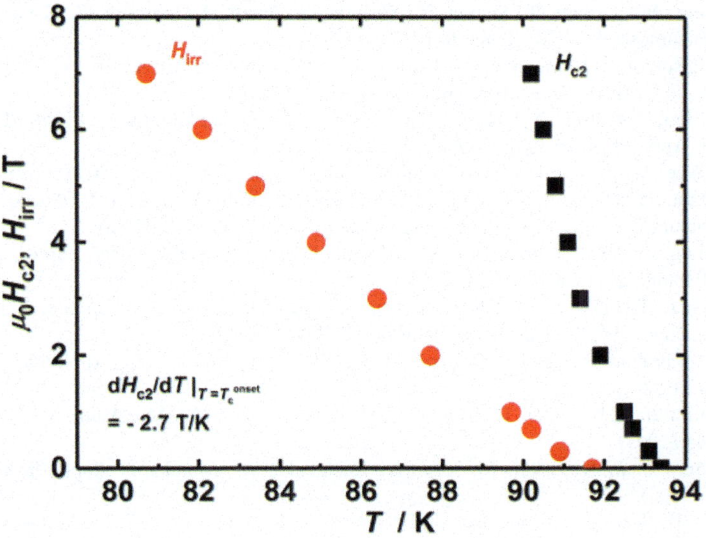

Fig. 8 Temperature dependence of upper critical field and irreversibility field of $(Nd_{0.33}Eu_{0.33}Gd_{0.33})Ba_2Cu_3O_y$ films deposited under following conditions: $t_{dep} = 30$ min, $T_{dep} = 800$ °C, $P_{dep} = 600$ mTorr and post-annealed at 500 °C for 15 h in 200 sccm O_2

4 Summary and Perspective

We successfully fabricated dense and c-axis-oriented $(Nd_{0.33}Eu_{0.33}Gd_{0.33})Ba_2Cu_3O_y$ (NEG-123)+ Gd_2BaCuO_5 (Gd-211) (NEG-123) superconducting films on $SrTiO_3$-substrate by pulsed laser deposition ablating OCMG targets. We carried out optimization of the deposition parameters and our best thin film of NEG123, showing $T_c =$ 92.6 K with transition width of ~2–3 K., has the following optimized processing parameters: substrate temperature: 800 °C; deposition time: 30 min; in-situ O_2 pressure: 600 mTorr; distance between substrate and target: 24 mm; post-annealing temperature: 500 °C; post-annealing duration: 15 h. The estimated values of critical fields are $H_{c2}(0) = 173$ T and H_{irr} (77 K) = 9 T for our optimized thin films. Next efforts will be focused on the improvement of H_{irr} and on the achievement of large critical current in external magnetic fields.

Acknowledgment Authors thank Prof. Marco Truccato (Physics Department, University of Torino, Italy) for performing SEM analysis.

References

1. P. Mele, K. Matsumoto, T. Horide, A. Ichinose, Y. Yoshida, M. Mukaida, S. Horii, R. Kita, *Supercond. Sci. Technol.* **21**, 032002 (2008)
2. P. Mele, R. Guzman, J. Gazquez, X. Obradors, T. Puig, S. Saini, Y. Yoshida, M. Mukaida, A. Ichinose, K. Matsumoto, M.I. Adam, *Supercond. Sci. Technol.* **28**, 024006 (2015)

3. P. Mele, M.I. Adam, T. Suzuki, Y. Yoshida, S. Awaji, A. Ichinose, S. Saini, A.K. Jha, K. Matsumoto, Sci. Adv. Mater. **6**, 1042 (2017)
4. K. Matsumoto, D. Takahara, T. Horide, A. Ichinose, S. Horii, Y. Yoshida, M. Mukaida, K. Osamura, IEEE Trans. Appl. Supercond. **15**, 2719 (2005)
5. M. Miura, Y. Yoshida, Y. Ichino, Y. Takai, K. Matsumoto, A. Ichinose, S. Horii, M. Mukaida, Jpn. J. Appl. Phys. **45**, L11 (2006)
6. S.H. Wee, A. Goyal, Y.L. Zuev, C. Cantoni, Appl. Phys. Express **1**, 111702 (2009)
7. J.L. MacManus-Driscoll, B. Maiorov, J. Durrell, S. Foltyn, Q. XJia, L. Civale, H. Wang, A. Kursumovic, D.E. Peterson, Supercond. Sci. Technol. **19**, S55 (2006)
8. T.J. Haugan, T.A. Campbell, N.A. Pierce, M.F. Locke, I. Maartense, P.N. Barnes, Supercond. Sci. Technol. **21**, 025014 (2008)
9. A. Xu, L. Delgado, N.N. Khatri, Y. Liu, V. Selvamanickam, D. Abraimov, J. Jarozynski, F. Kametani, D.C. Larbalestier, APL Mater. **2**, 064111 (2014)
10. M. Muralidhar, N. Sakai, M. Jirsa, M. Murakami, I. Hirabayashi, Appl. Phys. Lett. **92**, 162512 (2008)
11. M. Muralidhar, N. Sakai, T. Nishiyama, M. Jirsa, T. Machi, M. Murakami, Appl. Phys. Lett. **82**, 943 (2003)
12. M. Muralidhar, N. Sakai, M. Jirsa, N. Koshizuka, M. Murakami, Appl. Phys. Lett. **83**, 5005 (2003)
13. C. Cai, B. Holzapfel, J. Haenisch, L. Fernandez, L. Schultz, Appl. Phys. Lett. **84**, 377 (2004)
14. C. Cai, B. Holzapfel, J. Haenisch, L. Fernandez, L. Schultz, Phys. Rev. B **39**, 104531 (2004)
15. K. Inoue, Y. Yoshida, Y. Ichino, Y. Takai, K. Matsumoto, A. Ichinose, M. Mukaida, S. Horii, Physica C **468**, 1611 (2008)
16. R.-K. Ko, S.-H. Bae, M.-J. Jung, S.-H. Jang, K.-J. Song, C. Park, M.-H. Sohn, S.-I. Kang, S.-S. Oh, D.W. Ha, H.-S. Kim, Y.-C. Kim, IEEE Trans. Appl. Supercond. **19**, 3136 (2009)
17. N.R. Werthamer, E. Helfand, P.C. Hohenberg, Phys. Rev. **147**, 295 (1966)
18. D.C. Larbalestier, A. Gurevich, D.M. Feldmann, A. Polyanskii, Nature **414**, 368 (2001)

The Infiltration Growth Process of Single Grain $YBa_2Cu_3O_y$ Bulk Superconductor: Optimization and Effect of Liquid Phase Mass and Its Day-to-Day Life Applications

Sushma Miryala and Masato Murakami

1 Introduction

Applications of superconducting bulk magnets range from magnetron sputtering, non-contact rotating mechanics, transport systems, space ships and naval carriers to magnetic drug delivery system (MDDS) and portable magnetic resonance imaging (MRI) [1–6]. The use of high temperature superconductors such as Y-123 bulk, allows that all of these applications can operate at or below the liquid nitrogen boiling point. Recent advances in fabrication techniques such as top-seeded melt-grown (TSMG) for bulk Y123 and $REBa_2Cu_3O_y$ (RE: rare earth elements; Nd, Sm, Eu, Gd, NEG) superconductors have brought usage of superconducting permanent magnets into reality. Continuous technological advancements in development of novel Nd-123/MgO film seed crystals, Mg-doped Nd-123 generic seed crystals, and optimization of melt-growth techniques made it possible to produce large-size high-quality single-grain superconducting bulk and large-scale production of moderately sized Y-123 or Gd-123 bulks via batch-processing [7, 8]. Single-domain nature and high performance are the target parameters for these superconducting materials in various industrial applications. Later, it was confirmed that batch-processed bulk samples exhibit required properties (by means of Trapped Field measurements) and can be idealized for practical applications. Previous study suggests that Y-123 or RE-123 bulk superconducting permanent super-magnets have the ability to reach trapped fields (at low temperatures) by an order of magnitude higher than best

S. Miryala (✉)
Superconducting Materials Laboratory, Shibaura Institute of Technology (SIT), Tokyo, Japan

School of Engineering and Applied Science, Princeton University, Princeton, NJ, United States

M. Murakami
Superconducting Materials Laboratory, Shibaura Institute of Technology (SIT), Tokyo, Japan

A. G. Roca et al. (eds.), *Surfaces and Interfaces of Metal Oxide Thin Films, Multilayers, Nanoparticles and Nano-composites*,
https://doi.org/10.1007/978-3-030-74073-3_11

traditional permanent magnets [6, 9]. In order to put to use of these materials in practical applications, a further improvement in critical current density as well as mechanical properties is necessary so that it can be operated within a system around liquid nitrogen temperature (77 K) to reduce cost. Critical current density can be improved in low and high magnetic fields via improving vortex pinning [10–13]. Tiny submicron size ranged RE-211 particles can act as effective pinning centers in this system. However, it is challenging to control their size and growth during high temperature synthesis. On the other hand, in order to keep them refined and small, one can systematically use low cost cerium oxide [14]. This resulted in dramatic improvement of critical current density (J_c) and this improvement was due to 211 secondary phase growth restrictions created by minute quantity of cerium oxide (~1 wt%) [15]. As of now, the melt textured synthesis technique is well developed as well as optimized to maximize single grain growth and performance. Hence, current bulk Y-123 material is commercially available in several countries such as Japan, German, and Czech, etc. Unfortunately, melt textured growth technique obtains several drawbacks such as shrinkage in volume, non-uniform distribution of secondary phase particles (211 phases) in 123 matrix, pores, cracks (in case of slight improper mixing, etc.), formation of large size 211 secondary phase in 123 matrix (must use 211 refiners such as ceria, etc., for good performance as mentioned before).

In order to tackle some of these issues, a new synthesis technique infiltration-growth (IG) was developed and proved to be successful [11, 16–18]. Although the process looks reliable and doable, it needs much optimization and drawback analysis. Up to date, large single-grain bulk superconductors using IG have not been produced. One of the issues is that growth being terminated when the liquid stops to infiltrate the entire Y-211 green body or the liquid is insufficient because of losses after turning into liquid. To overcome this problem, we have proposed utilizing a special liquid phase source as mixture of Yb-123 (low peritactice temperature) and liquid ($Ba_3Cu_5O_8$) as a liquid source. This Yb-123 phase in mixture is to support its structure and to ensure complete liquid transfer at low temperatures as compared to Y-123. This enabled to produce a high quality large size, single-domain bulk superconductor.

To improve the performance of IG processed bulk Y-123 material, we firstly produced bulk Y-123 material grown by different liquid phase (LP) sources, i.e. Yb-123+liquid (1:1; 1:1.2; and 1:1.3), as a liquid source by isothermal growth and studied its superconducting transition temperature (T_c) and critical current density (J_c) at 77 K. Magnetization results confirmed that all samples were of good quality and critical current density value was around 39,000 A/cm^2 (in self-field at 77 K, *H*//*c*-axis). Through the usage of isothermal results, we have produced single grain Y-123 by setting a slow cooling process combined with a new class of liquid phase. Although most of the parameters seem to be fine with slow-cooled material, the obtained trapped field was low in value around 0.12 T (20 mm Y-123 at 77 K, 1:1 Yb-123). However, the use of a liquid rich combination about 1:1.3 (by infiltrating more liquid into Y-211 body during IG Process) leads to trapped three-times better compared to material produced liquid phase source i.e. Yb-123+liquid (1:1) as a liquid source.

2 Experimental Details

Y_2BaCuO_5 powders were prepared starting from commercial powders of Y_2O_3, $BaCO_3$ and CuO. The powder was mixed in the nominal composition of Y_2BaCuO_5 and calcined three times at 840 °C, 870 °C and 900 °C for 15 h each. This is to ensure simultaneously that powders of BaO_2 and CuO were mixed in a nominal composition of $Ba_3Cu_5O_8$ (Y-035). The first batch aimed to analyze the liquid phase effect on superconducting performance at different liquid phase (LP) sources, i.e. Yb-123+liquid (1:1; 1:1.2; and 1:1.3) as a liquid source was selected. Pre-forms were made out of Y211 and liquid phase using hydraulic press. The pre-forms of Y211 were placed on top of the liquid phase, then an Nd-123 melt-textured seed was placed at center of the top surface of Y-211 in order to grow Y-123 grains. This technique is a combination of cold top-seeded method and infiltration growth (IG) (see Fig. 1). In addition, Y_2O_3 and MgO plates were used under preform to suppress liquid loss and to enable spontaneous Y-123 nucleation. Eventually, all samples have been produced by isothermally heated at a fixed growth temperature of 995 °C for 50 h. The second batch production of bulk Y-123 material indicating the mixture of Yb-123+Ba_3Cu_5O (1:1) was selected as the liquid phase based on the isothermal results. The arrangement of IG experiment was similar to the isothermal process reported in Fig. 1. Later, the whole setup was subjected to custom made optimized

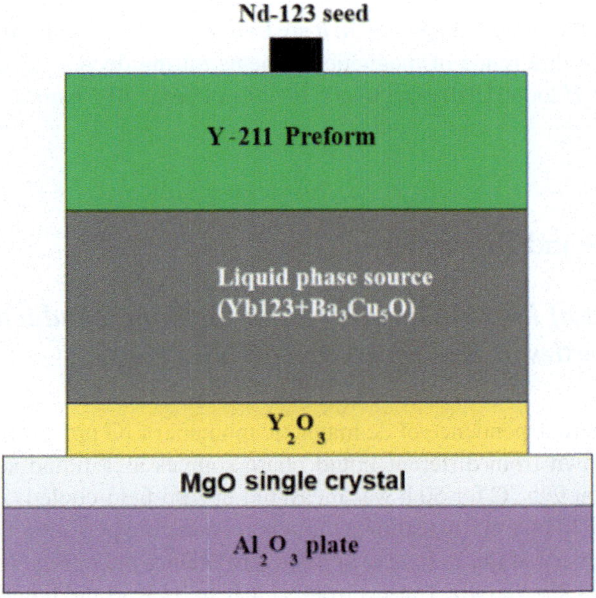

Fig. 1 A typical arrangement used in the IG experiment. A source of liquid phase mixture of Yb-123+ $Ba_3Cu_5O_y$ (1:1, 1:1.2, 1:1.3) is supported on a chemically inert, impervious substrate. The pre-form fabricated of Y-211 powder is placed on top of it. A *c*-axis Nd-123 crystal seed is placed in the center of the preform. (Reproduced with modifications from [18])

thermal cycle. The sample was heated to 880 °C at the rate of 200 °C/h and held there for 10 min. Then, the sample was heated for 50 min to 1060 °C. Then, the temperature was decreased to 1005 °C for 50 min and slowly cooled to 980 °C at a cooling rate of 0.5 °C/h. Finally, the temperature was lowered with a cooling rate of 200 °C/h to 100 °C followed by furnace cooling to room temperature. The melt-textured samples were annealed at 450 °C for 150 h in flowing pure O_2 gas that will oxygenate samples and make them into superconductors. The samples with 20 mm in diameter and 7 mm in thickness were produced. The third batch production of bulk Y-123 material has mixture of Yb-123+Ba_3Cu_5O (1:1.3) and was used as a liquid phase; whereas, the processing details were similar to (1:1) liquid phase and used Y-123 sample as cooling rate was changed 1005 °C to 980 °C at a cooling rate of 0.25 °C/h.

Superconducting properties and field trapping capability of these bulk samples were studied at 77 K. Trapped field experiments in field cooling (FC) process with the aid of hall probe. Small test specimens for magnetization measurements with the size of 1.5 × 1.5 × 0.5 mm^3 were cut from bulk samples. Measurements of critical temperature (T_c) and magnetization hysteresis (M-H_a) loops were done at 77 K from −1 to +5 T using a commercial SQUID magnetometer (Quantum Design, model MPMS5). Critical currents were calculated using extended Bean's critical state model formula for a rectangular sample given by

$$J_c = 20\Delta m / \left[a^2 d\left(b - a/3\right)\right], \quad (1)$$

where d is the sample thickness, a, b are transversal dimensions, $b \geq a$, all in mm and Δm is the difference of magnetic moments during decreasing and increasing field in the M-H loop [19], respectively, in units of emu=10^{-3} Am^2. J_c then results in A/cm^2 units.

3 Results and Discussion

3.1 *Effect of Liquid Phase Mass for the Superconducting Properties of IG Processed Bulk $YBa_2Cu_3O_y$*

The temperature dependence of dc magnetic moment of IG processed Y-123 superconductor grown from different liquid phase sources as a liquid source, isothermally grown at 995 °C for 50 h was measured in zero-field-cooled (ZFC) state and in magnetic field of 1 mT to estimate the superconducting performance of the material. The summary of the T_c (onset) and T_c (zero) values are in Fig. 2.

All samples showed a T_c (onset) from 91.1 K to 92 K as the liquid mixture goes from 1:1.3 to 1:1, whereas T_c (zero) values are around 90.1 K, which is similar to the earlier reports. [11] To determine the influence of the varying LP masses on critical current density (J_c) for IG processed Y-123 material, the magnetization

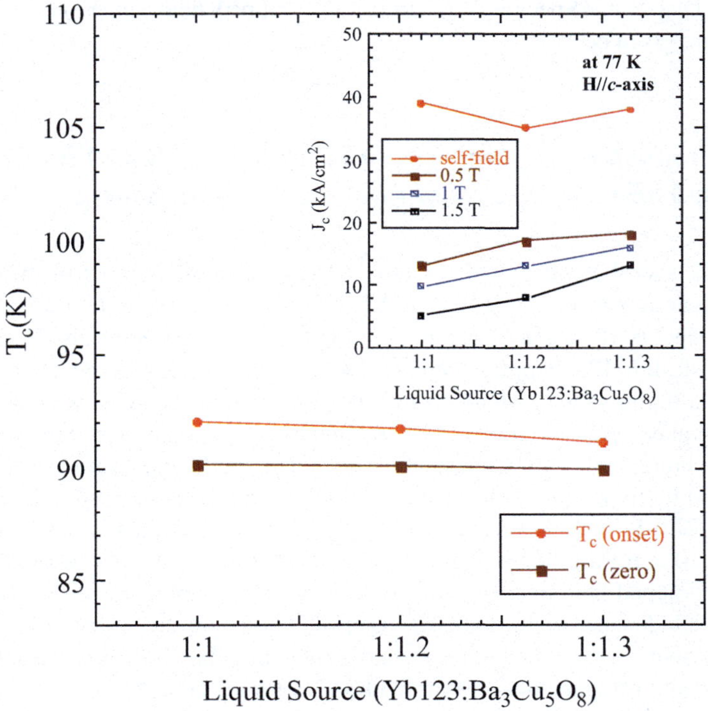

Fig. 2 The plot of the liquid phase mixture of Yb-123+ $Ba_3Cu_5O_y$ (1:1, 1:1.2 and 1:1.3) versus T_c (onset) and T_c (zero) for IG processed Y-123 material. Inset present the critical current density (J_c) values at self-field, 0.5 T, 1 T, and 1.5 T at 77 K for *H*//*c*-axis

hysteresis loops (MHL) were measured at 77 K along *H*//c-axis using a commercial SQUID magnetometer, and estimated the critical current performance using the Bean's critical current state model and the results are summarized in Fig. 2 (see inset). It is remarkable that (i) the highest remnant critical current density values of 39 kA/cm^2, 34 kA/cm^2, and 39 kA/cm^2 were achieved in the samples with the LP masses 1:1, 1:1.2 and 1:1.3 respectively, (ii) the medium field critical current performance was slightly improved in LP masses 1:1.2 and 1:1.3 as compared to the LP mass of 1:1, (iii) the highest critical current values of around 16 kA/cm^2 and 13 kA/cm^2 were recorded in 1 and 1.5 T for the sample with LP mass of 1:1.3. The self-field and low field critical currents probably influenced on δl-pinning controlled by Y-211 secondary phase particle quantity and its distribution. [11] The present results indicate that the sample grown with that LP mass of 1:1.3 have a better performance in terms of medium field critical currents and also somehow more LP mass 1:1.3 will help the melt growth process. Based on the superconducting performance of LP masses, LP masses of 1:1 and 1:1.3 were chosen and produced the IG processed bulk Y-123 in 20 mm size with a slow-cooling process and studied the superconducting transition temperature, critical current performance at 77 K, microstructure,

Energy Dispersive Spectroscopy (EDS) and trapped field values at 77 K. More details are given below.

3.2 Production of Single Grain IG Processed Bulk $YBa_2Cu_3O_y$ by $YbBa_2Cu_3O_y$+Liquid Phase as a Liquid Source (1:1)

Figure 3a inset shows the top surface of the top-seeded infiltration growth-processed Y-123 materials produced using the Yb-123+liquid (1:1) as a liquid source. The final product is 20 mm in diameter. Four growth facet lines were visible on the top surface of all Y-123 samples prepared by IG process utilizing the Yb-123+liquid (1:1) as a liquid source, which clearly indicated that the samples were grown into single grains (Fig. 3a inset). The IG-process combined with the mixture of Yb-123+liquid (1:1) as a liquid source is *new*. The single grain samples were grown within 50 h. To see the performance of this new material, trapped field was measured at 77.3 K. Sample was cooled at 0.5 T from room temperature to 77.3 K and kept for 10 min. After removing the applied field, waited for a period of of 30 s and then the trapped field distribution was recorded by scanning the surface with Hall sensors at a distance of 1.2 mm. The trapped field results are shows in Fig. 3a. The field distribution has a simple peak shape, indicating that the sample is not broken. The maximum-trapped field was around 0.12 T for the 20 mm in diameter samples. The present trapped field values are low compared to earlier reports [11] and a refinement of secondary phase particles and a processing optimization of the mixture of Yb-123+liquid (1:1) as a liquid source are needed. The superconducting

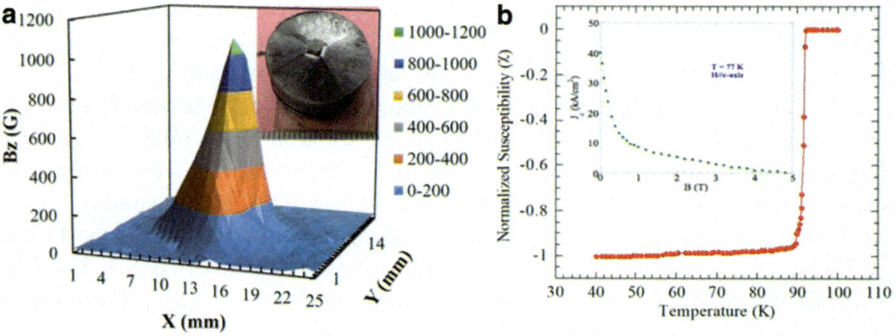

Fig. 3 (**a**). The trapped field distribution at 77.3 K for the bulk Y-123 superconductor produced by Top Seeded Infiltration Growth Process utilizing the mixture of Yb-123+liquid phase as a liquid source and inset shows the grown single grain bulk Y-123. (**b**). Superconducting transition temperature and field dependence of the super-current density at 77 K for a specimen cut 2 mm below the seed of the single-domain Y-123 bulk superconductors showed in inset. (Reproduced with modifications from [18])

performance, especially the superconducting transition temperature and critical current density at 77 K, was measured utilizing SQUID magnetometer. The results presented are in Fig. 3b.

A sharp superconducting transition with onset around 92.1 K and transition width of less than 1.9 K was observed, which indicates that the sample was of an excellent quality. The *M-H* curves were measured on the same sample at 77 K with the field applied parallel to the *c*-axis and critical current density, J_c, was calculated from the M-H curves using the extended Bean model formula, as shown in inset Fig. 3b. The sample showed J_c at 77 K, in self-field and 1 T of about 39,000 A/cm^2 and 9,750 A/cm^2, respectively (see inset of Fig. 3b). Note that dispersion of Y-211 secondary phase particles in the final Y-123 matrix has a great advantage on critical current density. This is due to the fact that Y-123/Y-211 interface will act as pinning centers for vortices in superconducting state and thus will help to increase critical current density especially at self-field as well as low magnetic fields [20, 21]. Therefore, for a given volume fraction of Y-211 particles, the Y-123/Y-211 interface is maximized when Y-211 secondary phase particles are small. To further improve J_c of IG processed Y-123, it is very important to control Y-211 preform before melt-growth experiment.

We also tried to observe uniformity of Y-123 material produced by mixture of Yb-123+liquid (1:1) as a liquid source by Energy Dispersive Spectroscopy (EDS) and results are presented in Fig. 4.

It is clear that Y, Ba, Cu and O are uniformly distributed in Y-123 matrix. The present results show that uniform, single grain Y-123 material can be produced by utilizing a mixture of Yb-123+liquid (1:1) as a liquid source within 50 h. However, further optimization is needed to enhance trapped field. For this, bulk Y-123 samples are grown in *c-growth* direction. In this case, more liquid phase supply is quite crucial and we utilized a mixture of Yb-123+liquid (1:1.3) as a liquid source and adopted its cooling rate of 0.25 °C/h.

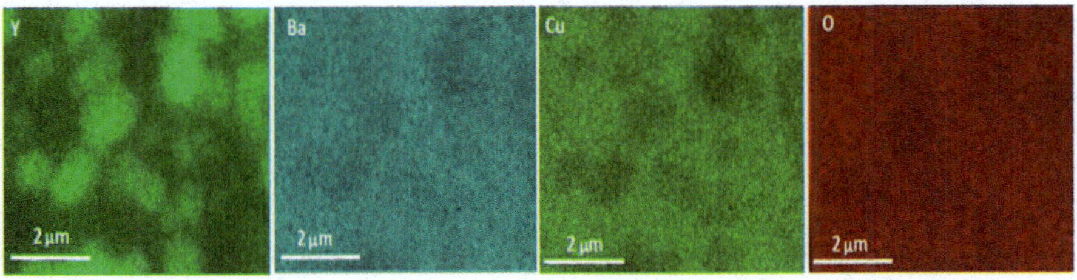

Fig. 4 Energy Dispersive Spectroscopy (EDS) results for the Y-123 superconductor produced by Top Seeded Infiltration Growth process utilizing the mixture of Yb-123+liquid phase as a liquid source. Note that Y, Ba, Cu, and O are uniformly distributed over the Y-123 matrix. (Reproduced with modifications from [18])

3.3 Production of Single Grain IG Processed Bulk $YBa_2Cu_3O_y$ by $YbBa_2Cu_3O_y$+Liquid Phase as a Liquid Source (1:1.3)

Figure 5b inset depicts top surface growth characteristics of as-grown Y-123 sample produced by the top-seeded infiltration growth using Yb-123+liquid (1:1.3) as a liquid source. The choice of Yb-123 was made because of its low melting temperature than Y-123. At the reaction, it immediately decomposes into a liquid and Yb211 providing more liquid into Y211 preform on the top. The melt processed Nd-123 seed crystal was used to produce single-grain Y-123 of 20 mm in diameter. This feature is observed due to the high melting temperature of Nd-123 compared to Y123. The visible four-growth facets on top surface and circumference of all Y-123 bulk samples prepared predict a *c*-axis growth. These growth features are quite

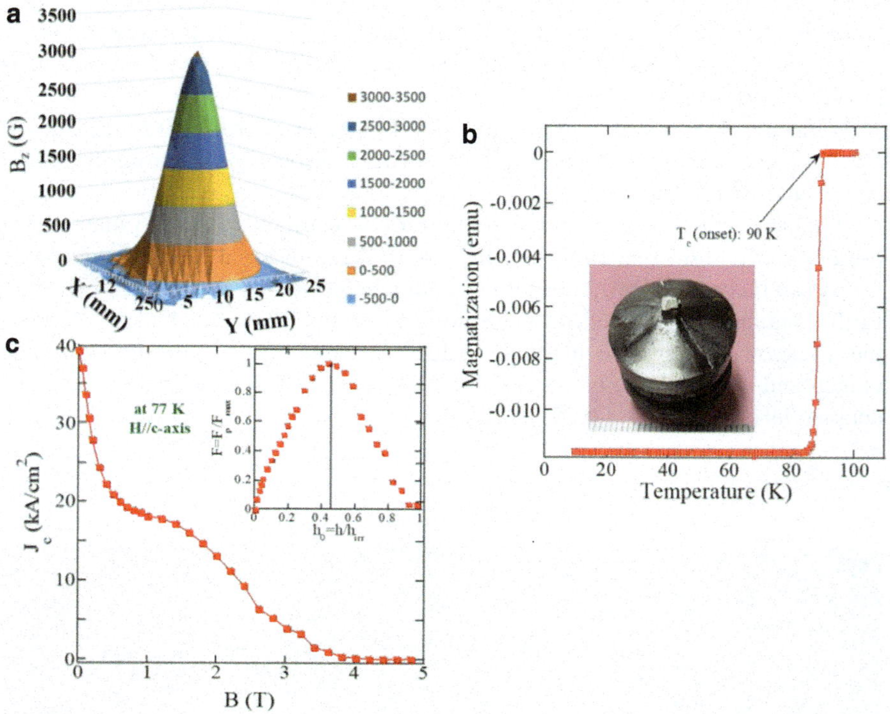

Fig. 5 The trapped field distribution at 77.3 K for the bulk Y-123 superconductor produced by Top Seeded Infiltration Growth Process utilizing the mixture of Yb-123+liquid phase as a liquid source (1:1.3). Note that the TF performance of Y-123 material with the LP of Yb-123+liquid phase (1:1.3) as a liquid source found to be superior 0.32 T (1.3 mm above the sample surface) as compared to the LP of Yb-123+liquid phase (1:1) as a liquid source (**a**). Note that the maximum trapped value recorded around 0.4 T when the Hall sensors at the distance of 0.3 mm. Superconducting transition and inset present the as grown bulk Y-123 (**b**) and field dependence of the super-current density at 77 K for a specimen cut 2 mm below the seed of the single-domain Y-123 (**c**)

common in well-grown melt processed bulk Y-123 or RE-123 materials produced by top seeding method [7, 8]. This reflects the fact that sample was grown into a single grain.

The trapped field has a single cone peak shape, indicating single crystalline nature and absence of weak links as any disturbance in matrix destroys smooth cone trapped field pattern. Trapped field of 0.32 T (at 77 K) was observed 1.3 mm above top center of bulk Y-123 material (see Fig. 5a). Note that the maximum-trapped field value was ***recorded around 0.4 T*** when Hall sensors were at a distance of 0.3 mm, i.e. when the probe is touching the surface of bulk and polymer was coating over Hall probe around 0.3 mm thick. The current measured trapped field (bulk made from Yb-123 and liquid in 1: 1.3 ratio) values are about three times to Fig. 3 where the mixture of Yb-123+liquid (1:1) was used as a liquid source. These results show that sufficient $Ba_3Cu_5O_8$ and appropriate cooling rate are crucial for a successful IG process because samples would grow properly along both *ab*- and *c*-growth directions. Experiments on optimization of the mixture of Yb-123+liquid as a liquid source and a refinement of secondary phase particles were made to improve the performance of Y-123 material.

Figure 5b shows the temperature dependence of magnetization in magnetic field of 1 mT for Y-123 material produced utilizing a mixture of Yb-123 and liquid in a ratio of 1:1.3 only as a liquid source. A sharp superconducting transition with T_c onset around 90 K and transition width (ΔT) of less than 1 K was observed, suggesting a good quality of the sample. Critical current density (J_c) of the specimen was calculated from M-H curves using extended Bean model formula. The sample showed J_c of about 40 kA/cm^2 and 13 kA/cm^2 at 77 K, in self-field and 2 T respectively (Fig. 5c). Although the self-field critical current density value of 1:1.3 mixture liquid phase is similar to that of Fig. 3 reported for Y-123 material produced with Yb-123+liquid (1:1) as a liquid source. However, the high-field critical current density values were improved and this hints that flux pinning mechanism might have changed. Usually, flux pinning at higher magnetic fields is mainly governed by nanometer size chemical substitutions, lattice mismatch defects, oxygen vacancies etc. [20]. Further, the micro-cracks and stacking faults were caused by the Y-211 secondary phase particles in Y-123 matrix and will also play a crucial role to improve high field critical current density. Hence, to understand the improved pinning at high magnetic fields, we had calculated volume pinning force density, F_p, normalized to the maximum value, $F_p/F_{p,\max}$, and plotted it against reduced magnetic field, h_0, (H/H_{irr}, H_{irr} being the irreversibility field, where the *M-H* Loop (MHL) closes). Note that H_{irr} was determined from MHLs utilizing a criterion of 100 A/cm^2. From inset of Fig. 5c, it is worth to note that the peak position lies at $h_0 = 0.45$, which clearly reveals the prevalence of δT_C pinning [22–24] over other pinning mechanisms. Similarly, recent reports on IG processed YGd-123 samples showed that the peak position h_{max} is around 0.4 and this clearly indicates that δT_c pinning is active in all samples due to variation in the distribution of YGd-211 (compositional fluctuations leading to fluctuations in transition temperature T_c) [25]. Furthermore, these values were more pronounced in GdDyBCO system and found that peak position moved from 0.41 to 0.44 for samples processed utilizing liquid phase

(Er-123+035) mass of 1 and 1.4 wt% [26]. These results prove that when liquid phase is comprised of excess $Ba_2Cu_3O_8$, the high-field pinning enhances and which in turn improves the performance in IG process. Hence, with optimized liquid phase mass ratio, it can significantly promote the pinning performance of Y-123.

The energy-dispersive spectroscopy (EDX) does not show any elemental variation of Y, Ba, Cu and O in the final product of bulk Y-123 material produced by mixture of Yb-123 + liquid (1:1.3) as a liquid source (see Fig. 6).

It is clear that one can produce uniform Y-123 material using liquid phase (LP) sources, i.e. Yb-123+liquid (1:1.3) as a liquid source which is quite beneficial to supply more liquid to grow samples in larger size. However, much precise optimization of liquid phase mixture composition, using a nanometer-size secondary phase Y-211 particles and improved growth along with *c*-axis will help in the production of high performance large IG processed Y-123 bulks and result in the enhancement of trapped field values as it is crucial for industrial applications.

3.4 Bulk $YBa_2Cu_3O_y$ and Its Use in Daily Applications

In recent years, significant advances have been made in fabricating single-grain, high-performance Y-123 pellets by means of the melt-growth process. Further, the processing was optimized and Y-123 batch is now processed with 30–40 mm material for several engineering applications. [7] As a result, 140 mm diameter melt textured superconductors are produced in SRL-ISTEC (Japan). On the other hand, reproducible processing was further developed in several commercial companies such as Nippon Steel (Japan), ATZ (Germany) or Can Superconductor (Czech Republic). The main advantage of bulk Y-123 material is that even small 25 bulk Y-123/Gd-123 single grain can trap magnetic field by order of magnitude higher than best hard ferromagnets [6, 9]. This will be more attractive for daily applications including transport, medical, space, power storage, and automobile and research applications. Nowadays, the performance has reached necessary level for industrial applications and we believe that bulk superconducting magnets will be promising in the near future when it comes to market of basic counter parts of compact NMR, drug delivery, water cleaning, wind power plant, and medical diagnosis systems, and/or motors based on trapped flux and flux concentration for large aircraft

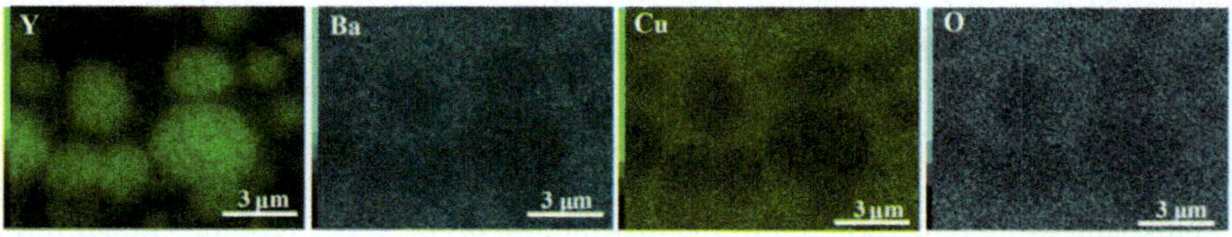

Fig. 6 Energy-dispersive spectroscopy (EDS) results for the Y-123 superconductor produced by top seeded infiltration growth process utilizing the mixture of Yb-123 + liquid phase (1:1.3) as a liquid source. Note that Y, Ba, Cu and O are uniformly distributed over the Y-123 matrix

propulsion [3, 4, 27–29]. The present review and recently reported results [18] had proven that a high-performance, new class of Y-123 material can scale up from laboratory conditions to industrial production by IG process that utilizes the mixture of Yb-123 and liquid as a liquid source.

4 Conclusion

In conclusion, we fruitfully grew single-grain bulk Y-123 material by IG process using a mixture of Yb-123 and liquid in 1:1.1 and 1:1.3 ratio as a liquid source. Magnetization results reveal a sharp superconducting transition of around 90 K width, less than 1 K. The J_c values were 40 kA/cm^2 (1:1.1) and 13 kA/cm^2, (1:1.3) at 77 K, *H//c*-axis, at self-field and 2 T. Flux pinning diagrams or scaling analysis of volume force showed peak position around h_o=0.45. The trapped field results confirmed single cone nature with improved performance as compared to Y-123 samples produced utilizing previous liquid phase (mixture of Yb-123 and liquid in the ratio of 1:1 as a liquid source). The present results indicate that mixture of Yb-123 and liquid in 1:1.3 ratio as a liquid source can aid in production of high performance single grain IG-processed Y-123 superconducting materials with improved trapped field properties. Furthermore, there is room for further optimization of liquid phase amount to use this material in medical, transport, and research applications.

References

1. M. Murakami, *Supercond. Sci. Technol.* **13**, R1 (1992)
2. T. Ohara, H. Kumakura, H. Wada, *Phys C* **13**, 1272 (2001)
3. S. Nishijima, S.I. Takeda, F. Mishima, Y. Tabata, M. Yamamoto, J.I. Joh, H. Iseki, Y. Muragaki, A. Sasaki, K. Jun, N. Saho, *IEEE Trans. Appl. Supercond.* **18**(**2**), 874 (2008)
4. T. Nakamura, D. Tamada, Y. Yanagi, Y. Itoh, T. Nemoto, H. Utumi, K. Kose, *J. Magn. Reson.* **259**, 68 (2015)
5. D. Zhou, M. Izumi, M. Miki, B. Felder, T. Ida, M. Kitano, *Supercond. Sci. Technol.* **25**, 103001 (2012)
6. J.H. Durrell, M.D. Ainslie, D. Zhou, P. Vanderbemden, T. Bradshaw, S. Speller, M. Filipenko, D.A. Cardwell, *Supercond. Sci. Technol.* **31**, 103501 (2018)
7. H.T. Ren, L. Xiao, Y.L. Jiao, M.H. Zheng, *Physica C* **412-414**, 597 (2004)
8. M. Muralidhar, M. Tomita, K. Suzuki, M. Jirsa, Y. Fukumoto, A. Ishihara, *Supercond. Sci. Technol.* **23**, 045033 (2010)
9. M. Tomita, M. Murakami, *Nature* **421**, 517 (2003)
10. M. Muralidhar, N. Sakai, N. Chikumoto, N. Jirsa, T. Machi, N. Nishiyama, Y. Wu, M. Murakami, *Phys. Rev. Lett.* **89**, 237001 (2002)
11. K. Nakazato, M. Muralidhar, M.R. Koblischka, M. Murakami, *Cryogenics* **63**, 129 (2014)
12. M. Santosh, *Acta Phys. Pol. A* **126**, 808 (2014)
13. P. Diko, C. Wende, D. Litzkendorf, T. Klupsch, W. Gawalek, *Supercond. Sci. Technol.* **11**, 49 (1998)

14. K. Chan-Joong, K.-B. Kim, H.-W. Park, T.-H. Sung, I.-H. Kuk, G.-W. Hong, *Supercond. Sci. Technol.* **9**, 76 (1996)
15. N. Vilalta, F. Sandiumenge, S. Piñol, X. Obradors, *J. Mater. Res.* **12**, 38–46 (1997)
16. E. Sudhakar Reddy, T. Rajasekharan, *Supercond. Sci. Technol.* **11**, 523 (1998)
17. A. Mahmood, B.-H. Jun, Y.H. Han, C.-J. Kim, *Supercond. Sci. Technol.* **23**, 065005 (2010)
18. M. Sushma, M. Murakami, *J. Supercond. Nov. Magn.* **31**, 2291 (2018)
19. D.X. Chen, R.B. Goldfarb, *J. Appl. Phys.* **66**, 2489 (1989)
20. D.F. Lee, V. Selvamanickam, K. Salama, Physica C **202**, 83 (1992)
21. P. Diko, V.R. Todt, D.J. Miller, K.C. Goretta, *Physica C* **278**, 192 (1997)
22. A.K. Pradhan, K. Kuroda, B. Chen, N. Koshizuka, *Phys. Rev. B* **58**, 9498 (1998)
23. M. Werner, F.M. Sauerzopf, H.W. Weber, A. Wisniewski, *Phys. Rev. B* **61**, 14795 (2000)
24. F.M. Sauerzopf, H.P. Wiesinger, H.W. Weber, G.W. Crabtree, *Phys. Rev. B* **51**, 6002 (1995)
25. D. Dhruba, M. Muralidhar, M.S. Ramachandra Rao, M. Murakami, *Supercond. Sci. Technol.* **30**, 105015 (2017)
26. K.N. Pavan, M. Muralidhar, T. Kento, M. Jirsa, M. Murakami, *J. Appl. Phys.* **125**, 093907 (2019)
27. H. Hayashi, K. Tsutsumi, N. Saho, N. Nishizima, K. Asano, *Physica C* **392–396**, 745 (2003)
28. M. Santosh, M.R. Koblischka, *Eur. J. Phys. Educ.* **5**, 1 (2014)
29. C.A. Luongo, J. Masson, T. Nam, D. Mavris, H.D. Kim, G.V. Brown, M.W. David Hall, *IEEE Trans. Appl. Supercond.* **19**, 1055 (2009)

Part IV
Nanostructured Metal Oxide Materials for Thermoelectricity and High-Frequency Performance

High-Frequency Soft Magnetic Properties of Nano-Granular Cobalt-(Metal-Oxide, Metal-Nitride) Thin Films with Perpendicular Magnetic Anisotropy

Hanae Kijima-Aoki

1 Introduction

Wireless technology provides rapid developments for worldwide communication by maximizing existing ultra-high frequency (UHF) communication bands (300 MHz–3 GHz). Today, the communication bands of highly integrated electric devices have shifted toward super-high frequency (SHF) frequencies. To achieve enhanced efficiencies and decreased footprints, high-frequency soft magnetic thin films have been intensively studied for their utility in electromagnetic applications such as magnetic sensors, inductors, and noise suppressors. The main problem for the practical application of these films is their decreased performance at high GHz frequencies and limited device design based on their anisotropic properties. To overcome such problems, the high-frequency soft magnetic properties of thin film materials can enable as follows: (1) increased ferromagnetic resonance frequency, (2) suppressed eddy current loss, and (3) multi/omnidirectional performance in-plane.

This chapter presents the omnidirectional soft magnetic performance up to 10 GHz frequency achieved by using low-core loss nano-granular cobalt-(metal-oxide) and cobalt-(metal-nitride) thin films with perpendicular magnetic anisotropy.

Ferromagnetic Resonance Frequency at GHz Frequency The ferromagnetic resonance phenomenon indicates the limits of the effective permeability, or the resonance peak itself that can be applicable to the high-frequency tunable device or electromagnetic noise filter. The ferromagnetic resonance frequencies of bulk materials and thin films are expressed as

H. Kijima-Aoki (✉)
Creative Interdisciplinary Research Division, Tohoku University, Sendai, Japan
e-mail: kijima@fris.tohoku.ac.jp

A. G. Roca et al. (eds.), *Surfaces and Interfaces of Metal Oxide Thin Films, Multilayers, Nanoparticles and Nano-composites*,
https://doi.org/10.1007/978-3-030-74073-3_12

$$f_r = \frac{\gamma}{2\pi} H_k \left(\text{for bulk material}\right) \tag{1}$$

$$f_r = \frac{\gamma}{2\pi} \sqrt{4\pi M_s H_k} \left(\text{for thin film shaped material}\right) \tag{2}$$

where $4\pi M_s$ is saturation magnetization (G), H_k is magnetic anisotropy field (Oe), and γ is gyro-magnetic ratio. The relative permeability (μ_r) is defined as the slope of the initial magnetization curve:

$$\mu_r = {4\pi M_s}/{H_k} + 1 \tag{3}$$

where increasing uniaxial magnetic anisotropy H_k is one strategy to extend f_r up to GHz.

As Landau-Lifshitz-Gilbert (LLG) theory[1] manifests high-frequency soft magnetic performance, in-plane uniaxial magnetic anisotropy is a sophisticated method for thin films to obtain both high f_r and μ_r. Because f_r appropriately increases not only H_k but also M_s, many researchers have reported that high-M_s films with in-plane uniaxial magnetic anisotropy show high μ_r about a few hundred up to a few GHz for the hard magnetization axis in-plane [2]. With properties such as large contributions to M_s and zero magnetostriction, amorphous alloyed thin films with rich magnetic metal composition have been researched and achieve both high μ_r and high f_r [3]. However, the high μ_r of in-plane anisotropic soft magnetic films do not exist over a few GHz. As frequency increases, the μ_r degrades because of eddy current loss in a metal rich conductive film.

Reduction of Eddy Current Loss at GHz Frequency Other than ferromagnetic resonance absorption, the μ_r values of thin films decrease by eddy current loss at GHz frequencies. Due to the eddy current of conductive magnetic films, effective RF magnetic fields only penetrate in the vicinity of the surface, which is called the skin depth effect [4]. The magnetic field strength at skin thickness, *s,* can be reduced to 1/e.

$$s = \frac{1}{2\pi\sqrt{{\mu f}/{\rho}}} \tag{4}$$

where μ is permeability, f is frequency, and ρ is electrical resistivity. For conductive soft magnetic sheet, s is known to be a few tens to a hundred nm at 1 GHz.

Thickness limitation due to eddy current loss is successfully overcome by insulating ferrite composite films [5], thin multilayer stacking with insulating layers [6], and scratch surfaces [7]. However, when the permeability amplitude increases, loss components at GHz frequency increase, and vice versa.

Omnidirectional μ-f Performance In-Plane While ferrite polymer materials offer excellent reductions in eddy current loss [5], it is difficult to decrease loss components (μ'') up to GHz frequencies. Its loss components are not small because anisotropic dispersion makes the width of the ferromagnetic resonance peak of μ'' broader. Uniaxial magnetic anisotropy makes the sharp ferromagnetic resonance peak, which greatly suppresses the loss components up to f_r. [8] When analyzing the demagnetization field, it is very reasonable to use in-plane uniaxial magnetic anisotropy for thin films. There are several origins of in-plane uniaxial magnetic anisotropy in soft magnetic films: field-induced anisotropy [9], magneto-elastic anisotropy [10], and shape anisotropy by patterning film [11]. However, based on application, in-plane uniaxial magnetic anisotropy has a directional limitation for one axis (Fig. 1a). To overcome directional limitations, crossed magnetic anisotropy has been proposed (Fig. 1b) [12]. A multilayer structure with stacks of thin magnetic layers with anisotropic axes perpendicular to each other, successfully work in biaxial directions in-plane and can decrease the eddy current loss as well.

Our motivation for more flexible soft magnetic material design for omnidirectional μ-f property and in-plane isotropic film with perpendicular H_k is proposed as shown in Fig. 1c. An advantage of perpendicular H_k is that the film plane is always along the hard magnetization axis, resulting in omnidirectional high μ-f performance in the film plane. In this chapter, the overview of our recent work on in-plane isotropic nano-granular cobalt- (metal-oxide, metal-nitride) films with perpendicular H_k will be presented [13–16].

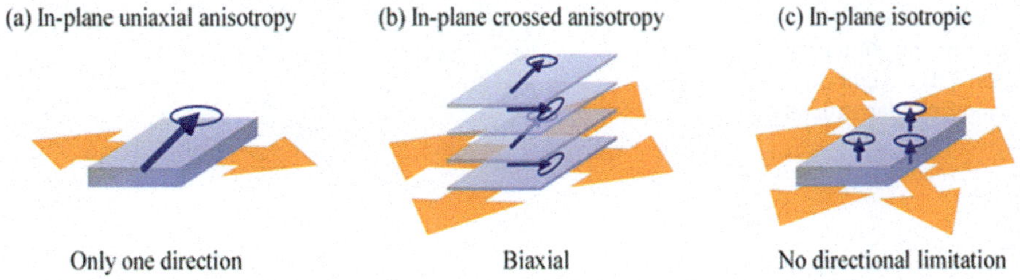

Fig. 1 Use of plane magnetic anisotropy (perpendicular H_k). (**a**) in-plane uniaxial magnetic anisotropy film, (**b**) in-plane crossed magnetic anisotropy film, and (**c**) in-plane isotropic film (with perpendicular H_k)

2 Results and Discussion

2.1 Synthesis and Structure of Metal-Insulator Nano-Granular Thin Films

Synthesis of nano-granular film via sputtering: Figure 2a describes the representative structure of nano-granular thin films produced by sputtering. Nano-granular structures are shown in Fig. 2, where magnetic metal nano-granules of a few nm in diameter are evenly dispersed in a non-magnetic matrix. Metal-oxide, metal-nitride, and metal-fluoride systems are preferable for the matrix when required for soft magnetic performance. Nano-granular structures were first found by *Karamon et al.* [17] of iron boron–nitride system films produced by a reactive sputtering process with nitrogen gas. Nano-granular films were grown on various sorts of substrate by any one of the sputtering composite targets, co-sputtering, or tandem sputtering both magnetic metal and ceramic targets [18]. In any of these methods, the nano-granules and surrounding matrix were both self-organized during the non-equilibrium sputtering process because phase separation between magnetic metal nano-granules and the ceramic (insulating) matrix was thermodynamically enhanced. For example, for the Co-Al_2O_3 system nano-granular film (Fig. 2b): metal particles and non-metal particles were sputtered first, reached the substrate surface, and precipitated. Then, metal-oxides with high heats of formation (or high melting temperature) segregated and became uniformly distributed clusters. Finally, subsequent Co segregation is hampered by Al_2O_3 clusters, which were then self-organized into nano-granules. Tandem sputtering is well-known as a way to control the granule size distribution and enable anisotropy in the in-plane structure along the rotational arc direction.

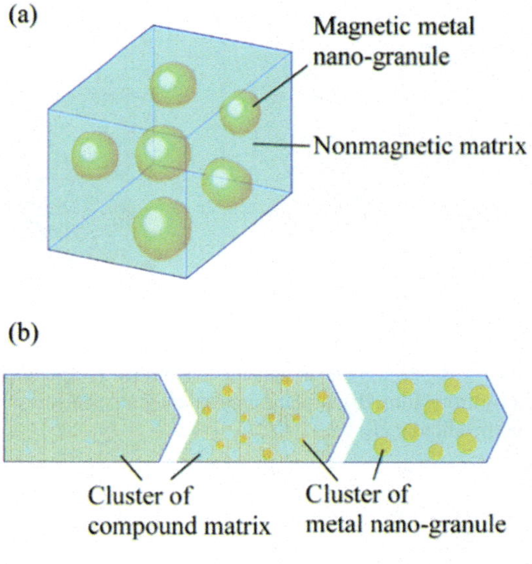

Fig. 2 (**a**) Schematic illustration of metal-insulator type nano-granular film. (**b**) self-organization process from left to right of nano-granular structure

Kinetic Phase Separation of Metal-Insulator Nano-Granular Films To obtain a metal-insulator nano-granular structure, the difference in heat of formation energy ($|\Delta H_{298K}|$) between magnetic particle material and matrix materials plays a key role because $|\Delta H_{298K}|$ should be large enough to phase separate. Figure 3 shows the heat of formation energy of oxide (circle), nitride (diamond), and fluoride (triangle) compounds. Compounds having higher ($|\Delta H_{298K}|$) are more easily to form than those with small ($|\Delta H_{298K}|$). Magnetic metal (Fe and Co) compounds are less stable than other compounds and can easily form single-phase granules. For instance, the well-known nano-granular Co-Al_2O_3 system films have the largest difference of heat of formation ($|\Delta H_{298K}|$) ~1400 kJ/mol.

In the case of Co-based nano-granular system films, metallurgical phase separation must be included since the magnetic anisotropy of cobalt depends on its crystal structure. It is known that the Co of as-deposited nano-granular films prepared by the sputtering process exhibits an amorphous phase or fcc crystal phase instead of hcp-phase [19, 20]. It is suggested that sputtering deposition proceeds under

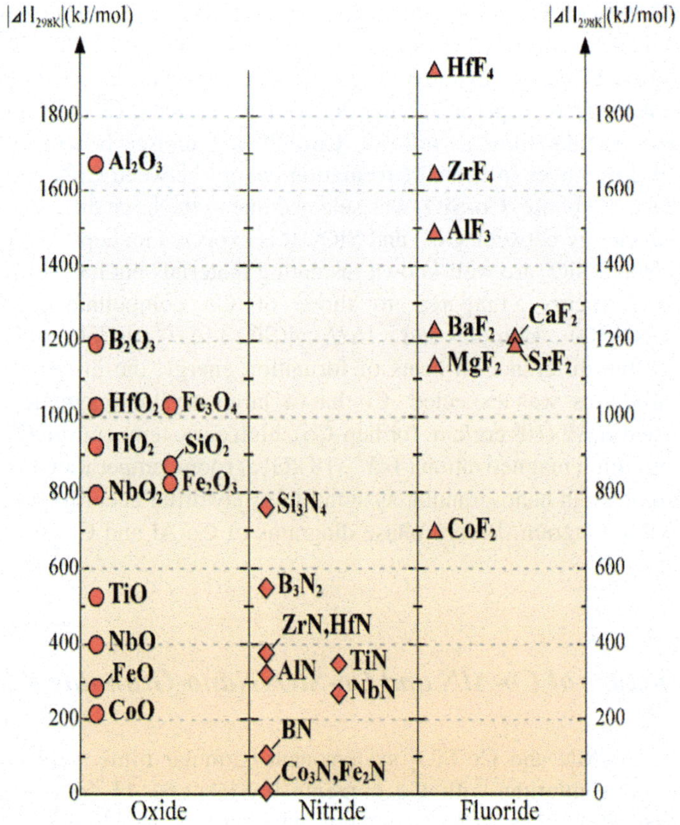

Fig. 3 Heat of formation of nominal oxide (circle), nitride (diamond), and fluoride (triangle) compounds

non-equilibrium conditions in the same manner as liquid quenching, and then a quasi-stable phase appears. In addition, stabilization of the fcc-Co phase in nano-sized Co particles (< 20 nm) synthesized by sputtering has been explained from free energy calculations of crystals [21]. The existence of cubic or amorphous Co phase in sputtered film is preferable for non-dispersive in-plane magnetic anisotropy (H_k) and good soft magnetic performance [22]. Since the hcp-phase made the film hard magnetic like nonvolatile perpendicular recording media, then, unselected to high-frequency soft magnetic film. So, the effect of the hcp-Co phase on soft magnetic properties is not clear.

Although the reasoning is mentioned above, is it impossible to obtain hcp-Co granules by sputtering. Given that a small amount of matrix is decomposed and alloyed with Co granules, it is indicative that the phase separation can be accounted for in the Co-metal binary alloy system as well. From equilibrium binary phase diagrams, the temperature range of the stable hcp-Co phase increased when Co was alloyed with specific metals. It has been reported that the hcp-phase has been successively grown on Co-metal (metal = Cr [23], Mo [23], and W [24]) alloyed thin film systems, so-called magnetic induced phase separation, for applications of perpendicular magnetic recording media. Also, nano-granular thin films with perpendicular magnetic anisotropy have been reported on CoCrPt-SiO_2 films [25].

Aiming to achieve soft magnetic nano-granular film with perpendicular H_k and investigate the effects of perpendicular H_k, we have selected Co-AlN and Co-SiO_2 system nano-granular films as follows. **Co-AlN** was chosen because of the relatively small differences in heats of formation energy between CoN and AlN. It is expected a fcc-Co phase. **Co-SiO_2** was selected due to the large differences in heats of formation energy between CoO and SiO_2. It is expected an hcp-Co phase.

Both AlN and SiO_2 are well-known insulating materials but have different heats of formation energy compared to those of Co compounds: $|\Delta H_{298K}|$(Al)-$|\Delta H_{298K}|$(CoN)~300 kJ/mol and $|\Delta H_{298K}|$(SiO_2)-$|\Delta H_{298K}|$(CoO)~650 kJ/mol. Referring to the difference in heats of formation energy, the difference in nano-granular structures was expected. Co has a large crystal magnetic anisotropy (K_1~4×10^6 and K_2~2×10^6 erg/cm^3 for hcp-Co), high curie temperature (T_c~1400 K), and high saturation magnetization ($4\pi M_s$~18 kG) at room temperature [4]. The crystal structure of Co in nano-granular system films may differ according to the binary Co-alloy phase diagram. Binary phase diagrams of Co-Al and Co-Si are given in Fig. 4 [26].

2.2 *Structure of Co-AlN and Co-SiO_2 Nano-Granular Films*

A series of Co-AlN and Co-SiO_2 system nano-granular films were prepared by RF-magnetron sputtering with Co contents 0–81% and 43–56%, respectively. Detailed experimental procedures are presented in [13–16]. Figure 5 shows the cross-sectional TEM images of (a) Co_{47}–$(AlN)_{53}$ and (b) Co_{52}–$(SiO_2)_{48}$ films as typical nano-granular structures. The Co_{47}–$(AlN)_{53}$ film shows spherical Co granules

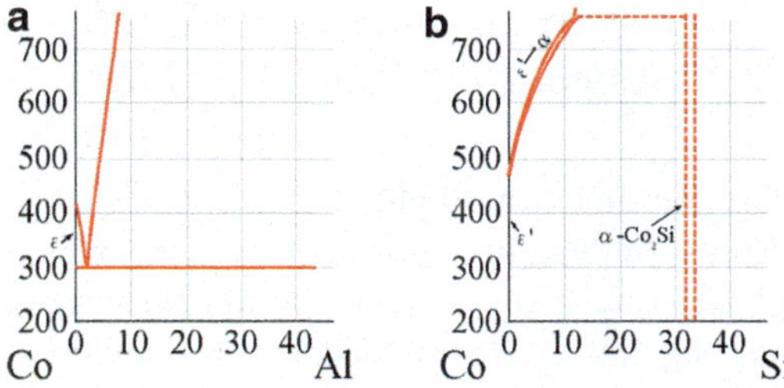

Fig. 4 Binary phase diagrams of (**a**) Co-Al and (**b**) Co-Si [26]

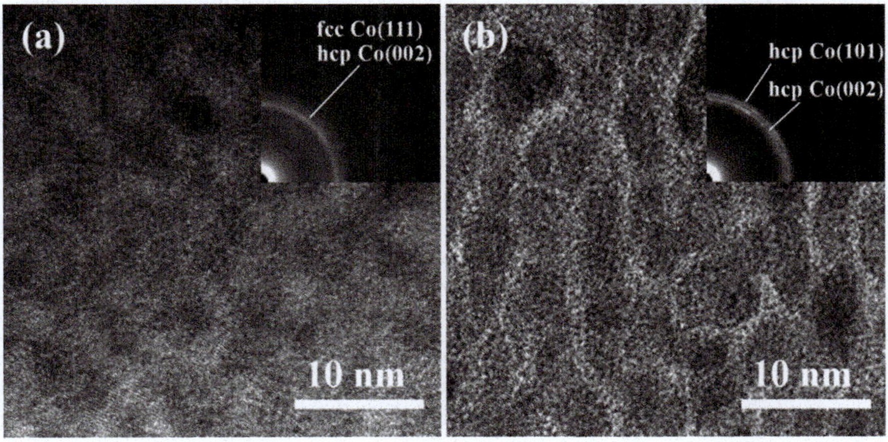

Fig. 5 Cross-sectional TEM image of (**a**) Co_{47}-AlN_{53} and (**b**) Co_{52}-$(SiO_2)_{48}$ nano-granular films. (Reprinted from [36])

with 4–5 nm embedded in the AlN amorphous matrix. The Co_{56}–$(SiO_2)_{44}$ film shows ellipsoidal Co granules that are 7–8 nm in length and 3–5 nm in diameter aligned perpendicularly to the film plane. Inter-granule thickness between ellipsoidal Co granules is about a few nm. The boundary between Co granules (dark area) and insulating matrix (bright area) of the Co-SiO_2 film is clearer than the Co-AlN boundary. This indicates that the thickness of the intermediate phase between the Co granule and AlN matrix is thicker than that of the Co-SiO_2 film because SiO_2 more easily forms than AlN due to the large difference of heat of formation energy (Fig. 3). Co granules of Co-AlN films are interconnected with each other through intermediate boundary layers. The diffraction profiles of Co-SiO_2 films were indexed as hcp-Co, whereas that of the Co-AlN film was probably fcc-Co. The difference in crystal phase of Co granules is consistent with the Co-Al and Co-Si alloyed systems (Fig. 4). Thus, perpendicularly aligned Co granules might be explained by magnetically separated hcp-Co and the intergranular phase grown perpendicular to the film plane.

2.3 Electric Properties of Co (Metal-Oxide or Metal-Nitride) System Nano-Granular Films

The abovementioned phase separation (2.2 and 2.3) strongly affects the conduction process of tunneling current through metal granules. Interconnection of neighboring metal granules often occurs due to imperfections in the matrix-granule interface and has negative effects on electrical resistivity. The electrical resistivities (ρ) of films were measured by conventional four-probe methods. Figure 6 shows the ρ dependence on Co contents of typical Co-(metal-oxide or metal-nitride) system nano-granular films [14, 16, and 27]. The ρ of matrix compounds, crystallinity, and interfacial separation strongly affect resistivity. The order of the ρ value of each system film mainly depends on the ρ value of the insulator matrix, which depends on the band gap energy E_g (e.g., E_g of AlN, Al_2O_3, and SiO_2 is 5.74, 9.5, and 11 eV, respectively [28]) The ρ of Co-SiO_2 is 1000 times higher than that of Co-AlN with the same Co contents. The small ρ value of the Co-AlN system film is possibly attributed to both the relatively small band gap energy of AlN matrix and imperfect phase separation between the Co and AlN interfaces.

The ρ value exponentially decreases with increasing Co contents. The slope of ρ, which reflects the formation and density of metal nano-granules, changed near the percolation threshold *x*. Tunneling currents of Co nano-granules pass through insulating inter-granule matrices under *x*. Conduction is attributed to spin-dependent tunneling probability, so this composition range is significant for tunneling

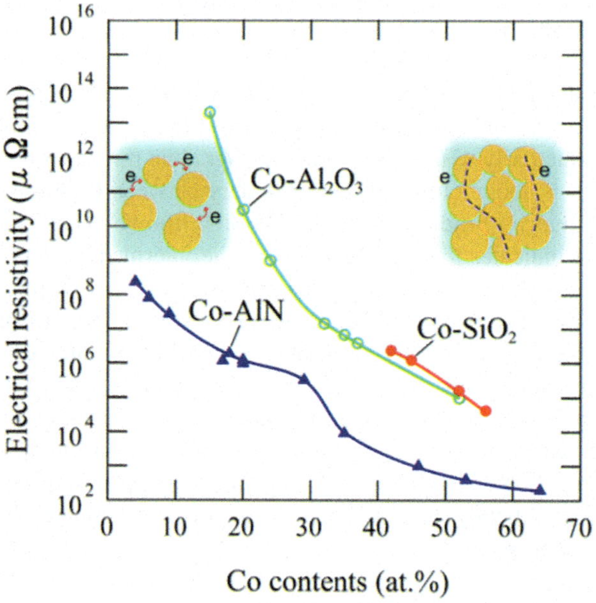

Fig. 6 Electrical resistivity (ρ) dependence on Co contents of Co-AlN [14], Co-Al_2O_3 [27], and Co-SiO_2 [16], nano-granular films

magnetoresistance properties of nano-granular films [29]. For nano-granular films with Co contents greater than x, metallic conduction is dominant among interconnected Co granules. By increasing the packing density of Co, granules will become magnetically coupled and exhibit excellent soft magnetic properties [30]. The network-like intergranular matrix is correlated to ρ, as seen in Fig. 5, where the nano-granular film with an oxide matrix has high resistivity up to high Co contents, which should decrease eddy current as defined in Sect. 4.

2.4 *Magnetic Properties of Nano-Granular Films with Perpendicular* $\mathbf{H}_k$

The magnetic properties of nano-granular films in the metallic composition range are well explained both by induced uniaxial magnetic anisotropy and random magneto anisotropy of nano-crystalline films [10, 31]. Coherent uniaxial magnetic anisotropy consists of the average random magnetic anisotropy of ferromagnetic metal grains located within the effective exchange length L_{ex}:

$$L_{ex} = \varphi_0 \sqrt{A/|K_1|} \tag{5}$$

where A is the exchange stiffness, K_1 is the local magnetic anisotropy constant, and φ_0 is a proportionality factor in the order of one. The magnetic domain width of conventional soft magnetic films with in-plane uniaxial H_k is a few tens to a hundred μm. The concept of magnetic domain design of in-plane isotropic high-frequency soft magnetic films is to segmentalize magnetic domains for orientation of perpendicular H_k.

2.5 *Magnetic Domains of Co-AlN and Co-SiO$_2$ Nano-Granular Films*

Magnetic domain observation was carried out by magnetic force microscopy [32]. Figure 7a shows the magnetic domain structure of the Co_{76}–$(AlN)_{24}$ film surface. Dark and bright contrast indicates the magnetic flux pointing up and down, respectively. Clear stripe domains of 400 nm in domain width were observed along the magnetically saturated axis of longitudinal direction. This striped domain structure has been reported on NiFe thin films with rotatable magnetic anisotropy, [33] a type of perpendicular magnetic anisotropy. The magnetic moment of the thin film is subject to strong demagnetization against the surface, attributed to shape anisotropy. For thin films with weak perpendicular anisotropy, the magnetic moments in the neighboring domains are potentially leaned toward each other, leading to a decrease of the demagnetization energy perpendicular to the film surface.

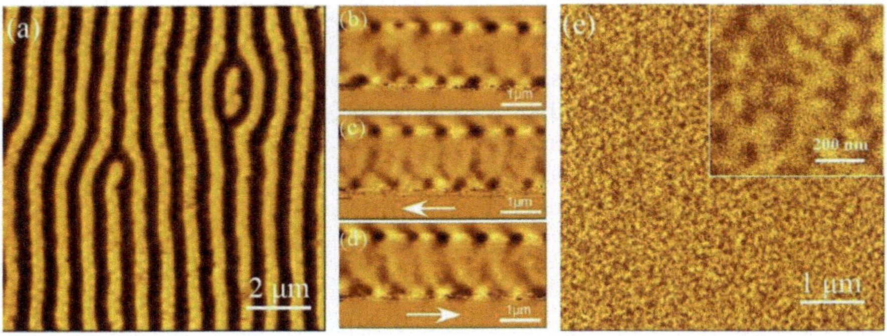

Fig. 7 Magnetic domain structure of the surface of (**a**)Co_{76}-$(AlN)_{24}$ and (d)Co_{56}-$(SiO_2)_{44}$ films. (**b**) Cross-sectional MFM image without zero-field and with external magnetic field (H_{ex}=100 Oe) (**c**) left and (**d**) right direction

Figure 7b is a cross-sectional magnetic force microscopy (MFM) image of the same film, which has been magnetically saturated perpendicular to the observed side. A couple of magnetic poles have appeared on the film surfaces and substrate interfaces and are alternately aligned in the film plane. The application of an external magnetic field (H_{ex}) of 100 Oe right (Fig. 7c) and left (Fig. 7d) side (indicated as white arrow) does not change polarity but does slightly affect magnetic flux in the domains. The magnetization of Co nano-granules in the same magnetic domain, within an exchange length, is manifested by the random anisotropy model. Under the small magnetic field, the mobility of internal magnetic moments is high whereas the stripe domain walls are pinned.

Figure 7e shows the magnetic domain structure of the surface of a 1.5 μm thick Co_{56}–$(SiO_2)_{44}$ film. Note that the width of the maze domains is smaller than 100 nm. This domain arrangement did not exhibit directional displacement, even after a large magnetic field was applied in every direction of the film plane. Such fine magnetic domain structures have been reported on amorphous Tb-Fe film with perpendicular anisotropy as well [34].

2.6 Magnetization Curves of Co-AlN and Co-SiO$_2$ Nano-Granular Films

Magnetization curves of (a)Co_{76}–$(AlN)_{24}$ and (b)Co_{52}–$(SiO_2)_{48}$ films are shown in Fig. 8. Each film shows the same magnetization curves regardless of the film plane direction; therefore, films have no uniaxial magnetic anisotropy in-plane. Both films have similar saturation magnetization values around 10 kG even at different Co contents in the film. Co-AlN system films show lower M_s values due to incomplete phase separation between Co granules and AlN matrix as can be seen in Fig. 5. For Co-AlN film, both perpendicular and in-plane magnetic anisotropy components were found. The perpendicular H_k components, which is represented as in-plane

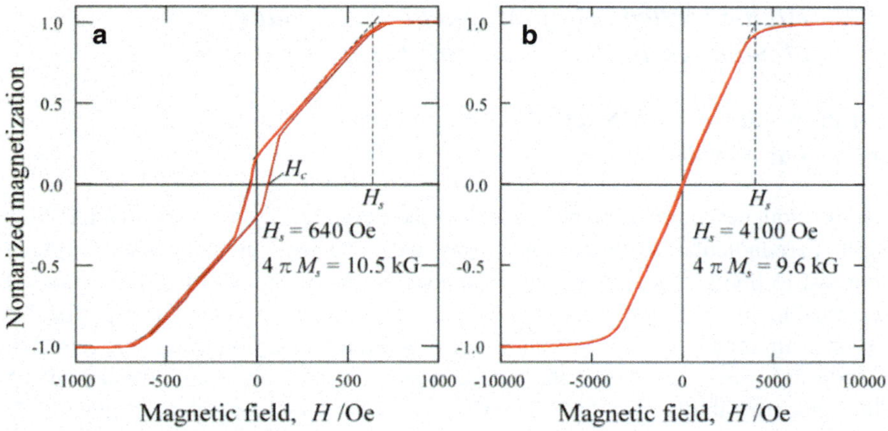

Fig. 8 In-plane magnetization curves (red: after 90-degree rotation in the plane) of (**a**) Co_{76}-$(AlN)_{24}$ and (**b**) Co_{56}-$(SiO_2)_{44}$ films

saturation magnetic field (H_s) in Fig. 8a), are the origin of texture-like structures constituting the Co granules chain arranged perpendicular to the plane [15]. The coercive field (H_c) of a few tens Oe is relatively large from that of in-plane soft magnetic films measured along the hard magnetization axis, indicating that the stripe domain wall (Fig. 7a) acts as a pinning site for the magnetic domain reversal process. We found that the initial magnetization process with an applied magnetic field less than H_c is reversible and independent of the magnetization direction [15].

In the magnetization process with small magnetic field excitation, rotation of magnetic moments predominantly occurs without domain wall movements, so that the hysteresis loss is extremely small.

The order of the saturation magnetic field of Co_{56}–$(SiO_2)_{44}$ films is close to that of perpendicular recording media, and which is 5–1000 times higher than that of conventional soft magnetic films with in-plane anisotropy. Such strong perpendicular magnetic anisotropy possibly originates from shape anisotropy of perpendicularly aligned ellipsoidal Co grains and crystal anisotropy of hcp-Co as shown in Fig. 5b). From the magnetization curves around zero magnetic field, the H_c in each direction is small due to the maze domain structure.

From the magnetic domains and magnetization curves of Co-AlN and Co-SiO_2 nano-granular films, the Co-AlN film with a striped domain has a smaller perpendicular H_k than Co-SiO_2 with fine maze domains.

Differences in static magnetic properties arise from the crystal structure, shape, and alignment of Co granules. Since the amplitude of perpendicular H_k closely relates ferromagnetic resonance frequency (f_r), high-frequency soft magnetic performances of Co-AlN film with weak perpendicular H_k, Co-SiO_2 film with large perpendicular H_k, and other application results for noise suppression are provided in the following section.

2.7 High-Frequency Soft Magnetic Performance of Nano-Granular Films with Perpendicular $\mathbf{H}_k$

μ-f Properties of Co-AlN &Co-SiO$_2$ Nano-Granular Films with Perpendicular H_k

High-frequency measurements related to the permeability of nano-granular films with perpendicular magnetic anisotropy were carried out using short-circuited microstrip line [35] and short-circuited coaxial line methods [36]. Applications of external magnetic field for background calibration were performed after permeability measurements, so that the magnetically saturated direction (magnetic remanent condition) was stored before the measurement. Figure 7 shows the frequency profiles on permeability μ' and μ'' of (a) Co_{76}–$(AlN)_{24}$ film with weak perpendicular H_k and (b) Co_{52}–$(SiO_2)_{48}$ films with strong perpendicular H_k. The RF magnetic field H_{rf} was applied in parallel (x: red) and perpendicular (y: blue) to the magnetically saturated direction (stripe direction for Co-AlN film).

Different μ-f profiles of Co-AlN film were obtained when the H_{rf} direction changed. For H_{rf} parallel to the magnetically saturated axis (stripe domain length direction), $\mu(x)$ was constant up to 3 GHz, and then a broad peak of μ'' appeared at a ferromagnetic resonance frequency (f_r) of 10 GHz. For H_{rf} perpendicular to the magnetically saturated axis, a sharp peak of μ'' appeared at f_r of 2.4 GHz. Consistent μ values were supported from initial magnetization curves [14], but the film does not have in-plane uniaxial magnetic anisotropy, and the magnetic stripe domain partly contributes to ferromagnetic resonance behavior. The tendency of f_r of $\mu''(x)$ and $\mu''(y)$ appearing is approximate to that reported by Vukadinovic, N [37]. The higher f_r value of $\mu''(x)$ than $\mu''(y)$ is explained by increased demagnetization components under a striped domain configuration. Since the magnetic charge is located at the domain wall by excitation of H_{rf}, the demagnetization factor preferably increases the effective perpendicular H_k via the smit-Wejin effect [38] (Fig. 9).

The frequency profiles of μ for Co-SiO$_2$ films do not change whether the application of H_{rf} changes from perpendicular or parallel to the magnetically saturated direction of the film plane due to the fine magnetic domain structure (Fig. 7e), and by the negligibly small magnetic remanence and coercivity in the film plane (Fig. 8b). The μ'', including magnetic anisotropy dispersion and eddy current loss, is negligibly small up to 5 GHz, even with a relatively thick film. Coherent magnetization precession locally occurs in segmentized magnetic domains. There are clear absorption peaks in μ'' at the magnetic resonance frequency $f_r = 13$ GHz. The μ' value of a well-known high-frequency (>10 GHz) soft magnetic material, such as NiZn ferrite, is less than 2 above 5 GHz because of the increased μ'' [39]. μ' and f_r of Co-SiO$_2$ films with large H_k are consistent with the estimated values using 1 and 3 equations. The omnidirectional high-frequency (>5 GHz) performance of the Co–SiO$_2$ films was improved for both high values of μ' and low values of μ''. Co-based highly-resistive nano-granular films with large perpendicular H_k have great potential for

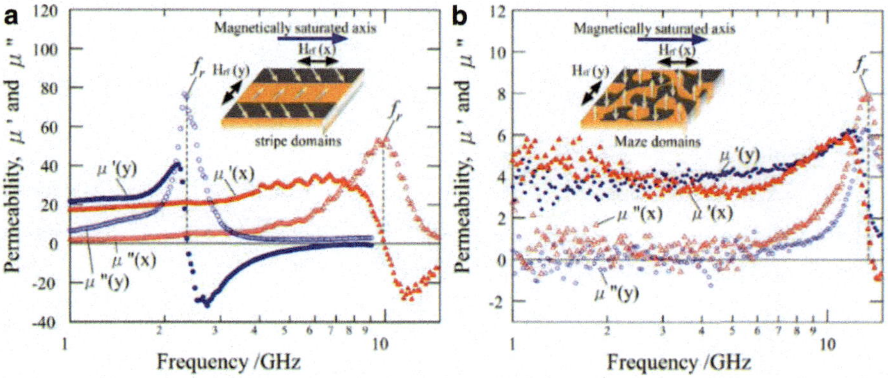

Fig. 9 Frequency profiles of μ' (solid symbols) and μ''(open symbols) of (**a**) Co_{76}–$(AlN)_{24}$ film with weak perpendicular H_k and (**b**) Co_{52}-$(SiO_2)_{48}$ films with strong perpendicular H_k. The μ-f was measured parallel(x) and perpendicular(y) to magnetically saturated direction in the film plane. (Reprinted from [15])

their applications as thin film inductors and antennae operating at high GHz frequencies.

2.8 *Omnidirectional* μ-f *Properties and Noise Suppression Performance of CoFe-AlN/AlN Multilayer Films with Perpendicular* H_k

In the applicational viewpoint for noise filters at GHz frequencies, the electromagnetic noise attenuation of thin magnetic films on integrated circuits was investigated [40]. In analog-digital integrated circuit, both conduction noise through signal line and near electromagnetic field noise at GHz frequency are crucial problems. Since intricate signal lines are arranged horizontally and vertically in-plane, the omnidirectional high noise suppression effect of magnetic thin films has potential application. Figure 10a presents a schematic illustration for noise suppression effect of in-plane isotropic films with perpendicular H_k. The electromagnetic noise from near-field and in-line was decreased by low frequency magnetization precession (μ) and high-frequency ferromagnetic resonance absorption around f_r of the magnetic film. The noise suppression affects both near-field intensity and in-line power was evaluated by comparing with and without magnetic films on the signal line. Figure 10b shows the frequency dependence of near-field intensity (left) and in-line power ratio (right) of $(CoFe\text{-}AlN)_n/(AlN)_{n+1}$ multilayer films for $n = 1, 3, 9, 18$, and 27. In-line power losses obtained without film were used for background subtraction.

For $(CoFe\text{-}AlN)_n/(AlN)_{n+1}$ multilayer films ($n = 3, 9, 18$, and 27), the attenuation of the near-field intensity gradually increased above 1 GHz, where the

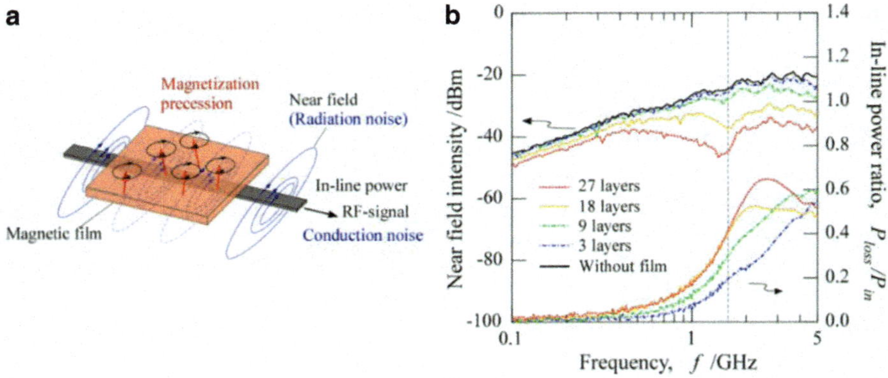

Fig. 10 (**a**) Schematic illustration for noise suppression effect of in-plane isotropic film with perpendicular H_k. (**b**) Frequency dependence of near-field intensity (left) and in-line power ratio (right) of (CoFe-AlN)$_n$/(AlN)$_{n+1}$ multilayer films for $n = 1, 3, 9, 18$, and 27. (Adapted from [40])

concentration of eddy current losses can be a prominent factor for near-field suppression. The highest attenuation in near-field intensity was observed around 1.7 GHz of the f_r. near-field intensity suppression was attributed to both ferromagnetic resonance absorption and eddy current losses. The in-line power losses exponentially increased with increasing frequency above 1 GHz [41]. This increase was more pronounced in films with large n values compared to those with small n values, which stems from the eddy current loss caused by the conductive magnetic film on the transmission line. Center points, at which a broad $P_{\rm loss}/P_{\rm in}$ peak appeared, shifted to frequencies exceeding f_r, consistent with the previously reported demagnetization effect [42]. Beyond a few gigahertz, this shift depends on the sample size and microstrip line shape. For in-plane isotropic (CoFe-AlN)$_n$/(AlN)$_{n+1}$ multilayer films, frequency-dependent noise suppression is possible in every direction in the film plane.

3 Conclusions

For omnidirectional high-frequency soft magnetic performance up to GHz frequencies, Co-AlN and Co-SiO_2 nano-granular films with various perpendicular anisotropies and magnetic nano-granules embedded in metal-oxide or metal-nitride matrices were prepared by sputtering.

The low-ρ Co-AlN film consists of spherical Co nano-granules embedded in an amorphous AlN matrix. Co-AlN films show both in-plane and out-of-plane magnetic anisotropy components and exhibit conventional striped domain structures. They show constant μ of around 20 up to f_r values of 2.4 GHz and 10 GHz under H_{rf} applications perpendicular and parallel to the magnetically saturated direction, respectively.

A high-ρ Co-SiO_2 film consists of perpendicularly aligned ellipsoidal hcp-Co nano-granules with an amorphous SiO_2 matrix. Large perpendicular anisotropy was successfully realized in Co-SiO_2 film. The Co-SiO_2 film shows a constant μ of four with small loss components up to 5 GHz and a high f_r of 13 GHz. The frequency dependence of the Co-SiO_2 film does not change regardless of the direction of applied $H_{rf.}$

Co-based high resistive nano-granular films with perpendicular H_k have achieved omnidirectional high-frequency soft magnetic performance and provide great potential for their applications as thin film inductors and antennae operating at high GHz frequencies.

Acknowledgments I would like to express my great appreciation to Dr. H. Masumoto (Tohoku Univ.) for his dedication as mentor for whole research. The author appreciates the support of Dr. T. Miyazaki (Tohoku Univ.) in performing TEM observations. I am also grateful to Dr. T. Yamaoka (SII NanoTechnology Inc.) for his taking MFM observations. Dr S. Takeda (KEYCOM Co.) provided us refined high-frequency measurements up to GHz frequencies and helpful perspectives. My thanks are also extended to Dr. S. Ohnuma (Denjiken), Dr. Y. Endo (Tohoku Univ.), M. Yamaguchi (Tohoku Univ.), and Dr. Y. Shimada (Tohoku Univ.) for their contribution in providing fruitful discussion and helpful advice.

References

1. C. Kittel, On the theory of ferromagnetic resonance absorption. *Phys. Rev.* ***73***(2), 155 (1948)
2. S. Ohnuma, H. Fujimori, T. Masumoto, X.Y. Xiong, D.H. Ping, K. Hono, FeCo–Zr–O nanogranular soft-magnetic thin films with a high magnetic flux density. *Appl. Phys. Lett.* ***82***(6), 946–948 (2003)
3. H. Fujimori, N.S. Kazama, K. Hirose, J. Zhang, H. Morita, I. Sato, H. Sugawara, Magnetostriction of Co-base amorphous alloys and high frequency permeability in their sputtered thin films. *J. Appl. Phys.* ***55***(6), 1769–1774 (1984)
4. R. M. Bozorth, *Ferromagnetism* (1993), p. 264, 568, and 771
5. A. Verma, A.K. Saxena, D.C. Dube, Microwave permittivity and permeability of ferrite–polymer thick films. *J. Magn. Magn. Mater.* ***263***(1–2), 228–234 (2003)
6. M. Munakata, M. Yagi, Y. Shimada, High-resistive (CoFeB)-(SiO/sub 2/) soft magnetic amorphous film for micro-cores in a few MHz range. *IEEE Trans. Magn.* ***35***(5), 3430–3432 (1999)
7. S.P. Morgan Jr., Effect of surface roughness on eddy current losses at microwave frequencies. *J. Appl. Phys.* ***20***(4), 352–362 (1949)
8. S.X. Wang, N.X. Sun, M. Yamaguchi, S. Yabukami, Properties of a new soft magnetic material. *Nature* ***407***(6801), 150–151 (2000)
9. G.S. Cargill III, T. Mizoguchi, Dipolar mechanisms for magnetic anisotropy in amorphous ferrimagnetic alloys. *J. Appl. Phys.* ***49***(3), 1753–1755 (1978)
10. G. Herzer, Anisotropies in soft magnetic nanocrystalline alloys. *J. Magn. Magn. Mater.* ***294***(2), 99–106 (2005)
11. G. Perrin, J.C. Peuzin, O. Acher, Control of the resonance frequency of soft ferromagnetic amorphous thin films by strip patterning. *J. Appl. Phys.* ***81***(8), 5166–5168 (1997)
12. M. Frommberger, C. Schmutz, M. Tewes, J. McCord, W. Hartung, R. Losehand, E. Quandt, Integration of crossed anisotropy magnetic core into toroidal thin-film inductors. *IEEE Trans. Microwave Theory Tech.* ***53***(6), 2096–2100 (2005)

13. H. Kijima, S. Ohnuma, H. Masumoto, High-frequency soft magnetic properties of isotropic Co-Al-N films. *IEEE Trans. Magn.* ***47***(10), 3928–3931 (2011)
14. H. Kijima, Y. Zhang, N. Kobayashi, S. Ohnuma, H. Masumoto, High frequency soft magnetic performance on magnetically isotropic Co-Al-N films in external bias field. *IEEE Trans. Magn.* ***48***(11), 2910–2913 (2012)
15. H. Kijima, S. Ohnuma, H. Masumoto, Effect of perpendicular magnetic anisotropy on the high frequency soft magnetic properties of Co-Al-N films. *J. Magn. Soc. Jpn.* ***36***(4), 287–292 (2012)
16. H. Kijima-Aoki, S. Takeda, S. Ohnuma, H. Masumoto, High-frequency soft magnetic properties of Co–SiO 2 nanogranular films with large out-of-plane magnetic anisotropy. *IEEE Magn. Lett.* ***9***, 1–5 (2018)
17. H. Karamon, T. Masumoto, Y. Makino, Magnetic and electrical properties of Fe-B-N amorphous films. *J. Appl. Phys.* ***57***(8), 3527–3532 (1985)
18. N. Kobayashi, S. Ohnuma, T. Masumoto, H. Fujimori, Tunnel-type magnetoresistance in matel-nonmetal granular filmes prepared by tandem deposition method. *J. Magn. Soc. Jpn.* ***23***(1_2), 76–78 (1999)
19. D. Kumar, A. Gupta, Evolution of structural and magnetic properties of sputtered nanocrystalline Co thin films with thermal annealing. *J. Magn. Magn. Mater.* ***308***(2), 318–324 (2007)
20. B.X. Gu, H. Wang, Structure and magnetic properties of sputtered FCC Co (1 1 1) films grown on a glass substrate. *J. Magn. Magn. Mater.* ***187***(1), 47–50 (1998)
21. O. Kitakami, H. Sato, Y. Shimada, F. Sato, M. Tanaka, Size effect on the crystal phase of cobalt fine particles. *Phys. Rev. B* ***56***(21), 13849 (1997)
22. S. Ohnuma, H. Fujimori, S. Mitani, T. Masumoto, High-frequency magnetic properties in metal–nonmetal granular films. *J. Appl. Phys.* ***79***(8), 5130–5135 (1996)
23. K. Oikawa, G.W. Qin, M. Sato, O. Kitakami, Y. Shimada, J. Sato, K. Fukamichi, K. Ishida, Magnetically induced phase separation and magnetic properties of Co–Mo hexagonal-close-packed structure thin films. *Appl. Phys. Lett.* ***83***(5), 966–968 (2003)
24. J.J. Wang, Y. Tan, C.M. Liu, O. Kitakami, Crystal structures and magnetic properties of epitaxial Co–W perpendicular films. *J. Magn. Magn. Mater.* ***334***, 119–123 (2013)
25. H. Uwazumi, K. Enomoto, Y. Sakai, S. Takenoiri, T. Oikawa, S. Watanabe, CoPtCr–SiO_2 granular media for high-density perpendicular recording. *IEEE Trans. Magn.* **39**, 1914 (2003)
26. M. Hansen, K. Anderko, in *Metallurgy and Metallurgical Engineering Series*. (Mc Graw-Hil Book Company, 1958), p. 51, New York, USA
27. H. Fujimori, S. Mitani, S. Ohnuma, Tunnel-type GMR in Co-Al-O insulated granular system—Its oxygen-concentration dependence. *J. Magn. Magn. Mater.* ***156***(1–3), 311–314 (1996)
28. W.H. Strehlow, E.L. Cook, Compilation of energy band gaps in elemental and binary compound semiconductors and insulators. *J. Phys. Chem. Ref. Data* ***2***(1), 163–200 (1973)
29. H. Fujimori, S. Mitani, S. Ohnuma, Tunnel-type GMR in metal-nonmetal granular alloy thin films. *Mater. Sci. Eng. B* ***31***(1–2), 219–223 (1995)
30. M. Ohnuma, K. Hono, H. Onodera, S. Ohnuma, H. Fujimori, J.S. Pedersen, Microstructures and magnetic properties of Co–Al–O granular thin films. *J. Appl. Phys.* ***87***(2), 817–823 (2000)
31. S. Flohrer, G. Herzer, Random and uniform anisotropy in soft magnetic nanocrystalline alloys. *J. Magn. Magn. Mater.* ***322***(9–12), 1511–1514 (2010)
32. T. Yamaoka, K. Watanabe, Y. Shirakawabe, K. Chinone, E. Saitoh, M. Tanaka, H. Miyajima, Applications of high-resolution MFM system with low-moment probe in a vacuum. *IEEE Trans. Magn.* **41**(10), 3733–3735 (2005)
33. N. Saito, H. Fujiwara, Y. Sugita, A new type of magnetic domain structure in negative magnetostriction Ni-Fe films. *J. Phys. Soc. Jpn.* **19**(7), 1116–1125 (1964)
34. C. Bathany, M. Le Romancer, J.N. Armstrong, H.D. Chopra, Morphogenesis of maze-like magnetic domains. *Phys. Rev. B* ***82***(18), 184411 (2010)
35. S. Takeda, T. Hotchi, S. Motomura, H. Suzuki, *J. Magn. Soc. Jpn.* ***39***, 227–231 (2015)
36. S. Takeda, H. Kijima-Aoki, H. Masumoto, H. Suzuki, Permeability measurement up to 30 GHz of a magnetically isotropic thin film using a short-circuited coaxial line. *J. Magn. Soc. Jpn.* ***43***(5), 91–98 (2019)

37. N. Vukadinovic, M. Labrune, D. Pain, Magnetic excitations in a weak-stripe-domain structure: A 2D dynamic micromagnetic approach. *Phys. Rev. Lett.* ***85***(13), 2817 (2000)
38. J. Smit H.P.J. Wijn, Ferrites, in *Philips Technical Library* (1959), p. 83, N.V. Philips Gloeilampenfabrieken, Eindhoven, The Netherlands
39. H.J. Kwon, J.Y. Shin, J.H. Oh, The microwave absorbing and resonance phenomena of Y-type hexagonal ferrite microwave absorbers. *J. Appl. Phys.* **75**, 6109 (1994)
40. Y. Shimada, J. Ma, T. Ito, K. Yanagi, Y. Endo, S. Muroga, M. Yamaguchi, Performance of crossed anisotropy multilayered CoZrNb films as IC chip level electromagnetic noise suppressor. *IEEE Trans. Magn.* ***50***(11), 1–4 (2014)
41. H. Kijima, S. Ohnuma, H. Masumoto, Y. Shimada, Y. Endo, M. Yamaguchi, High noise suppression using magnetically isotropic (CoFe-AlN)/(AlN) multilayer films. *J. Appl. Phys.* ***117***(17), 17E514 (2015)
42. S. Muroga, Y. Endo, Y. Mitsuzuka, Y. Shimada, M. Yamaguchi, Estimation of peak frequency of loss in noise suppressor using demagnetizing factor. *IEEE Trans. Magn.* **47**, 300–303 (2010)

Theoretical and Experimental Surveys of Doped Thermoelectric Na_xCoO_2

M. Hussein N. Assadi

1 Introduction

Sodium cobaltate (Na_xCoO_2), is a dopable oxide which is stable for a wide range of Na concentrations $0.3 \leq x \leq 1.0$ [1]. The magnetic, structural and thermoelectric characteristics of Na_xCoO_2 depend critically on x, resulting in a rich magnetic and structural phase diagrams [2–4], including a relatively high thermoelectric figure of merit (ZT) at temperatures around 800 K [5]. As shown in Fig. 1, the Na_xCoO_2 lattice comprises of stacked alternating Na and edge-sharing CoO_2 octahedral layers that constitute a quasi-two-dimensional system. In Na-deficient ($x < 1$) Na_xCoO_2, each Na vacancy introduces a hole that oxidises a Co^{3+} ion to Co^{4+}. As a result, Na_xCoO_2 exhibits p-type conductivity and its carrier concentration (n) is inversely related to x. Moreover, the coexistence of Co^{3+} and Co^{4+} ions in the CoO_2 layers generates a competitive Seebeck potential through spin entropy flow [6, 7]. For instance, the Seebeck coefficient (S) of polycrystalline $Na_{0.75}CoO_2$ at 800 K is within the range of ~132 – ~143 μV K^{-1}, depending on the preparation protocol [8–10]. Furthermore, in the Na layers, the mobile Na^+ ions for which their motion is schematically shown with arrows in Fig. 2a, diffusively scatter the phonons [11, 12], decreasing the lattice thermal conductivity (κ_L) [1, 13–16] to ~0.01 W cm^{-1} K^{-1} at ~1000 K [17]. ZnO, for comparison, has a κ_L value of ~1.25 W cm^{-1} K^{-1} at ~1000 K [18].

As shown in Fig. 1, the lattice structure of the fully sodiated Na_1CoO_2 has hexagonal symmetry with space group number 194. The primitive cell of the Na_1CoO_2 has a chemical composition of $Na_2Co_2O_4$, in which the ions are arranged in two alternating Na and CoO_2 layers. The close-packed oxygen ions stack as

M. H. N. Assadi (✉)
School of Materials Science and Engineering, UNSW Sydney, Sydney, NSW, Australia
e-mail: h.assadi.2008@ieee.org

A. G. Roca et al. (eds.), *Surfaces and Interfaces of Metal Oxide Thin Films, Multilayers, Nanoparticles and Nano-composites*,
https://doi.org/10.1007/978-3-030-74073-3_13

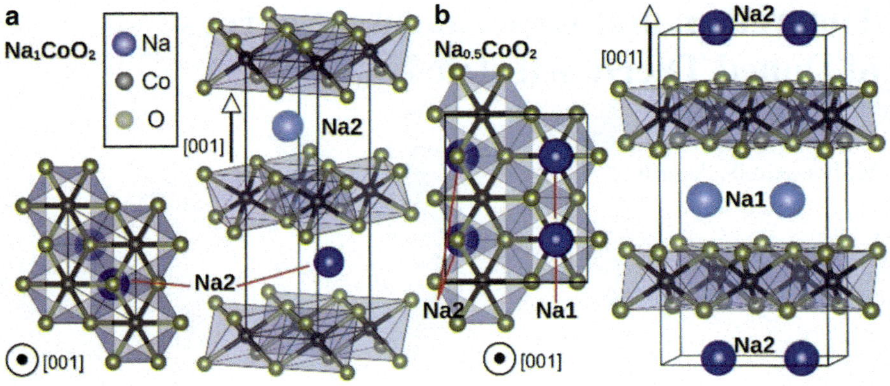

Fig. 1 (**a**) The top and side view of the of $P6_3/mmc$ primitive cell of $P2$ Na_1CoO_2. (**b**) The top and side view of the supercell of $P2$ $Na_{0.5}CoO_2$. Co and O ions occupy the Wyckoff $2a$ and $4f$ sites, respectively. In Na_1CoO_2, all Na ions occupy the Wyckoff $2c$ sites which share basal coordinates with O and commonly referred to as Na2 sites. In Na-deficient compounds, all Na sites are partially occupied. Some Na ions also move to $2b$ sites which share basal coordinates with Co which are also known as Na1 site

ABBA-ABBA, commonly referred to as *P*2 stacking, in which the oxygen ions on each side of a Na layer face each other directly. This stacking creates two types of prismatic sites for the Na ions; Na1 and Na2, defined by the location of the Na ions relative to the O ions (Fig. 1). The oxygen prism around the Na1 site shares a face with a CoO6 octahedron in both direction along *Z*, while the oxygen prism around the Na2 site only shares the edges with adjacent CoO6 octahedra [19]. In Na_1CoO_2, all Na ions occupy the Na2 site. As the Na concentration decreases, more and more Na ions move to the Na1 site. The creeping of Na ions to Na1 sites minimises the electrostatic repulsion among the Na^+ ions in the Na layer. Furthermore, as shown in Table 1, the lattice parameters of Na_xCoO_2 vary with Na concentration. As x decreases, a also decreases, while c generally increases.

There are a few common synthesis techniques for preparing sodium cobaltate. The fully sodiated compound Na_1CoO_2 can be prepared by dissolving stoichiometric quantities of sodium acetate CH_3COONa and cobalt tartrate $C_4H_4O_6Co$ in ethanol with a gelling agent, then drying and calcinating the gel, followed by annealing it at 650 °C [23]. Polycrystalline samples of Na-deficient Na_xCoO_2 ($x < 1$) can be prepared by employing the conventional solid-state reaction technique. In this method, a powder mixture of Na_2CO_3 and Co_3O_4 with the desired stoichiometric molar ratios are directly placed into a furnace for preheating at 850 °C to minimise Na evaporation and loss, followed by sintering for 16 h in air. The resulting powder is then pulverised, pelletised, and sintered again at 870 °C for 20 h [24]. Alternatively, the Co_3O_4 and Na_2CO_3 powders can be fabricated into ceramic pellets by the spark plasma sintering method. In this method, the powder usually placed in a graphite die and pressed at 50 MPa at 700 °C for 5 min. Spark plasma sintering produces highly dense and highly textured ceramic samples with nanotextures that are beneficial for

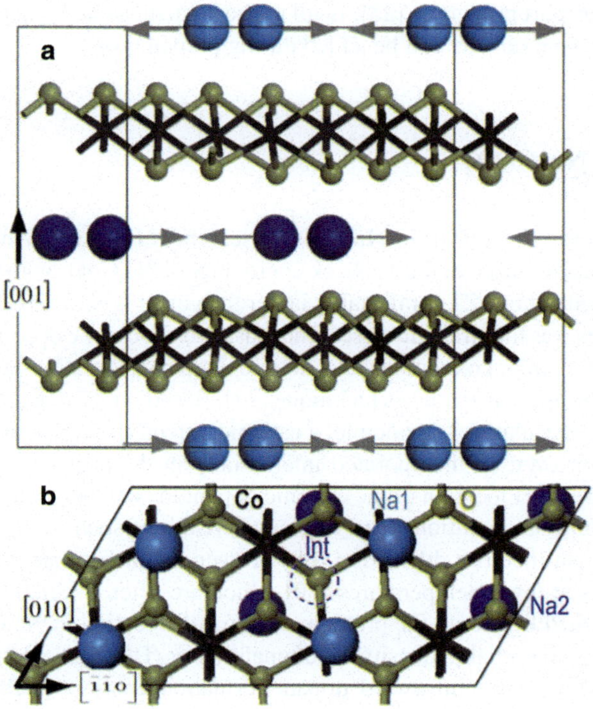

Fig. 2 A side (upper panels) and top (lower panels) preview of the $Na_{0.50}CoO_2$ structure constructed by using a $4a \times 2a \times 2c$ Na_1CoO_2 supercell as a representative of the Na_xCoO_2 structures. In $Na_{0.50}CoO_2$ compound, one-half of the Na ions occupy $2b$ (Na1) sites which share basal coordinates with Co, and the other half of the Na ions occupy $2c$ (Na2) sites which share basal coordinates with O. The ratio of Na1/Na2 and the arrangement of the occupied Na1 and Na2 sites depend on Na concentration [39]. Adapted with permission from Ref. [40]. Copyright Elsevier 2015

Table 1 The measured and calculated structural properties of Na_xCoO_2

Compound	Na_1CoO_2	$Na_{0.75}CoO_2$	$Na_{0.5}CoO_2$
a (Å) (Theoretical)	2.906[a]	2.890[a]	2.834[a]
c (Å) (Theoretical)	10.566[a]	10.899[a]	11.171[a]
a (Å) (experimental)	2.8829[b]	2.8329[c]	2.81508[b]
c (Å) (experimental)	10.4927[b]	10.990[c]	11.1296[b]
Unit cell dimension	$1a \times 1a \times 1c$	$1a \times \sqrt{3}a \times 1c$	$1a \times 2\sqrt{3}a \times 1c$
Na1/Na2	0[d,e]	0.5[d,e]	1[d,e]

[a][20]
[b][21]
[c][9]
[d][22]
[e][19]

the carrier transport properties [25]. In all of these methods, a dopant in the form of oxide or carbonate powder can be added during preparations.

2 Doped Na_xCoO_2

Given the promising low κ_L in Na_xCoO_2, researchers have extensively examined various dopants to improve the Seebeck coefficient of Na_xCoO_2 with the unfulfilled ambition of achieving a *ZT* comfortably larger than one [8, 9, 17, 26–36]. One of the main obstacles was the strategies researchers used to choose the dopants, e.g. mainly based on mere nominal oxidation state; atomic mass considerations and the solubility limits of the applied synthesis technique [37]. Based on these factors, the doping site is usually speculated. In principle, a cationic dopant can replace either a Na ion or a Co ion, or even be incorporated interstitially in the host lattice of Na_xCoO_2. Identifying the exact location of the cationic dopant in Na_xCoO_2's host lattice under a specific synthesis condition requires characterisations such as X-ray absorption spectroscopy and neutron diffraction [38] or even supplementary magnetic phase measurements at lower temperatures [30] which are often omitted. An alternative tool that can shed light onto the problem of dopant's location in Na_xCoO_2, albeit at a lower cost, is the reliable density functional theory (DFT) calculations. DFT can provide us with a straightforward dopant formation energy which indicates the energy required to place a dopant in the host lattice. By comparing the formation energies of a specific dopant at different lattice sites, called here *configurations*, one can identify the most stable dopant's location in the host lattice of Na_xCoO_2. In this section, we first survey the available experimental reports on the dopants' effect on the thermoelectric performance of Na_xCoO_2. We then briefly survey the theoretical works in this field. By comparing these two lines of inquery, we then recommend precise strategies for future research in Na_xCoO_2 based thermoelectric materials.

Figure 3a shows the Seebeck coefficient (*S*) while Fig. 3b shows the electrical resistivity (ρ) at $T = {\sim}800$ K for Na_xCoO_2 doped with a wide variety of dopants as reported in the literature. In Fig. 3, each circle represents an experimental report for a specific dopant. The circles are arranged based on the dopant's atomic number. The colour of each circle represents the dopant concentration while the size of the circle represents the sodium concentration in the host Na_xCoO_2 lattice. After a close inspection, we notice that the largest *S* is obtained for late transition elements dopants like Cu and Ni for which the Seebeck coefficient exceeds ~320 $\mu V\ K^{-1}$, followed by noble metals dopants such as Ag and Au for which the $S = {\sim}200\ \mu V\ K^{-1}$ and the rare-earth dopants such as Dy and Sm for which the $S = {\sim}180\ \mu V\ K^{-1}$. Furthermore, large *S* values are achieved when the Na concentration is 50% as higher Na concentrations ($x > 0.5$) persistently produce lower *S* values. Finally, the lowest ρ is achieved for dopant concentrations of ~2.5% or lower as larger dopant concentrations increase the resistivity (ρ), occasionally by orders of magnitude.

To obtain an atomistic understanding based on these experimental results, we cross-examined the formation energy of a wide range of dopants calculated with

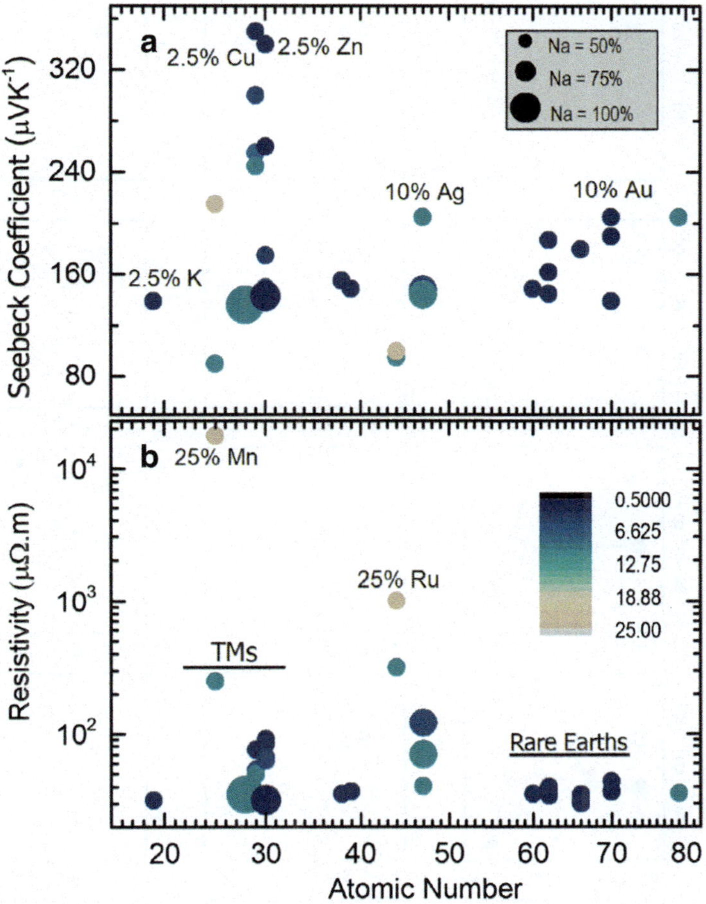

Fig. 3 The Seebeck coefficient (S) and the resistivity (ρ) of doped Na_xCoO_2 as reported in the literature at $T = \sim 800$ K presented in (**a**) and (**b**), respectively. The radius of the circles proportionally correlates to the Na concentration as demonstrated in the legend of (**a**). The shade of the circles represents the concentration of the dopants as expressed in the legend of (**b**). The data are taken from the following articles: K [17], Mn [30], Ni [31]. Cu [32], Zn [8, 33], Sr and Y [17], Ru [30], Ag [9, 27], Nd [17], Sm [17, 26], Dy [26], Yb [17, 26], Au [27]. (Reproduced with permission from Ref. [40]. Copyright 2015 Elsevier)

DFT presented in Fig. 4. Each panel in Fig. 4 shows the formation energy (E^f) of a dopant as a function of the incorporation sites, or configurations *Na1*, *Na2*, *Int* and *Co*. The first two configurations represent a dopant replacing an Na ion from Na1 and Na2 sites, respectively (for details see the caption of Fig. 2). Subscript Int refers to a configuration in which the dopant is incorporated interstitially in the Na later, and subscript Co refers to configurations in which the dopant replaces a Co ion from the Na_xCoO_2 host lattice. The dopant formation energy (E^f) was calculated based on the standard formula [41]:

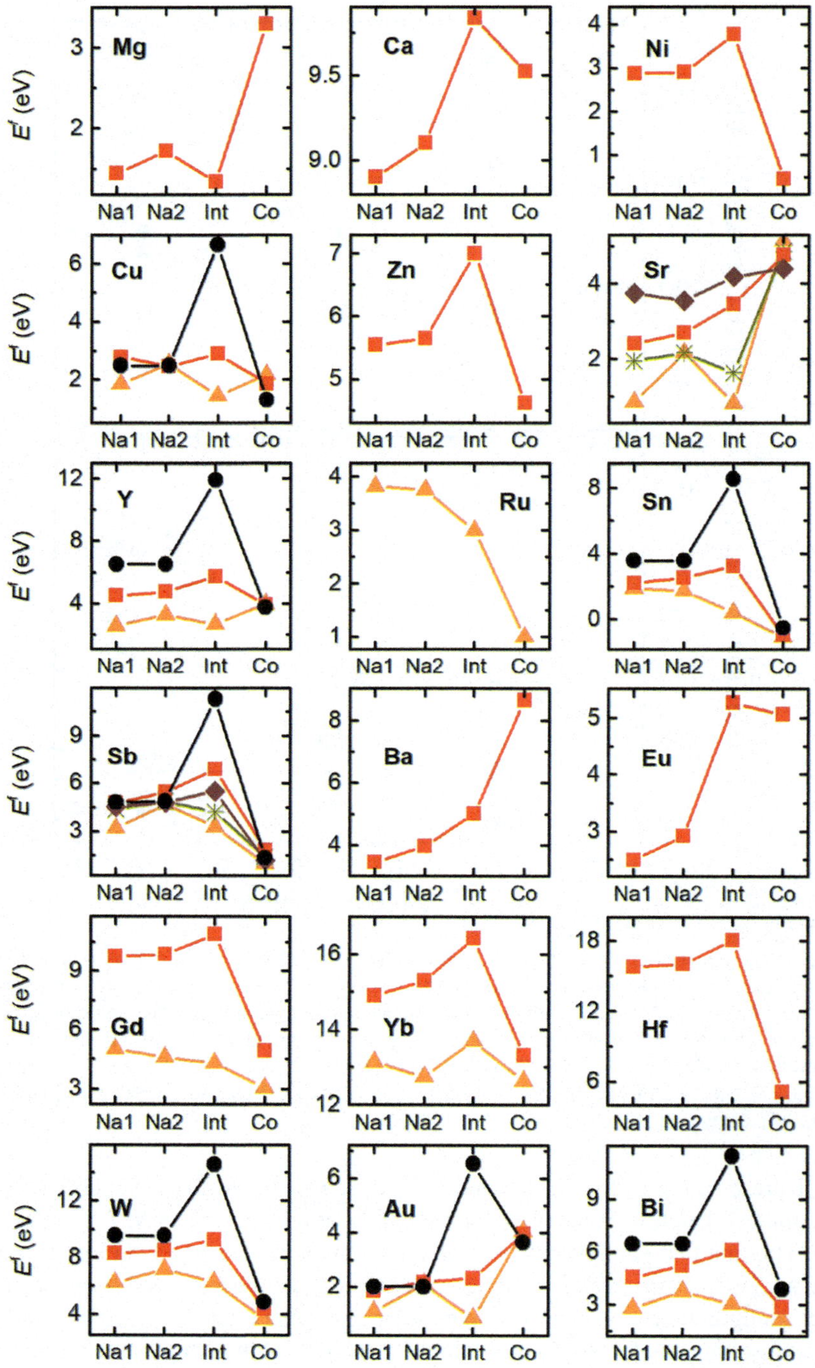

Fig. 4 The formation energy of various dopants in Na_xCoO_2 calculated for an oxygen-rich environment. The upright triangles in orange, the star signs in dark yellow, the solid squares in red, the diamonds in brown and the circles in black symbols correspond to x values of 0.50, 0.625, 0.75, 0.875 and 1.00, respectively in Na_xCO_2. The values were copied from the following articles: Mg [24], Ca and Ba [43], Ni, Zn, Eu [44], Cu, Y, Sn, W, Au, Bi [40], Ru [45], Sr [46], Sb [20, 46, 47], Gd, Yb [48], Hf [49]

$$E^f = E^t\left(\mathrm{Na}_x\mathrm{CoO}_2 : \mathrm{M}\right) + \mu_\alpha - E^t\left(\mathrm{Na}_x\mathrm{CoO}_2\right) - \mu_\mathrm{M}, \tag{1}$$

in which, $E^t(Na_xCoO_2{:}M)$ is the total DFT energy of the Na_xCoO_2 supercell containing dopant M, and $E^t(Na_xCoO_2)$ is the total energy of the pristine Na_xCoO_2 supercell. μ_α and μ_M are the chemical potentials of the removed and added elements, respectively. The chemical potentials used in the calculations presented in Fig. 4 pertain to the oxygen-rich environment [42], which is the case for most synthesis techniques such as powder sintering in air. In all presented E^fs, one dopant was placed in the $4a \times 2a \times 2c$ Na_xCoO_2 supercell, resulting in a dopant concentration of ~6.25%. More details regarding the DFT settings of each calculation are provided in the cited literature for each dopant.

By examining Fig. 4, we first notice that dopants' radius is not a good predictor of its incorporation site. For example, Mg^{2+} has a radius of 0.57 Å close to that of Co^{3+} (0.55 Å) in octahedral coordination. Mg, however, is more stable at Na layer as an interstitial dopant. The radius of Gd^{3+},0.94 Å, on the other hand, is closer to that of Na^+ (1.02 Å), but it is more stable when replacing Co for a wide range of Na concentrations. Second, for the dopants that we have formation energy for multiple x values, we notice that some dopants such as Sr and Au tend to be stable in the Na layer (either substitutionally or interstitially) for a wide range of Na concentrations (x). In contrast, some other dopants like Bi, W, Sb and Sn, Gd and Y are more stable when substituting for Co. The incorporation site of dopants such as Cu and Y, nonetheless, depends strongly on Na concentration (x). Cu dopants are the most stable in Cu_{Co} configuration in Na_1CoO_2 and $Na_{0.75}CoO_2$, and in Cu_{Int} configuration in $Na_{0.50}CoO_2$. Y dopants follow the same trend. Furthermore, dopant like Mg, Ca, Ba and Eu are more stable in the Na layer in $Na_{0.75}CoO_2$, while Ni, Zn, Ru and Hf are more stable when replacing a Co in $Na_{0.75}CoO_2$. Similarly, Ru is also stable when replacing Co in $Na_{0.50}CoO_2$. In the case of Mg_{Na}, we could directly verify Mg's local chemical environment using Raman spectroscopy validating the theoretical results [24], confirming the DFT prediction.

By comparing the experimental results with the theoretical calculations, we can identify some trends correlating those dopants that produce high S values with their location in the lattice site predicted by density functional calculations. The first category is those dopant residing in the Na layer: configurations Na1, Na2 and Int. This category includes Mg [24], Ca [50], Cu [32], and Au [27]. The second category is made of those dopants that replace a Co and have an oxidation state of +4 such as Ru [51], Ni [31]. Now let's examine the how these two categories of dopants enhance the thermoelectric performance of Na_xCoO_2 in terms of Seebeck coefficient (S), carrier concentration (n), resistivity (ρ) and power factor ($PF = S^2/\rho$).

The high-temperature Seebeck coefficient (S) in Na_xCoO_2 can be explained by Koshibae's equation [52, 53], which is based on a modified form of Heikes formula [54, 55], which itself is obtained by solving a Hubbard model at an infinite temperature [56]:

$$S(T\to\infty)=-\frac{k_B}{e}\ln\left[\frac{g(\mathrm{Co}^{4+})}{g(\mathrm{Co}^{3+})}\frac{n_{\mathrm{Co}^{4+}}}{n_{\mathrm{Co}^{3+}}}\right]. \tag{2}$$

Here k_B is the Boltzmann constant, e is the electronic unit charge, g is the different possible ways in which electrons can be arranged in the orbitals of Co^{3+} and Co^{4+} ions, and n is the concentration of a given Co species. g is the product of spin degeneracy (g_s) and orbital (g_o) degeneracy: $g = g_s \bullet g_o$, in which $g_s = 2\zeta + 1$ where ζ is the ions' total spin number and g_o is the number of valid permutations for distributing the electrons across its orbitals:

$$g_o=\frac{3!}{n^{\uparrow}!\left(3-n^{\uparrow}\right)!}\bullet\frac{3!}{n^{\downarrow}!\left(3-n^{\downarrow}\right)!}, \tag{3}$$

in which $n^{\uparrow}$ and $n_{\downarrow}$ are the number of spin-up and spin-down electrons, respectively. Equation 3 is valid for those transition metal ions under octahedral coordination in which only t_{2g} band is either fully or partially filled such as Co^{3+} and Co^{4+}. Figure 5 schematically shows how electronic conduction through hopping in Na_xCoO_2 results in a spin entropy flow in the opposite direction.

Assuming that Co ions are in low spin state in Na_xCoO_2 ($\zeta = 0$ for Co^{3+} and $\zeta = ½$ for Co^{4+}), we obtain $g(Co^{4+}) = 6$ and $g(Co^{3+}) = 1$. By substituting these values in Koshibae's formula for S, we obtain values of $S = 249\ \mu V\ K^{-1}$ for $Na_{0.75}CoO_2$ and

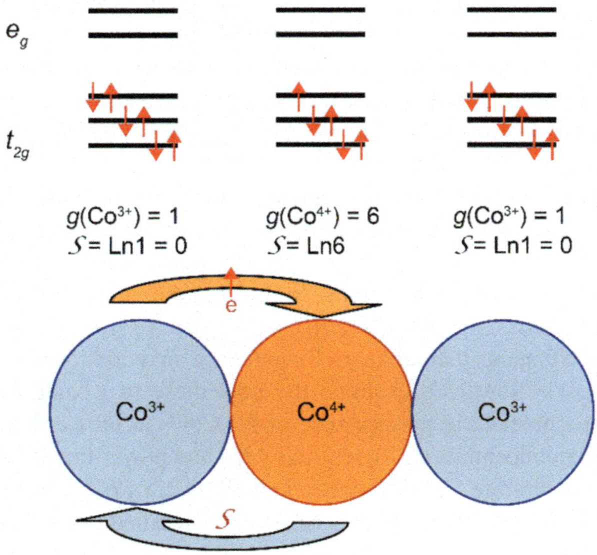

Fig. 5 The schematics of spin entropy flow in Na_xCoO_2. S and g are the entropy and the electronic degeneracy per site, respectively. (Reproduced with permission from Ref. [48]. Copyright 2020 Institute of Physics)

$S = 154\ \mu V\ K^{-1}$ for $Na_{0.50}CoO_2$. Note the S has a reverse relationship with the concentration of Co^{4+}. Moreover, we can approximate the resistivity by assuming that the carrier mobility (μ) remains $1.0\ cm^2\ V^{-1}\ s^{-1}$ for doped Na_xCoO_2 [57] according to $\rho = (e \bullet n \bullet \mu)^{-1}$. Here, n is carrier concentration and equals to the concentration of Co^{4+}, as there is a hole for every Co^{4+} in Na_xCoO_2. The constant mobility approximation in Na_xCoO_2 is conservative as heavier and more positively charged dopants in the Na layer generally improve carrier mobility modestly [40]. This approximation gives a resistivity value of $9.706 \times 10^{-4}\ \Omega$ cm for $Na_{0.75}CoO_2$ and $4.853 \times 10^{-4}\ \Omega$ cm for $Na_{0.50}CoO_2$. The power factor, which is a parabolic function of S and an inverse function of ρ (PF = S^2/ρ), comes out as 6.392 mW $m^{-1}\ K^{-2}$ for $Na_{0.75}CoO_2$ and 4.912 mW $m^{-1}\ K^{-2}$ in $Na_{0.50}CoO_2$. These values are well in agreement with measurements in single-crystal [58, 59] and epitaxial thin film Na_xCoO_2 [60].

3 Dopants at Na Site

When a cationic dopant is located in the Na layer, it introduces one or several electrons that partially recombine with the holes of the host Na_xCoO_2, reducing the concentration of Co^{4+} and the carrier concentration (n). For instance, a dopant with +3 oxidation state that replaces an Na, such as Eu^{3+}_{Na1}, introduces two new electrons. Likewise, a dopant with +2 oxidation state that is interstitially incorporated in the Na layer, such as Mg^{2+}_{Int}, also introduces two electrons. As marked with red and blue arrows in Fig. 6a, the newly introduced electrons (and the consequent reduction in Co^{4+} concentration) increase S to $322\ \mu V\ K^{-1}$ in electron-doped $Na_{0.75}CoO_2$, and to $198\ \mu V\ K^{-1}$ in electron-doped $Na_{0.50}CoO_2$. The reduction in Co^{4+} concentration also decreases the carrier concentration (n) and, therefore, increases the resistivity (ρ). As marked with red and blue arrows in Fig. 6b, in $Na_{0.75}CoO_2$ and $Na_{0.50}CoO_2$ doped with two electrons, ρ increases to $1.941 \times 10^{-3}\ \Omega$ cm and $6.471 \times 10^{-4}\ \Omega$ cm, respectively. The power factor for $Na_{0.75}CoO_2$ and $Na_{0.50}CoO_2$, as a result, comes to 5.344 mW $m^{-1}\ K^{-2}$ and 6.085 mW $m^{-1}\ K^{-2}$, respectively. In the case of three electrons doping, such as Y^{3+}_{Int}, the changes in S, ρ, follow a similar trend, though more drastic; S rises to $388\ \mu V\ K^{-1}$ in $Na_{0.75}CoO_2$ and to $222\ \mu V\ K^{-1}$ in $Na_{0.50}CoO_2$; ρ rises to $3.882 \times 10^{-3}\ \Omega$ cm in $Na_{0.75}CoO_2$ and to $7.765 \times 10^{-4}\ \Omega$ cm in $Na_{0.5}CoO_2$. PF, consequently, becomes 3.873 mW $m^{-1}\ K^{-2}$ in $Na_{0.75}CoO_2$ and 6.367 mW $m^{-1}\ K^{-2}$ in $Na_{0.50}CoO_2$. We see that electron doping via dopants that are located in Na layer increases both S and ρ of which the first is the desired outcome but the second is not. These two changes oppose one another in determining PF; only in $Na_{0.50}CoO_2$, electron doping improves the overall PF. See Tables 2 and 3 for more details.

Fig. 6 (**a**) Seebeck coefficient (*S*) based on Koshibae's equation, (**b**) resistivity (*ρ*) and (**c**) the power factor (PF) in Na_xCoO_2 as a function of Co^{4+} concentration. Green and pink arrows indicate the change in *S* for cobalt site doping. The blue and red arrows indicate the change in *S*, *ρ* and PF for Na site doping. (Partially adapted from Ref. [48]. Copyright 2020 Institute of Physics)

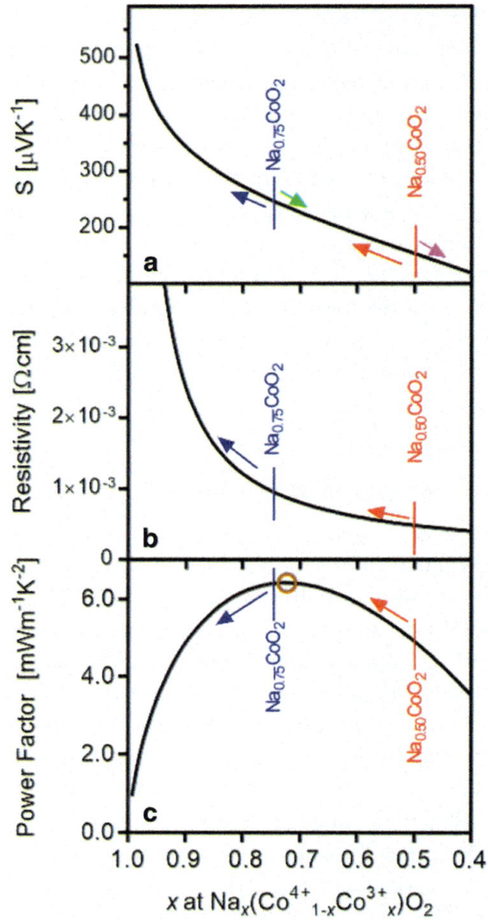

Table 2 Tabulated transport properties calculated for electron-doped $Na_{0.75}CoO_2$ at a dopant concentration of M/Co = ~6.25%

Number of doped electrons	Example	n [cm^{-3}]	ρ [Ω cm]	S [μV K^{-1}]	PF [mW m^{-1} K^{-2}]
0	Pristine	6.431×10^{21}	9.706×10^{-4}	249	6.392
1	Cu^{2+}_{Na1}	4.823×10^{21}	1.294×10^{-3}	280	6.091
2	Mg^{2+}_{Int}	3.215×10^{21}	1.941×10^{-3}	322	5.344
3	Gd^{3+}_{Int}	1.608×10^{21}	3.882×10^{-3}	388	3.873

4 Dopants at Co Site

A dopant with oxidation state +3 substituting for Co^{3+}, such as Gd^{3+}_{Co} does not change the carrier (hole) concentration as the Co^{4+} ions remain unaltered. However, such dopants change the dynamics of spin entropy flow. The electric conduction and the

Table 3 Tabulated transport properties calculated for electron-doped $Na_{0.50}CoO_2$ at a dopant concentration of M/Co = ~6.25%

Number of doped electrons	Example	n [cm^{-3}]	ρ [Ω cm]	S [μV K^{-1}]	PF [mW m^{-1} K^{-2}]
0	Pristine	1.286×10^{22}	4.853×10^{-4}	154	4.912
1	Sr^{2+}_{Na1}	1.125×10^{22}	5.546×10^{-4}	176	5.589
2	Cu^{2+}_{Int}	9.646×10^{21}	6.471×10^{-4}	198	6.085
3	Y^{3+}_{Int}	8.038×10^{21}	7.765×10^{-4}	222	6.367

resulting spin entropy flow is facilitated by the electrons hopping from a full t^6_{2g} states of a Co^{3+} ion to a singly vacant t^5_{2g} states of a Co^{4+}. A dopant like Gd^{3+}_{Co}, in practice, reduces the Co^{3+} sites available for conduction, increasing the effective concentration of Co^{4+} among the available sites. Taking the reduced number of Co^{3+} sites into account, as marked with green and pink arrows in Fig. 6a, according to Koshibae's equation, S is reduced to 241 μV K^{-1} for $Na_{0.75}CoO_2$ and to 142 μV K^{-1} for $Na_{0.50}CoO_2$. The carrier concentration, on the other hand, is not altered by Co side doping as the concentration of hole bearing Co^{4+} does not change by isovalent doping at Co^{3+} site. The net result of such Co^{3+} site doping is a decrease in the power factor to 6.013 in $Na_{0.75}CoO_2$ mW m^{-1} K^{-2} and to 4.207 mW m^{-1} K^{-2} in $Na_{0.50}CoO_2$. Based on the trend shown in Fig. 6b, doing cationic dopants with +3 oxidation state reduces S and PF for all possible x values and is detrimental to the thermoelectric performance of Na_xCoO_2.

A dopant with oxidation state +4 substituting for Co^{4+}, such as Ru^{4+}_{Co} and Ni^{4+}_{Co}, reduces the number of the Co^{4+} sites available for spin entropy flow, reducing the hole concentration as the holes initially borne on Co^{4+} are now borne on the dopant instead and may not be readily mobile within the Co ions' network. According to eq. (2), the decrease in the concentration of Co^{4+} increases S to 274 μV K^{-1} in $Na_{0.75}CoO_2$ and to 165 μV K^{-1} in $Na_{0.50}CoO_2$. Accordingly, resistivity rises to 1.294×10^{-3} Ω cm in $Na_{0.75}CoO_2$ and to 5.546×10^{-4} Ω cm in $Na_{0.50}CoO_2$. Given the amount of the rise in both S and ρ, the PF comes to 5.796 mW m^{-1} K^{-2} in $Na_{0.75}CoO_2$ and to 4.963 mW m^{-1} K^{-2} in $Na_{0.50}CoO_2$. Similar to the case of electron doping in the Na layer, here, we see a minor enhancement in the PF of $Na_{0.5}CoO_2$ but a reduction of PF in $Na_{0.75}CoO_2$ with respect to the values of the undoped compounds shown in the second rows of Tables 2 and 3.

5 Summary

We finalise this chapter by summarising the presented arguments in the following points: (a) Common guesswork based on the dopant's radius or atomic number in determining the doping site in Na_xCoO_2 is not reliable. More stringent characterisations or complementary density functional calculations are required; (b) Cationic doping at Na later, generally, introduces extra electrons that recombine with holes of Na_xCoO_2. This decrease in itinerant carriers increases the Seebeck coefficient and the resistivity. These two effects oppose each other in achieving the ideal thermoelectric material. Only in $Na_{0.50}CoO_2$ the increase in the Seebeck coefficient is significant enough to negate the accompanying increase in resistivity; (c) For dopants replacing Co, only those with an oxidation state of +4 improve the Seebeck coefficient. In this case, the increase in the resistivity is less prominent than the case of Na layer doping. The power factor, nonetheless, still only improves in doped $Na_{0.50}CoO_2$; (d) As marked with a brown circle in Fig. 6c, the power factor of the pristine $Na_{0.75}CoO_2$ is already near the maximum theoretical value attainable in Na_xCoO_2. As a result, doping, and therefore any change in carrier concentrations, tends to move S, ρ and PF away to less desirable values. Carrier doping, thus, seems to be a viable strategy only in $Na_{0.50}CoO_2$. Finally, we would like to draw the attention of the readers that the DFT results presented here were valid under oxygen-rich environment. In some cases, like Gd^{3+} and Yb^{3+}, the dopants are more stable in the Na layer instead of replacing a Co under oxygen-poor environment [48]. As a consequence, for these dopants, by controlling the synthesis environment, one can change its role from detrimental to favourable in the thermoelectric performance of Na_xCoO_2.

References

1. M. Roger, D.J.P. Morris, D.A. Tennant, M.J. Gutmann, J.P. Goff, J.U. Hoffmann, et al., Patterning of sodium ions and the control of electrons in sodium cobaltate. Nature **445**(7128), 631–634 (2007). https://doi.org/10.1038/nature05531
2. P. Mendels, D. Bono, J. Bobroff, G. Collin, D. Colson, N. Blanchard, et al., Cascade of bulk magnetic phase transitions in Na_xCoO_2 as studied by muon spin rotation. Phys. Rev. Lett. **94**(13), 136403 (2005). https://doi.org/10.1103/PhysRevLett.94.136403
3. R. Berthelot, D. Carlier, C. Delmas, Electrochemical investigation of the P2–Na_xCoO_2 phase diagram. Nat. Mater. **10**(1), 74–80 (2011). https://doi.org/10.1038/nmat2920
4. E. Vera, B. Alcántar-Vázquez, Y. Duan, H. Pfeiffer, Bifunctional application of sodium cobaltate as a catalyst and captor through CO oxidation and subsequent CO_2 chemisorption processes. RSC Adv. **6**(3), 2162–2170 (2016). https://doi.org/10.1039/C5RA22749F
5. J.W. Fergus, Oxide materials for high temperature thermoelectric energy conversion. J. Eur. Ceram. Soc. **32**(3), 525–540 (2012). https://doi.org/10.1016/j.jeurceramsoc.2011.10.007
6. I. Terasaki, Y. Sasago, K. Uchinokura, Large thermoelectric power in $NaCo_2O_4$ single crystals. Phys. Rev. B **56**(20), R12685–R126R7 (1997). https://doi.org/10.1103/PhysRevB.56.R12685
7. Y. Wang, N.S. Rogado, R. Cava, N. Ong, Spin entropy as the likely source of enhanced thermopower in Na_xCo2O_4. Nature **423**(6938), 425–428 (2003). https://doi.org/10.1038/nature01639
8. P.H. Tsai, T.S. Zhang, R. Donelson, T.T. Tan, S. Li, Power factor enhancement in Zn-doped $Na_{0.8}CoO_2$. J. Alloy Compd. **509**(16), 5183–5186 (2011). https://doi.org/10.1016/j.jallcom.2011.02.045
9. T. Seetawan, V. Amornkitbamrung, T. Burinprakhon, S. Maensiri, K. Kurosaki, H. Muta, et al., Thermoelectric power and electrical resistivity of Ag-doped $Na_{1.5}Co_2O_4$. J. Alloys Compd. **407**(1), 314–317 (2006). https://doi.org/10.1016/j.jallcom.2005.06.032
10. A. Nag, V. Shubha, Oxide thermoelectric materials: a structure–property relationship. J. Electron. Mater. **43**(4), 962–977 (2014). https://doi.org/10.1007/s11664-014-3024-6
11. M.L. Foo, Y. Wang, S. Watauchi, H. Zandbergen, T. He, R. Cava, et al., Charge ordering, commensurability, and metallicity in the phase diagram of the layered Na_xCoO_2. Phys. Rev. Lett. **92**(24), 247001 (2004). https://doi.org/10.1103/PhysRevLett.92.247001

12. M. Weller, A. Sacchetti, H.R. Ott, K. Mattenberger, B. Batlogg, Melting of the Na layers in solid $Na_{0.8}CoO_2$. Phys. Rev. Lett. **102**(5), 056401 (2009). https://doi.org/10.1103/PhysRevLett.102.056401
13. G. Mahan, J. Sofo, The best thermoelectric. Proc. Natl. Acad. Sci. U. S. A. **93**(15), 7436–7439 (1996). https://doi.org/10.1073/pnas.93.15.7436
14. G.J. Snyder, E.S. Toberer, Complex thermoelectric materials. Nat. Mater. **7**(2), 105–114 (2008). https://doi.org/10.1038/nmat2090
15. K. Koumoto, I. Terasaki, R. Funahashi, Complex oxide materials for potential thermoelectric applications. MRS Bull. **31**(03), 206–210 (2006). https://doi.org/10.1557/mrs2006.46
16. D.J. Voneshen, K. Refson, E. Borissenko, M. Krisch, A. Bosak, A. Piovano, et al., Suppression of thermal conductivity by rattling modes in thermoelectric sodium cobaltate. Nat. Mater. **12**(11), 1028–1032 (2013). https://doi.org/10.1038/nmat3739
17. T. Nagira, M. Ito, S. Katsuyama, K. Majima, H. Nagai, Thermoelectric properties of $(Na_{1-y}M_y)_xCo_2O_4$ (M=K, Sr, Y, Nd, Sm and Yb; y=0.01 similar to 0.35). J. Alloy Compd. **348**(1–2), 263–269 (2003). https://doi.org/10.1016/s0925-8388(02)00799-5
18. U. Ozgur, X. Gu, S. Chevtchenko, J. Spradlin, S.J. Cho, H. Morkoc, et al., Thermal conductivity of bulk ZnO after different thermal treatments. J. Electron. Mater. **35**(4), 550–555 (2006). https://doi.org/10.1007/s11664-006-0098-9
19. Y.S. Meng, Y. Hinuma, G. Ceder, An investigation of the sodium patterning in Na_xCoO_2 (0.5< x< 1) by density functional theory methods. J. Chem. Phys. **128**, 104708 (2008). https://doi.org/10.1063/1.2839292
20. M.H.N. Assadi, S. Li, R.K. Zheng, S.P. Ringer, A.B. Yu, Magnetic, electrochemical and thermoelectric properties of P2-$Na_x(Co_{7/8}Sb_{1/8})O_2$. Chem. Phys. Lett. **687**, 233–237 (2017). https://doi.org/10.1016/j.cplett.2017.09.026
21. Q. Huang, M.L. Foo, R.A. Pascal, J.W. Lynn, B.H. Toby, T. He, et al., Coupling between electronic and structural degrees of freedom in the triangular lattice conductor Na_xCoO_2. Phys. Rev. B **70**(18), 184110 (2004). https://doi.org/10.1103/PhysRevB.70.184110
22. P.H. Zhang, R.B. Capaz, M.L. Cohen, S.G. Louie, Theory of sodium ordering in Na_xCoO_2. Phys. Rev. B **71**(15), 153102 (2005). https://doi.org/10.1103/PhysRevB.71.153102
23. N.K. Samin, R. Roshidah, N. Kamarudin, N. Kamarulzaman, Synthesis and battery studies of sodium cobalt oxides, $NaCoO_2$ cathodes. Adv. Mater. Res. **545**, 185–189 (2012). https://doi.org/10.4028/www.scientific.net/AMR.545.185
24. P.-H. Tsai, M.H.N. Assadi, T. Zhang, C. Ulrich, T.T. Tan, R. Donelson, et al., Immobilisation of Na ions for substantial power factor enhancement: site-specific defect engineering in $Na_{0.8}CoO_2$. J. Phys. Chem. C **116**(6), 4324–4329 (2012). https://doi.org/10.1021/jp209343v
25. W. Zhang, P. Liu, Y. Wang, K. Zhu, G. Tai, J. Liu, et al., Textured Na_xCoO_2 ceramics sintered from hydrothermal platelet nanocrystals: growth mechanism and transport properties. J. Electron. Mater. **47**(7), 4070–4077 (2018). https://doi.org/10.1007/s11664-018-6296-4
26. T. Nagira, M. Ito, S. Hara, Effect of partial substitutions of rare-earth metals for Na-site on the thermoelectric properties of $Na_xCo_2O_4$ prepared by the polymerised complex method. Mater. Trans. **45**(4), 1339–1345 (2004). https://doi.org/10.2320/matertrans.45.1339
27. M. Ito, D. Furumoto, Effects of noble metal addition on microstructure and thermoelectric properties of $Na_xCo_2O_4$. J. Alloy Compd. **450**(1), 494–498 (2008). https://doi.org/10.1016/j.jallcom.2006.11.032
28. E. Ermawan, S. Poertadji, Thermoelectric properties enhancement of $NaCo_2O_4$ through ca atom partial substitution on Na atom. Adv. Mater. Res. **1123**, 140–144 (2015). https://doi.org/10.4028/www.scientific.net/AMR.1123.140
29. P. Mandal, Anomalous transport properties of co-site impurity doped Na_xCoO_2. J. Appl. Phys. **104**(6), 063902 (2008). https://doi.org/10.1063/1.2978212
30. S. Li, R. Funahashi, I. Matsubara, S. Sodeoka, Magnetic and thermoelectric properties of $NaCo_{2-x}M_xO_4$ (M = Mn, Ru). Mater. Res. Bull. **35**(14), 2371–2378 (2000). https://doi.org/10.1016/S0025-5408(00)00441-4

31. L. Wang, M. Wang, D. Zhao, Thermoelectric properties of c-axis oriented Ni-substituted $NaCoO_2$ thermoelectric oxide by the citric acid complex method. J. Alloy Compd. **471**(1), 519–523 (2009). https://doi.org/10.1016/j.jallcom.2008.04.013
32. K. Park, K.U. Jang, H.C. Kwon, J.G. Kim, W.S. Cho, Influence of partial substitution of cu for co on the thermoelectric properties of $NaCo_2O_4$. J. Alloy Compd. **419**(1), 213–219 (2006). https://doi.org/10.1016/j.jallcom.2005.08.081
33. K. Park, J.H. Lee, Enhanced thermoelectric properties of $NaCo_2O_4$ by adding ZnO. Mater. Lett. **62**(15), 2366–2368 (2008). https://doi.org/10.1016/j.matlet.2007.11.090
34. A.I. Klyndyuk, N.S. Krasutskaya, E.A. Chizhova, L.E. Evseeva, S.A. Tanaeva, Synthesis and properties of $Na_{0.55}Co_{0.9}M_{0.1}O_2$ (M = Sc, Ti, Cr–Zn, Mo, W, Pb, bi) solid solutions. Glas. Phys. Chem. **42**(1), 100–107 (2016). https://doi.org/10.1134/s1087659616010053
35. M.O. Erdal, M. Koyuncu, M.L. Aksu, I. Uslu, S. Koçyiğit, Thermoelectric properties of nickel and boron co-substituted $NaCo_2O_4$ prepared by electrospinning technique. Nano. Hybrid. Compos. **19**, 34–45 (2018). https://doi.org/10.4028/www.scientific.net/NHC.19.34
36. D.K. Aswal, R. Basu, A. Singh, Key issues in development of thermoelectric power generators: High figure-of-merit materials and their highly conducting interfaces with metallic interconnects. Energy Convers. Manag. **114**, 50–67 (2016). https://doi.org/10.1016/j.enconman.2016.01.065
37. X. Zhang, L.-D. Zhao, Thermoelectric materials: energy conversion between heat and electricity. J. Mater. **1**(2), 92–105 (2015). https://doi.org/10.1016/j.jmat.2015.01.001
38. J.A. Alonso, M.J. Martínez-Lope, A. Aguadero, L. Daza, Neutron powder diffraction as a characterisation tool of solid oxide fuel cell materials. Prog. Solid State Chem. **36**(1), 134–150 (2008). https://doi.org/10.1016/j.progsolidstchem.2007.03.004
39. M.H.N. Assadi, H. Katayama-Yoshida, Interplay between magnetism and Na concentration in Na_xCoO_2. Funct. Mater. Lett. **08**(03), 1540016 (2015). https://doi.org/10.1142/S1793604715400160
40. M.H.N. Assadi, H. Katayama-Yoshida, Restoration of long range order of Na ions in Na_xCoO_2 at high temperatures by sodium site doping. Comput. Mater. Sci. **109**, 308–311 (2015). https://doi.org/10.1016/j.commatsci.2015.07.043
41. C. Freysoldt, B. Grabowski, T. Hickel, J. Neugebauer, G. Kresse, A. Janotti, et al., First-principles calculations for point defects in solids. Rev. Mod. Phys. **86**(1), 253–305 (2014). https://doi.org/10.1103/RevModPhys.86.253
42. K. Reuter, M. Scheffler, Composition, structure, and stability of RuO_2 (110) as a function of oxygen pressure. Phys. Rev. B **65**(3), 035406 (2001). https://doi.org/10.1103/PhysRevB.65.035406
43. M.H.N. Assadi, H. Katayama-Yoshida, Sodium cobaltate engineered with alkaline earth metal doping for waste energy harvesting; a theoretical study. Energy Procedia **75**, 3259–3264 (2015). https://doi.org/10.1016/j.egypro.2015.07.697
44. M.H.N. Assadi, S. Li, A.B. Yu, Selecting the suitable dopants: Electronic structures of transition metal and rare earth doped thermoelectric sodium cobaltate. RSC Adv. **3**(5), 1442–1449 (2013). https://doi.org/10.1039/c2ra22514j
45. M.H.N. Assadi, H. Katayama-Yoshida, Magnetism and Spin Entropy in Ru Doped $Na_{0.5}CoO_2$. Phys. Chem. Chem. Phys. **19**, 23425–23430 (2017). https://doi.org/10.1039/C7CP03752J
46. M.H.N. Assadi, H. Katayama-Yoshida, Dopant incorporation site in sodium cobaltate's host lattice: a critical factor for thermoelectric performance. J. Phys. Condens. Matter **27**(17), 175504 (2015). https://doi.org/10.1088/0953-8984/27/17/175504
47. M.H.N. Assadi, M. Fronzi, P. Mele, Suppression of magnetism and Seebeck effect in $Na_{0.875}CoO_2$ induced by Sb_{Co} dopants. Mater. Renew Sust. Energy **9**, 5 (2019). https://doi.org/10.1007/s40243-020-0165-9
48. M.H.N. Assadi, Na site doping a pathway for enhanced thermoelectric performance in $Na_{1-x}CoO_2$, The case of Gd and Yb dopants. J. Phys. Condens. Matter. **32**, 125502 (2020). https://doi.org/10.1088/1361-648X/ab5bdb

49. M.H.N. Assadi, Hf doping at Co site for enhancing the thermoelectric performance in layered $Na_{0.75}CoO_2$. Mater Today Proc. (2020. in press). https://doi.org/10.1016/j.matpr.2020.01.349
50. Y. Ono, N. Kato, Y. Miyazaki, T. Kajitani, Transport properties of Ca-doped γ-Na_xCoO_2. J. Ceram Soc. Jpn **112**, S626–S6S8 (2004). https://doi.org/10.14852/jcersjsuppl.112.0.S626.0
51. P. Strobel, H. Muguerra, S. Hébert, E. Pachoud, C. Colin, M.-H. Julien, Effect of ruthenium substitution in layered sodium cobaltate Na_xCoO_2: synthesis, structural and physical properties. J. Solid State Chem. **182**(7), 1872–1878 (2009). https://doi.org/10.1016/j.jssc.2009.04.030
52. I. Terasaki, High-temperature oxide thermoelectrics. J. Appl. Phys. **110**(5), 053705 (2011). https://doi.org/10.1063/1.3626459
53. W. Koshibae, K. Tsutsui, S. Maekawa, Thermopower in cobalt oxides. Phys. Rev. B **62**(11), 6869–6872 (2000). https://doi.org/10.1103/PhysRevB.62.6869
54. R.R. Heikes, R.W. Ure, *Thermoelectricity: Science and Engineering* (Interscience Publishers, New York, 1961)
55. P.M. Chaikin, G. Beni, Thermopower in the correlated hopping regime. Phys. Rev. B **13**(2), 647–651 (1976). https://doi.org/10.1103/PhysRevB.13.647
56. S. Mukerjee, Thermopower of the Hubbard model: Effects of multiple orbitals and magnetic fields in the atomic limit. Phys. Rev. B **72**(19), 195109 (2005). https://doi.org/10.1103/PhysRevB.72.195109
57. P. Brinks, G. Rijnders, M. Huijben, Size effects on thermoelectric behavior of ultrathin Na_xCoO_2 films. Appl. Phys. Lett. **105**(19), 193902 (2014). https://doi.org/10.1063/1.4901447
58. K. Fujita, T. Mochida, K. Nakamura, High-temperature thermoelectric properties of $Na_xCoO_{2-\delta}$ single crystals. Jp J. Appl. Phys. **40**(1–7), 4644–4647 (2001). https://doi.org/10.1143/jjap.40.4644
59. C. Thinaharan, D.K. Aswal, A. Singh, S. Bhattacharya, N. Joshi, S.K. Gupta, et al., Growth and morphology of the single crystals of thermoelectric oxide material Na_xCoO_2. Cryst. Res. Technol. **39**(7), 572–576 (2004). https://doi.org/10.1002/crat.200310226
60. L. Yu, L. Gu, Y. Wang, P.X. Zhang, H.U. Habermeier, Epitaxial layered cobaltite Na_xCoO_2 thin films grown on planar and vicinal cut substrates. J. Cryst. Growth **328**(1), 34–38 (2011). https://doi.org/10.1016/j.jcrysgro.2011.06.033

Index

A. G. Roca et al. (eds.), *Surfaces and Interfaces of Metal Oxide Thin Films, Multilayers, Nanoparticles and Nano-composites*,
https://doi.org/10.1007/978-3-030-74073-3

T

U

V

The manufacturer's authorised representative in the EU is Springer Nature Customer Service Centre GmbH, Europaplatz 3, 69115 Heidelberg, Germany. If you have any concerns regarding our products, please contact ProductSafety@springernature.com

Printed and bound by CPI Group (UK) Ltd, Croydon, CR0 4YY
30/09/2026
02225593-0006